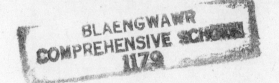

REBECCA'S CHILDREN

REBECCA'S CHILDREN

A Study Of Rural Society, Crime, And Protest

DAVID J. V. JONES

CLARENDON PRESS · OXFORD
1989

Oxford University Press, Walton Street, Oxford OX2 6DP
Oxford New York Toronto
Delhi Bombay Calcutta Madras Karachi
Petaling Jaya Singapore Hong Kong Tokyo
Nairobi Dar es Salaam Cape Town
Melbourne Auckland
and associated companies in
Berlin Ibadan

Oxford is a trade mark of Oxford University Press

Published in the United States
by Oxford University Press, New York

© David Jones 1989

All rights reserved. No part of this publication may be reproduced, stored in a retrieval system, or transmitted, in any form or by any means, electronic, mechanical, photocopying, recording, or otherwise, without the prior permission of Oxford University Press

British Library Cataloguing in Publication Data
Jones, David, 1941–
Rebecca's children: a study of rural society, crime, and protest.
1. Wales. Social change, 1837–1901
I. Title 942.9081
ISBN 0–19–820099–4

Library of Congress Cataloging in Publication Data
(data available)

Typeset by Hope Services, Abingdon
Printed and bound in
Great Britain by Biddles Ltd.
Guildford and King's Lynn

Preface

On the 150th anniversary of the outbreak of the Rebecca riots it is time to look again at this remarkable event. With the passing of time more material comes to light and our views of a familiar story gradually change. I have been able to examine certain newspaper, estate, and legal records, as well as the papers of Home Secretary Sir James Graham and Major-General George Brown which other scholars of Rebecca have not, to my knowledge, consulted. This has given me a great deal of new factual information on the riots, especially on the disturbances which were concerned with matters other than turnpikes. I am now clearer in my own mind why newspapers paid the movement so much attention, why the Queen was so concerned about it, and why it produced such a late but comprehensive military and police response. The Rebecca riots, which hold such a prominent position in Welsh history, were larger than we thought and less respectable, and for a time rivalled Irish affairs as the chief topic of debate in Westminster and the country. Sir Robert Peel, the Prime Minister during most of the period, was as bemused as anyone by the size and success of this great rural protest movement, not least because it sprang from a people who, 'speaking of them generally, [are] novices in agitation and systematic outrage'. The commission of enquiry which he launched never fully explained this strange dichotomy, though it initiated a chain of reactions which were of the utmost importance in our history.

My own interest in the story was rekindled some years ago when I was working on a project on crime and protest in the nineteenth century. It seemed to me that there was a case for seeing the Rebecca riots in the wider context of illegal activities, and in a longer chronological perspective, not least because strands of the movement were present before and after the middle decades of the century. However, the real spur to writing was comments by two very different characters, Queen

Victoria and the Chartist thinker Bronterre O'Brien. In their way, their comments were representative of two views of disturbances in rural Wales, the one contemporary and the other historical. Her Majesty, whose lip trembled when she spoke of the riots in her speech to Parliament in the autumn of 1843, asked Prime Minister Peel to explain how a rising could occur in that part of her empire which, she had been assured, was unrivalled for its loyalty and peaceful nature and where turnpikes had been in existence for many years. Bronterre O'Brien, writing for a wider audience, refused to worry about such matters. He was deeply suspicious of the movement, and asked rhetorically whether 'middle-class farmers', on horseback and in women's clothes, had ideas in their heads. Was it likely, the Chartist continued, that such a reactionary movement had any importance in the march of history? I hope that the complex answers to these questions will bring the reader closer to an understanding of the movement and the society out of which it grew.

The riots are a part of the history of every Welsh man and woman. Until recent times we lived on the land, and the process of removal from it has been a painful one. At the time of this story thousands of people were marching out of one past and into another. Of course, they ultimately became different people, but the lines between rural, urban, and industrial in the mid-nineteenth century were more blurred than we imagine. Ideas and beliefs, of Painite radicalism, Liberal Dissent, and much else, ran across geographical boundaries. Whether one lived in Llandysul or Merthyr there were patterns of language, thought, and action which were familiar. The reader will find that there were people who moved to and fro across south Wales and, more bewilderingly, from one type of protest to another. Historians, like enclosure commissioners, are unhappy with such behaviour. It has become customary, prompted no doubt by the Merthyr and Newport risings, to see Welsh industrial society as more criminal, violent, and politically advanced than the backwaters of south-west Wales, but even in these areas of life the town–country divide was not as wide as some have suggested. A traveller in 1850 would have met with soldiers and uniformed policemen west of Carmarthen as well as north of Cardiff.

Preface

This was not a situation that contemporaries welcomed. It was an uncomfortable reminder of a turbulent past. Some people in the mid-Victorian years preferred to forget that the Rebecca riots had actually happened. One writer, in the *Red Dragon* of 1883, who was daily familiar with 'the quiet placid look' of a few elderly farmers, had to remind himself that these had been 'in the full vigour of their youth . . . the daring children of Rebecca'. The first historian of the movement, Henry Tobit Evans, also found it difficult to make the connection. 'When reading the accounts of these wonderful riots', he wrote at the turn of the century, 'one is easily persuaded that it is an account of passionate Ireland . . . rather than of quiet, peaceful Wales.' From the days of Thomas Phillips and Henry Richard onwards people have interpreted these riots in a way that reflected their notions of the march of Welsh history and the spirit of the people. Most interpretations, including the latest revisionist accounts of rural history, have played down the ideological and violent aspects of Rebeccaism. Outside the large towns, Welsh society in the first half of the nineteenth century is still seen as relatively peaceful and harmonious, its people stirred only by poverty and outsiders.

The main purpose of this book is to examine the Rebecca movement in its own right, time, and place, and to give it due weight in the history of land and people in nineteenth-century Wales. I have made a particular study of the many aspects of Rebeccaism which have often been ignored or confined to footnotes, and of the public meetings that accompanied the nights of violence. The riots were more serious than we have been led to believe, and about much more than tollgates. Those who came to south Wales at this time with preconceived notions of a primitive people engaged in a limited exercise against toll charges were soon disabused. 'Any man who sets down the small farmers of South Wales as a parcel of ignorant clod-hoppers,' wrote the reporter of the Chartist *Northern Star*, 'for once in his life is wide of the mark.' He had been attending meetings in the region which, he came to realize, were without parallel in rural society. They have been sadly neglected by its historians.

To understand the speakers, the language, and the grievances at these meetings, and indeed to comprehend the movement generally, it was necessary to dig deep into rural society. The title of the book confirms that this is a study of Rebecca's children as much as of the Lady herself. It has not been an easy task. The ordinary people have to be recovered from census returns, parish registers, vestry minutes, court rolls, and other records, and their real opinions are not to be found conveniently within parliamentary papers and estate books. I have looked at aspects of rural society, such as family life, popular culture, criminal behaviour, and parish politics, which have usually been ignored by agrarian historians. If all this makes Rebeccaism more intelligible, it will have served its purpose. My chief regret is that I have not been able to research in depth the various forms of late nineteenth-century Rebeccaism about which we know so little. This will, I hope, be the subject of another book.

The reactions to the riots, by both the government and the people of south-west Wales, have been considered in some detail. They are fascinating for historians of both Wales and England. They show that Rebeccaism, like other popular movements of the time, was taken seriously, and that the response was comprehensive and calculated. One of the most interesting aspects of the story of Rebecca's children is the way in which a disturbed society was outwardly transformed into a region of comparative peace. In the process the memory and history of the riots were themselves changed, so that the modern writer has to deal with myths as well as voluminous evidence.

My debts in the writing of this book are many. I would like to pay tribute to those who have gone before, from Henry Tobit Evans to Pat Molloy, and especially to David Williams, the most thorough and elegant of scholars. Like every historian of rural Wales I have benefited enormously from three masterpieces: Alwyn Rees, *Life in the Welsh Countryside* (Cardiff, 1950), David Jenkins, *The Agricultural Community in South-West Wales at the Turn of the Twentieth Century* (Cardiff, 1971), and David Howell, *Land and People in Nineteenth-Century Wales* (London, 1977). David Howell, Diane Thomas, Audrey Philpin, and my wife Gwenda have helped with queries and translations. I am sure that it is a better book for the scholarship of Francis Jones, Geraint Jenkins, Richard Colyer,

Preface

Muriel Evans, and others, and always there has been the stimulation of comparative books by Gwyn Williams, Kenneth Morgan, John Davies, and Ieuan Gwynedd Jones.

Kind people have sent me references, and librarians at Edinburgh, Cambridge, London, Aberystwyth, and many other places have saved me wasted time. The staff of the Carmarthen Public Library and the Carmarthen and Haverfordwest Record Offices have been especially tolerant and helpful. My greatest debts are to the University College of Swansea, which gave me leave to research and write the book, to the British Academy, which financed several expeditions to London, Cambridge, and Aberystwyth, and to Ivon Asquith, Anthony Morris, and the Oxford University Press.

DAVID JONES

September 1988

Contents

ILLUSTRATIONS	xii
FIGURE	xii
MAPS	xii
TABLES	xii
ABBREVIATIONS	xiii
1. Rebecca's Country	1
2. Greater and Lesser Men	45
3. Poverty, Despair, and Crisis	99
4. Crime and Deviance	150
5. The Rebecca Riots	199
6. Rebecca the Redresser	256
7. A Slow Death	319
NOTES	378
SOURCES	411
INDEX	415

Illustrations

1. A Cardiganshire cottage in the late nineteenth century (Welsh Folk Museum, St Fagans)   102
2. Rebecca and Her Daughters (Punch, 1843)   122
3. The Welsh Rioters (Illustrated London News, 1843)   205
4. Great Meeting on Mynydd Sylen (Illustrated London News, 1843).   326

Figure

1. Indictable crime in Cardiganshire, Carmarthenshire, and Pembrokeshire, 1805–1850   151

Maps

1. South-west Wales in 1843   2
2. Attacks on tollgates, 1839–1844   202
3. Distribution of troops, October 1843–February 1844   237
4. Rebeccaism: attacks on persons and property, 1839–1844   261

Tables

1. Landownership in south-west Wales in the 1870s (major landowners)   46
2. Landownership in south-west Wales in the 1870s (lesser landowners)   46
3. People occupied on the land in 1851   64
4. Attacks on tollgates, tollhouses, bars, and chains, 1839–1844   201
5. Rebeccaism: attacks on persons and property, 1839–1844   260

Abbreviations

BCR	Brecknock, Cardigan, and Radnor
CGP	Carmarthen, Glamorgan, and Pembroke
CJ	*Carmarthen Journal*
CRO	Carmarthen Record Office
CUL	Cambridge University Library
HO	Home Office Letters and Papers
NLS	National Library of Scotland
NLW	National Library of Wales
PP	Parliamentary Papers
PRO	Public Record Office
W	*Welshman*

1
Rebecca's Country

It was said that in August 1843 one member of the 4th Light Dragoons, on asking a fellow soldier about their whereabouts, was told, 'this is Rebecca's country'. 'How do you know?' was the reply. 'Mountain, rain, and sheep as thin as yesterday's broth', groaned the old trooper. A few miles further on their march, as they reached half-way between Llandovery and Llandeilo, all doubt disappeared. On the hillside a sinister monument rose high into the morning sky. The detachment halted, and a party marched the short distance to look at the collection of three large rocks. Those who could reach checked the large inscriptions on the pillars: 'REBECCA AND HER DAUGHTER' and 'MISS CROMWELL'.[1] 'This is a dreadful place,' cried a sergeant, 'let us move on.' Within hours they were in Carmarthen, 'a dirty disease-ridden' rural metropolis, now heaving with soldiers. There the dragoons met two other detachments, the one bound for the bustling port of Aberystwyth and the other for the Wesleyan citadel of Haverfordwest.

On that day in the autumn of 1843 these three parties of soldiers traversed most of the land that had fallen into Rebecca's hands. It was a distinct and peculiar terrain, some 2,000 square miles in size, cut off from the rest of Wales to the north and east by Plynlimon and the Black Mountain, and embraced by the indented coastline and the short Ystwyth and Loughor rivers. At night, when the rain swept across the moors and upland reaches of the hinterland, the land had an eerie quality. Even in the early nineteenth century people died on the highest slopes, and there were sightings of ghosts and monsters. 'You cannot imagine a ride through a more romantic country',[2] exclaimed the first seeker after Rebecca, and all the newspaper correspondents who followed him above Carmarthen were struck by the beauty of the rolling countryside, by the turbulent evening skies, and by the white cottages lying like the last snow against the hills.

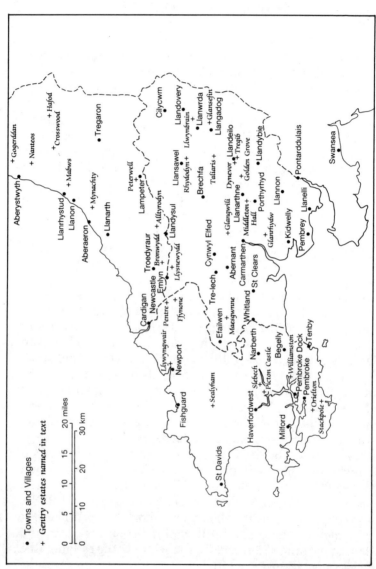

MAP 1. South-west Wales in 1843

The carriers and drovers, who criss-crossed this landscape, were always relieved to drop into the upland dales and deeper valley bottoms. The valleys to the west ran out of the crescent-shaped range of hills which divided the land between Tregaron in the north and Fishguard in the south. Contrary to first impressions, much of the land of Cardiganshire, Carmarthenshire, and Pembrokeshire was under 1,000 feet. Where the rivers Teifi, Cleddaus, Taf, Tywi, Gwendraeths, and Loughor cut their way to the sea, the countryside opened out into a mass of green vegetation and low marshland. The Tywi valley, down which the soldiers of the 4th Light Dragoons marched in August 1843, was already gaining a reputation for its fine green pastures and picturesque country houses. Southwards, along the coastal districts of Pembrokeshire and Carmarthenshire, the deep-sea inlets and warmer climate produced a 'sub-continent of rich farming and horticulture', while the more observant travellers also noticed the popularity of fishing and quarrying, and the newer interest in industry and leisure.

THE ECONOMY: TRADITION AND CHANGE

If south-west Wales was more varied in landscape and economy than was often thought, the dominance of agriculture as a way of life was undeniable. Since the later eighteenth century, and more especially during the heady days of the Napoleonic Wars, an increasing proportion (50–60 per cent) of the 1,400,000 acres had been cultivated. All except the highest and most barren land was being farmed, and two-thirds of the population was directly and indirectly involved in this effort. Farming in this humid, wet, and windy climate was largely dependent on geography. On the upland reaches, where some vegetation existed, livestock farming naturally predominated. Cardiganshire was synonymous with sheep rearing; on its mountainous unfenced walks shepherds tended the mixed flocks of small and wiry animals. In the worst of the winter weather some of them were brought down to lower grazing levels, only to be sold months later after lambing. It is now impossible to guess the annual number of sheep in the region, but there were several hundred thousand in Cardiganshire alone, many times the total head of cattle and horses reared on the land below them.

The second type of farming was to be found in northern Carmarthenshire, and along the middle and lower hill slopes of Rebecca's country. There, where the land was damp with peat and clay, mixed farming was the order of the day. Sheep were popular, but there was a higher ratio of cattle, pigs, and poultry. Up to a third of the cattle were kept for milk, for dairy products had a growing local market and Welsh butter and cheese were sent as far as Gloucester, Bristol, and London. Yet the real value of livestock was in its meat. Pembrokeshire Blacks and Glamorgan Browns were reared for three years or less, and then sold to drovers who took them, with sheep and pigs, to be fattened on the lush pastures across the English border. Most were sold in the late autumn, as the grass grew noticeably thinner. Winter feed was a constant problem for the Welsh hill farmers, and this was one reason why they increasingly harvested hay and a few root crops besides the regular cereals. Oats and, to a lesser extent, barley had for generations satisfied the dietary needs of man, but by the 1840s wheat, and even a few hundred acres of potatoes and turnips, could be seen in the fields around Lampeter and Carmarthen.

Where the soil was at its most fertile, along the rich lower valleys of the Aeron, Teifi, Taf, and Tywi, and along the coastal lowlands, the character of the farming changed again. Pasture gave way to arable, and small and closely observed experiments in commercial horticulture and afforestation. From Pembroke to Laugharne square patches of green beans and potatoes broke up the long cloth of ripening corn. Here sales of wheat and dairy produce, rather than regular sales of cattle, supplied the income necessary for rents and taxes. Clare Sewell Read and other botanical experts found much to admire in this coastal plain, although their praise was usually at the expense of other aspects of west-Walian agriculture.[3]

With very few exceptions, visitors to the Principality criticized the agrarian economy for being 'backward' and 'lacking capital and foresight', and it was true that the productivity of Welsh farming lagged well behind that of England and Scotland. Several investigators attributed the Rebecca riots to these economic shortcomings. The Welsh were blamed for under-using wastes and marshland, and, at the same time, attacked for overstocking pasture and overworking arable land. The recurring image was of a peasant never allowing himself or his land any rest, while the farm buildings collapsed, the hedges grew, and the water table rose. In this wet land, drainage schemes were few and, apart from the extensive use of lime, virtually no attempt was made to put back the nutrients washed out of the soil. The growth of root crops, the early panacea for all ills, was still in its infancy, and the use of modern machinery was so rare that one Welsh show of farm implements in 1845 drew gasps of amazement from spectators. Ploughing, harvesting, and threshing on the smallholdings were done by hand or with the assistance of oxen and horses. Only in a few places, such as the home farms of the Golden Grove and Taliaris estates in Carmarthenshire, were there real signs of approved agricultural progress. Landowners, agents, and large farmers tried to outline the scientific advantages of land consolidation and enclosure, and promised higher returns on new crop rotations and breeding schemes, but it was a slow business.

Much of the debate on the merits of Welsh agriculture took place in the agricultural societies that had sprung up since the late eighteenth century. These often began life in a small way, like the Pembroke Farmers' Club (1817–), their main functions being to put on shows and ploughing matches and to host discussions and air grievances. As the years passed, and especially in the first quarter of Victoria's reign, these clubs were bound together into district and county associations. Inevitably, they were dominated by a few prominent enthusiasts, men such as the Reverend Thomas Bowen of Troedyraur, a member of the Royal Agricultural Society, Abel Lewes Gower of Castle Malgwyn, 'the drain-cutter', and the earl of Cawdor, the crop rotator of Stackpole Court. Their speeches at the annual dinners mixed sound advice with promises of free guano and rewards for the best sows and long service. Some of their statements, and particularly those of visiting speakers, deeply wounded the beleaguered freeholders and tenant farmers. For many years the people of the south-west were told that they had about 'the worst husbandry in Britain'.[4]

It was, of course, impossible to make sensible judgements on the nature of Welsh agriculture so long as other models and standards were constantly used. In the three counties, in contrast to much of Britain, high farming formed a comparatively small part of the rural economy. In north-western Pembrokeshire, and in the favoured Carmarthenshire vales, there were large farms which had the capital, size, and skills to make the kind of progress which Clare Sewell Read and Lord Cawdor wanted. About Llandovery during the mid-nineteenth century the agricultural society lovingly documented the yearly increase in the size of their cattle, turnips, and cabbages. But this was, in general, a land of smallholdings, where people had neither the environment nor the resources to participate fully in the changes of the agricultural revolution. 'How far is the Welsh farmer of the present day wrong in taking all he can from the land with as little outlay as possible?', asked one of the tenants' spokesmen in the 1860s. 'Precious little . . .'[5]

Most farms in Rebecca's country were under 75 acres in size, and were run by the family with perhaps one or two helpers. This economic fact alone became the starting-point of the simplest explanation of rural discontent. 'The natural consequence of the smallness of the farms is the poverty of the farmers,' declared the reporter of the *Morning Herald*, '[and] to the poverty of the Welsh farmers . . . I attribute the present disturbances'.[6] Of the poverty, no one had any doubt; wills reveal that four-fifths of these farmers had virtually no capital outside the buildings in which they lived and the animals which they fed.[7] This only increased their natural conservatism, bred by history, custom, and a deep sense of fate. 'He [the Welsh peasant] jogs on the beaten path of his forefathers', said one of the newspaper correspondents of the period.[8] Setting the tone for generations, these writers dismissed his agriculture and his ambitions with the label 'subsistence farming'. This was correct in the sense that survival was the first priority of peasant life, but it ignored the fact that these men and women, apparently lost in the hills and language of west Wales, were an integral part of a larger economy. Each year, because of financial and other pressures, small farmers sold off much of their stock and dairy produce, and even, in some cases, crops as well.

The fortunes of both small- and large-scale farming in the region were bound up with the wider world of trade and industry. The rise in population, the growth of towns and ports, and the sudden development of heavy industry in the neighbouring counties increased the demand for foodstuffs. During the war years the largest farmers, and their wives, were said to have ridden through Carmarthen and Haverfordwest 'like little princes'. Dealers, agents, and drovers scoured the three counties, seeking private contracts with them, and profitable bargains were struck at the numerous markets and fairs. By the mid-nineteenth century the largest towns had market halls and exchanges serving the rural economy. Each town in the region had one or, as in the case of Carmarthen, two markets a week, where people bought and sold according to long-established rules. Tolls were demanded on stalls and sales, a source of annoyance which became more acute with the increase in trade and changes in the methods of doing business. Dealers tried to bypass such restrictions, even setting up rival marketing institutions on common and waste land, and gradually lords of the manor and burgesses began to concede 'free trade'.

Fairs, like markets, were an important element in the rural economy, but they tended to be larger and served a wider district. There were at least 250 fairs a year in the three counties, and in the busiest months of September and May there were five or six on the same day.[9] Places like Carmarthen, Cardigan, Lampeter, Narberth, Llanarthne, and Llanboidy hosted a large number of them, and some, such as the Whitsun fair at Newport, Pembrokeshire, were huge jamborees of delight. Dancing bears, raven-haired gypsies, blind ballad-singers, Irish pickpockets, and Cockney gamblers wandered amongst the booths, where entranced customers were viewing the wax replica of the latest murderer or gaping at Joseph Sewell, the young giant, and his three-foot friend. Meanwhile, during the two or three days of noise and drinking, serious commerce was in progress, and thousands of pounds changed hands. Thomas Jones of Llandovery and John Griffiths of Cilgerran were regular visitors, buying hundreds of cattle, horses, and pigs for the English markets, and butchers from Swansea and Llanelli put in smaller bids. Soon the roads were full of stock being driven to the southern counties of England, and of carts groaning under the weight of cheeses, bacon, and eggs. The names of those responsible for guiding these caravans were well known, and we shall meet them later in this book. Of course, not all the farm produce moved along the drovers' roads; some of the corn and dairy produce was transported around the coast by a flotilla of small ships. The people of Rebecca's country may have been 'inward-looking', with 'all their feelings . . . concentrated' in the parish, but their economic and social contacts kept them in touch with the wider world.[10]

By its very nature, the small and mixed farming of the region was said to have been less affected by the misfortunes that befell the more progressive and specialized agriculture of Britain during the nineteenth century. Yet the economy which we have been describing was undoubtedly sensitive to price movements. Towards the end of the eighteenth century, and for twelve years into the nineteenth, the price of cereals, store cattle, and dairy products was high, but during the post-war depression, especially in 1816–17, the demand for the products of the south-west collapsed. The result was long remembered, particularly in Cardiganshire. Animals were left unfed, people lived on nettle soup, and rents and other taxes went unpaid. 'The landed Proprietors in general [are] all very cruel to their tenants', complained David Davies, freeholder of Troedyraur, in March 1816, 'they . . . give them up to the attorneys.'[11] In parts of the region the level of bankruptcies and desertions produced a significant change in land ownership and occupation. Some parishes witnessed the first serious wave of farmers' migration to the towns and mining areas of the Principality, while braver souls hoped to create a better Wales across the seas.

The 1820s were years of mixed fortunes for Welsh farmers. Cereal prices never returned to the exceptionally high war levels, but for half the decade the demand for store cattle was good. The Reform crisis ushered in a short commercial panic and a longer period of low prices for most agricultural products; rent arrears piled up and farmers held protest meetings, though Captain Swing made only a shadowy appearance on this side of the border. The years 1834–7 had reasonably good harvests, but markets and fairs were soon overburdened, and once again George Goode and his fellow auctioneers were busy with sales of houses and stock. In 1836, when things were at their worst, the Narberth Agricultural Association called on Parliament to end the long depression in their industry, a plea that was to be repeated several times in the next decade.[12] For a while, in the late 1830s, it seemed that the bright future had arrived, and the south-western counties benefited from a welcome surge in prices. The expansion of the mining and metal industries of south Wales contributed to this agricultural prosperity, for more meat and corn were now carried eastwards to Lanelli, Swansea, Cardiff, Newport, and Bristol. It was no accident that Merthyr Tydfil, the centre of the iron industry, held a special fascination for the people of Cardiganshire, Carmarthenshire, and Pembrokeshire. 'On the prosperity of this place,' said Thomas Campbell Foster, *The Times* journalist, 'depends in great measure the agricultural prosperity of South Wales.'[13] As if to prove his point, there were already in 1840–1 the first signs in the mining districts of that catastrophic decline in demand which was to form the bleak economic context of the Rebecca story.

The comprehensive nature of the crisis of the early 1840s illustrated the interaction not just between the major agricultural and industrial economies but also between the various smaller trades of south-west Wales. Agriculture gave rise to a number of secondary industries, including the making of clothes, hats, stockings, and shoes. The stocking trade was dominated by women, often by single and widowed females. The business thrived along the Teifi, and in other northern parishes of Carmarthenshire and Pembrokeshire, with thousands of items sold annually at fairs in Llandovery, Tregaron, Lampeter, and Cardigan. Some were exported, but most were bought for home use. Cardiganshire farms also supplied much of the local demand for larger pieces of clothing. Wool was turned into garments by men and women working from home. Like most part-time rural industries, it thrived in the slack winter months. However, in the second quarter of the nineteenth century the trade became specialized, as carding and spinning mills were established around Felindre and Drefach, and in districts just to the north-west of Narberth and Carmarthen. The mills, like the weaving industry generally, were small-scale and often managed by the families of small farmers. Along with colonies of weavers in all the major towns of the area, these workers produced coarse cloth for 'the agricultural slaves of Britain and the Americas'.[14]

There was also a close relationship between agriculture, forestry, and quarrying. Great estates like Golden Grove and Rhydodyn had large areas of woodland, and in the war years especially their product was needed for ship-building. By the mid-nineteenth century large fir plantations were a prominent feature of the landscape; the earl of Lisburne, Edward Adams, and a host of smaller landowners planted hundreds of acres of new forest, to supply the expanding railway and mining industries. The bark of the oak was used in the tanning of leather; every year, for example, trees on the Nanteos estate in Cardiganshire and the Slebech estate of Baron de Rutzen were stripped by women and children. Wagons then took the bark to neighbouring tanneries, or it was shipped from Aberystwyth to Ireland, the Isle of Man, and more exotic parts.

Quarrying had long been part of the rural economy, but around the turn of the century it became increasingly important. The drive to improve the quality of a poor soil involved the extensive use of lime, for the southern half of the three counties was rich in limestone rock. The lime was burnt in kilns built along the coast and around the edges of the coalfield. During the lime-burning season, in May and June, hundreds of carts and small ships carried the lime from the neighbourhood of Llandybie, St Clears, and Llangendeirne. Another profitable undertaking was the quarrying of slate, for this, as the port records show, was in great demand throughout Britain. Landowners, and one or two contractors between Aberaeron and Fishguard, were to the fore in this trade, and not without their critics. In the lower Teifi valley, where more than a thousand people were said to be dependent, in full or part, on quarrying for their livelihood, there was mounting anger over the silting of the river from falling debris. Slate and stone had become precious commodities, required for the boom in new houses and hotels, and supplying the growing demand for gravel on the miles of turnpikes now being laid in the region. The quarries east of Carmarthen were busy in the Rebecca years, and stone-cutters and masons found it hard to reconcile self-interest with loyalty to 'the mistress of rebellion'.

Another important industry of the south-west was fishing, and here the link with agriculture was ancient and symbiotic. Small farmers near the coast had shares in the fishing boats, and some of them took to sea for part of the year. The waters from Cardigan to Swansea bays teamed with herring and mackerel, and places like Aberystwyth, Fishguard, and Haverfordwest had their own fishing fleets. In the early nineteenth century record catches of sea fish became commonplace and the business more professional. Boats and coracles also ran up the long estuaries of the Teifi, Nevern, Gwaun, Cleddau, and Tywi, sweeping for salmon and sewin, and further up still there was a collection of man-made weirs and pools where these and other fish were trapped. The types of wide and narrow nets used by the estuary men were as bewildering to the magistrates as they were to the fish, but until stocks became depleted in the mid-century, all regulations were either unnecessary or ignored. Altogether, several thousand families in the three counties were dependent on the fishing trade, and it gave a balance to the diet and economy of the coastal districts. It was always a hazardous enterprise, for besides the vagaries of the weather there was growing competition from English boats and local rivalries between, for instance, the Carmarthen and Ferryside men. The Carmarthen fishermen, huddled for generations on Tywiside and Kidwelly Fach, begged, drank, and fought their way through the winter months.[15]

If fishing enjoyed mixed fortunes in the second quarter of the nineteenth century, there were three areas of economic activity which made exceptional progress in the south-west: leisure, heavy industry, and transport. The first of these was regarded as perhaps the most important, for it brought status as well as wealth to the area. Upper- and middle-class visitors came to Wales to admire the romantic scenery of mountains and cliffs, while others found the water and air of the inland spas invigorating. Every summer and autumn the families of hundreds of local clergymen, retired army officers, and the lesser gentry relaxed for three or four months at Aberporth, Aberaeron, Llangrannog, Fishguard, Ferryside, Builth, and Llanwrtyd Wells. A few of these families had second homes in the towns, but most enjoyed the benefits of hotel accommodation, and soon forgot about the troubles of the countryside. Conversation on the terraces was about servants, the previous night's dramatics, and the latest musical offering, and, when the clouds lifted, the visitors strolled 'in Brighton fashion' down the parades and promenades.

It was a matter of pride in the early Victorian period that hundreds of English and Scottish families now made the long annual trip to south-west Wales, with Aberystwyth—'the Queen of towns'—and Tenby as their favourite haunts. Every week in high season sixty carriages a week arrived at the former place, seeking accommodation in the Belle Vue hotel or Gogerddan Arms. To cater for these special guests, Aberystwyth and other towns were rapidly improved, with good water and gas supplies as the main priority. New piers were built, and a range of rooms, halls, and baths provided. If the entertainment never quite equalled the quality of that of Bath or Cheltenham, it was strong on variety: regattas, race meetings, cricket and archery matches, brass bands and dances, concerts, picnic parties, and excursions by steamboat. There was also the moveable feast of theatrical groups, circuses, magic shows, German musicians, and, not far behind, the travelling surgeons and dentists.

Not everyone welcomed this intrusion into Welsh life. There were complaints from Haverfordwest Dissenters about the moral damage done by the theatre and female steeplechases, and a more permanent grouse that the interest in the leisure of the wealthy diverted attention from the real problems of country living. The irony, during the Rebecca riots, of gentlefolk searching for the wonders of geographical and archaeological Wales on Preseli mountain and being waited upon by sullen locals in European peasant dress was duly noted. So were the evenings at the theatre spent watching 'Twm Siôn Cati', 'The Farmer's Boy', and 'Rebecca and her Daughters', while the Metropolitan police and dragoons kept the streets quiet outside. At Quarter Sessions, turnpike trust meetings, and in the respectable press there were constant warnings that the slightest sign of violent behaviour would drive holiday-makers and capitalists from the region. Tenby was not affected by the riots, and Aberystwyth's corporation tried hard to give the same impression, being deeply conscious that the annual visitors were the foundation of its latest prosperity and a welcome stimulus to local agriculture, industry, and transport.

The development of heavy industry in south-west Wales was older than the exploitation of leisure. There had been lead and coal mines, and even a few iron forges, in the region for centuries, and processed ore had been sent as far as Ireland and Cornwall. However, after 1750, and again after the 1790s and the 1830s, the pace of activity suddenly increased. In the lead industry, which was located mainly in the hills of north Cardiganshire, and to the west of Llanwrtyd and the north-west of Carmarthen, the mid-1830s marked a recovery that was to last until 1870. Most mines were small; altogether in the mid-century some 2,000 people made a living out of cutting, carting, and shipping the lead. The biggest enterprises were in the hinterland of Aberystwyth, and, as their names suggested, they owed much to the energies of landowners like the Pryses of Gogerddan, the Powells of Nanteos, and the Lisburnes of Crosswood, while the same was true of the Rhandirmwyn mine in Carmarthenshire and the earl of Cawdor. Abel L. Gower of Castle Malgwyn, near Cardigan, a director of the Bank of England who had his fingers in many commercial endeavours, also worked a few small mines.

The capital required for lead-mining in remote areas was considerable, and most landowners leased the working to various companies and contractors. These often preferred to give the more skilled jobs to outsiders. Local labourers found wages lower than they had hoped and payments irregular. Every two or three years the Welsh miners broke their characteristic passivity and invaded the nearest town to take vengeance on Cornish families and the unscrupulous agents who employed them. Conditions of work in the lead mines were bad, droughts and floods interrupted their job, but employment was reasonably secure, except when, as in 1842, the bottom fell out of the market. For much of the Rebecca period lead-mining was in a profitable state, with new veins being discovered and new ventures promised, and this economic optimism was used to explain the comparative tranquillity of much of northern Cardiganshire at the time. 'Had that county no mines, and had those mines not been at work by English enterprise and English capital,' argued a loyal correspondent for the *Welshman*, 'we do not know whether Cardiganshire' would have escaped so lightly.[16]

The contribution of the iron, coal, and copper industries to the economic life of the three counties quickly surpassed that of lead. In the late eighteenth century the largest ironworks were those in the vicinity of Carmarthen and Kidwelly, which were associated with that remarkable native entrepreneur Robert Morgan. There were a number of smaller forges elsewhere, but these were mainly leased by the gentry to supply a local market. As coal replaced charcoal in the smelting of iron ore, so London merchant Alexander Raby and a few other non-Welsh pioneers erected ironworks in the coalfield to the north of Llanelli and Swansea, matching that of the Prices at Neath Abbey. In 1838 anthracite coal was first used in the manufacture of pig-iron, and very soon there were three major enterprises exploiting this process: the Gwendraeth, Trimsaran, and Brynamman works, each employing over a hundred men. Such was the optimism of these years that a dozen ironworks were planned, but by 1842 the market had collapsed, and employers like Charles Newman at Pontyberem, the English owner of the Gwendraeth works, survived partly on a secondary income from tally shops and farming.

In sharp contrast to the picture in Monmouthshire and Glamorgan, the iron industry in this part of the world always took second place to coal, and from the 1750s onwards the output of the latter rose sharply. The responsibility for extracting and selling the coal was shared between landowners like the Philippses of Picton Castle and the Mansels of Trimsaran and Stepney, local farmers and merchants, and immigrant businessmen. The arrival, at the turn of the century, of Richard Janion Nevill from Birmingham and later William Chambers from Kent signalled the triumph of the Llanelli coalfield. These two men of energy played an important role in all the industrial enterprises of the region, and with their moves to Llangennech Park and Llanelli House they joined, a little unsteadily, the second rank of landed society. As a result of this, and in other ways, they managed to achieve star billing in the drama of the Rebecca riots.

By the early 1840s there were almost 200 coal mines across

the lower half of Dyfed. They were situated in three main groups. The first group were located about Haverfordwest, Kilgetty, and Begelly on the western side of Carmarthen Bay. These employed about 800 people. The largest pits, owned by the Owens, Childs, Stokes, and other local notables, each had a labour force of about eighty men and thirty women, with many of the latter working underground. The geology of the area made these mines difficult and dangerous. The loss of forty lives at Hugh Owen's Landshipping colliery in 1844, when the overhead river broke through the roof, was one of the worst of many disasters by rock fall, the ingress of water, and firedamp. Productivity in this district was inevitably lower than in some other areas, and so were wages; in fact, the Pembrokeshire miners often had smallholdings or other forms of supplementary income. The coal which they cut was carried to Milford or neighbouring ports and thence to the west of England and further afield. For much of the early nineteenth century the trade in both coal and iron here and elsewhere on the western reaches of the south Wales coalfield grew impressively, but there were always seasonal fluctuations and, as happened in the early 1820s, the early 1830s, and, of course, in 1842–3, the occasional slump.

Much to its regret Milford at the end of the eighteenth century was replaced as the leading coal-exporting centre of the south-west by Llanelli. Half a century later there were fifty small and dangerously open pits almost encircling the town, producing bituminous coal for fuel. The Llangennech Coal Company was one of the main employers in the area, with particular interest in the eastern district that embraced Hendy and Pontarddulais. In 1844 a hundred of their men died in a mine two miles outside Llanelli. From that town another line of pits followed the coast westwards towards Kidwelly, and from there an arc of anthracite-coal workings took its direction from the river Gwendraeth Fawr.

The latter group of pits, which reached up to the Amman coalfield, were comparatively new, and consequently places like Pontyates, Pontyberem, Cwmmawr, and Llannon retained much of their agricultural past. Although the census returns show that the colliers' families were more dependent on industrial work, and smaller, than we have been told, nevertheless the personal links with their more numerous rural neighbours were close. At Cwmmawr John Thomas, the oldest and leading employer of the neighbourhood, was said to have been surrounded by a thousand people who took what they could, when they could, from mining, farm, and road work. The first occupation had the best financial rewards, but payments were poorly calculated, long in coming, and soon expended at the company shops. Nor were colliers unaware of the problems of demand, for there were times during the early 1840s when only small amounts of coal and culm from the pits of this district were taken to the ports of Pembrey, Bury Port, and Llanelli. In such a period the colliers were willing recruits to the cause of Rebecca. Finally, in this story of coal-mining, there were the pits in the valleys about Swansea and Neath. Perhaps half of their output, and that of other mines, was dispatched in ships with such appropriate names as *Farmers' Delight* and *Industry*.

The remainder of the coal was used locally either for domestic fuel or in the furnaces of other industries. Chief of these were tin, which was worked at Carmarthen, Neath, and a few other places, and especially copper, for the coast between Kidwelly and Aberavon was the home of British copper smelting. Copper ore was brought initially from the West Country, and later from Cuba and Chile. At Llanelli, where much of it arrived at the turn of the century, the ore was taken to the new smelting works on the perimeter of the town. There the Nevills and their colleagues used mountains of coal to melt the ore, and built 250-foot giraffe-like structures to carry the sulphurous smoke into the higher atmosphere. Ten miles away, on the east side of Swansea, the Vivians, Grenfells, and Williams built an even larger complex of copper works, employing over 1,500 skilled people. Industrial relations were poor; on at least half a dozen occasions in the early nineteenth century the male and female copper workers protested violently against their brutal employer-magistrates and their cheap Irish labour. Hours of work were long, conditions appalling, the smell of sulphur and urine all pervasive, and the environment almost totally destroyed. Fortunes were made rapidly, for during the middle decades of the nineteenth century the trade in copper was consistently good, and prices and wages held steady. But the change in import duties in 1842, and the industrial troubles in copper and tin during the next two years, caused exceptional anxiety, and added to the sense of a general economic crisis.[17]

The depression suffered by most of industry at the time of the Rebecca riots highlighted the contribution which non-agricultural employment made to the south-western counties. It loosened a corseted economy, provided landowners like the earl of Cawdor, Pryse Pryse, and Hugh Owen with an alternative income, brought capital, entrepreneurs, and workmen from England, Ireland, and Scotland, and gave new employment opportunities to native managers, tradesmen and work people. Above all, it gave some relief from the grinding poverty suffered by so many of the Welsh population. Occupational detail on known Rebeccaites shows that, where possible, small farmers and labourers supplemented their income from sources outside agriculture, and a few of these later moved to a more rewarding life in the mining districts. Without manufactures, argued John Jenkins, the south Wales free-trade lecturer at this time, the Welsh like the Irish would have been reduced to beggary, starvation, and rebellion.

Yet it would be wrong to exaggerate the industrial progress. Despite the vital technological innovations and the beginnings of a financial and communications infrastructure, mining and metal industries generated less than a fifth of the region's wealth in 1842, and gave employment to only a tenth of the working population. Moreover, much of this industry was concentrated in a few districts, notably those on the south Carmarthenshire–Glamorgan border, and even there observers noted the marked differences between the blackened industrial villages and the huge works towns of Merthyr Tydfil or Tredegar. Newspaper correspondents, on obligatory visits towards the end of the Rebecca troubles, claimed that in the three counties the industrial revolution was only just beginning. 'Her people are rural by work and instinct', suggested one writer. It was an exaggeration, but an illuminating comment.[18]

Earlier visitors to the south-west, in their analyses of the comparative backwardness of the region, were convinced that one of the main obstacles to progress was transport. It was slow, costly, and dangerous. The Reverend Ebenezer Morris, vicar at Llanelli, recalled in later life how he had needed God's guidance throughout his first Himalayan journey across Wales. As we have seen, the greatest hill walkers were the drovers, who took hundreds of animals every year along well-worn but difficult routes, via Tregaron, Lampeter, Llandysul, and other stopping places. The transport of other farm goods, and the products of the early slate, stone, lime, lead, coal, and iron industries, were placed in the hands of hundreds of carriers, many of whom were part-time farmers and craftsmen. Their pack horses, sledges, carts, and wagons inflicted terrible damage on the uneven road surfaces. Despite damning criticisms, communications did not change much in the eighteenth century; parishes, lords of the manor, and private individuals put the responsibility for improvement on one another. As a result, the lines of roads in the Principality bore only a passing resemblance to the needs of the changing economy. Speaking of those days in the 1890s, Major Price Lewes of Ciliau Aeron suddenly exclaimed: 'What a fearfully uncivilised condition this country must have been in then.'[19] He was right; roads were no better than muddy tracks, and the weary traveller had to negotiate sharp inclines and numerous rivers and streams.

Where it was feasible water transport provided fewer problems, and many of the first industrial sites were located near to river and coast. Shallow-bottomed vessels carried goods down river and into one of the forty estuaries along the south-west. The demand for more and larger vessels grew rapidly. By the 1840s the ship-building industry was at its height in Aberystwyth, New Quay, and Cardigan, and in the Royal Dockyard at Pembroke they were launching the biggest ships in the world. Every day in each of the main ports forty or fifty ships were loaded and unloaded. Indeed, for some of these places times were never quite as prosperous again. During a Carmarthen weekend, the arrival of fruit, ore, and coal from France, England, and Porthcawl was matched by the departure of oats for Cardiff, timber for Neath, and bark for Ireland. In addition, there were new types of vessels: the packets, with their letters and parcels, the pleasure boats waiting for customers, and the steamers *Phoenix*, *Hercules*, *Pilot*, and *Harriet* on their round trips between Aberystwyth, Milford, and Bristol. Even some of the soldiers used to subdue the rebellious Celts were transported in this way. It was a quick but risky means of transport, for Cardigan Bay and the coastal waters around the southern tip of Pembrokeshire were hard to negotiate in bad weather. In one of the worst storms ever, at the end of October 1843, scores of bodies were washed up on the rocks and sands. Almost before they stopped breathing, their goods were spirited away by smugglers and wreckers.

In retrospect, the 1840s marked a change in the fortunes of water transport, as people began to take advantage of the improvements in inland communications. These had gathered pace since the late eighteenth century, and were a response to demands from agriculture, industry, and wealthy travellers. The initiative in road building was taken by pioneering landowners or by enterprising parishes like Llangeler in Carmarthenshire, but the main agency of change was the turnpike trusts and the technical wizards whom they employed. The trusts were set up under Acts of Parliament to raise money for road improvements, with the expectation that a system of tolls would pay off all mortgages and creditors. The trustees, a cross-section of the wealthy and powerful, borrowed thousands of pounds from the Exchequer, from Lords Lisburne, Cawdor, and Dynevor, and from a host of smaller property owners, bankers, and businessmen.

Those who later defended these trusts claimed that, because of their size, independence, and income, they were able to transform the road system in a way that parish authorities never could have done. New roads were built, old ones widened and straightened, surfaces improved, hedges and ditches cleared, and bridges constructed and improved. The main roads linking Aberystwyth and Swansea, Llandovery and Pembroke, and Haverfordwest and Neath were brought up to a satisfactory standard, and there was special praise in the mid-century for the macadamed surface and condition of the roads about Carmarthen and on the lowest slopes of the Black Mountain. At long last new centres of industry and population had been linked up, and rivers, marshes, and estuaries crossed. Yet there were parts of the south-west, like northern Pembrokeshire, which were largely untouched by the turnpike revolution. There people still relied on the parish roads, which after 1835 were upgraded with money from the highway rate.

The thousand miles of turnpikes were built and repaired at a much greater cost. The major developments had been carried out by well-meaning people, who had run ahead of themselves in the early nineteenth century in an effort to promote their own districts. The future was heavily mortgaged, and more turnpike gates and higher tolls were needed to stave off bankruptcy.[20]

The other transport changes were closely related to the needs of heavy industry near the coast. Late in the eighteenth century a handful of short-distance canals were built, connecting coal mines to the other works and to the sea, and most of these were located in the south-eastern corner of Carmarthenshire. More followed in the first decades of the next century, so that, for example, by the late 1820s the collieries about Pontyberem and Fforest could send their black diamonds directly to Kidwelly, Pembrey, and Llanelli. Elsewhere there were experiments in wooden, and then iron, tramroads. Horsepower gave way slowly to steam power, but by 1842 there were railways radiating inland from Saundersfoot, Llanelli, and the other industrial ports. The Reverend Ebenezer Morris, riding northwards towards Llandeilo at this time, was delighted by the noisy and persistent companionship of the coal wagons; he found it hard to believe that this was the uninhabited countryside which he had first passed through only twenty-four years before.[21] As yet, however, 'railway fever' had not reached the rest of west Wales; in another decade railway companies would be vying with one another to pull the region into the modern world.

The coming of the railways, which so alarmed some of Morris's Dissenting colleagues, was an important social as well as economic event. It transformed people's perceptions of Rebecca's country. Before that time it was easy to ignore the area, or to see it in a simple and timeless way. The travel writers and artists who discovered Wales in the late eighteenth century, and the military officers and journalists who followed them, were struck by the apparent isolation and provincialism of its south-western region. They described the pre-eminence of its backward agriculture, and the lack of enterprise and commercial spirit amongst its primitive people. Yet their comparisons with the more remote parts of Ireland, Scotland,

and other European countries were deceiving. There were, as we have seen, not one but several economies in the south-west, each related to the others, each affected by external factors, and each in the process of change. Rebecca was the child of old and new economic forces.

SOCIETY: PEOPLE, PLACES, AND OCCUPATIONS

Economic change was intimately connected with social movements, the most impressive of which was the increase in population. For centuries south-west Wales had been renowned for the small number of its people, and for the unique nature of their settlement. This was not a land of large nucleated villages, like southern England; instead the inhabitants were scattered as if by a great wind. 'Everything conspires to disorder', commented a magistrate during the Rebecca troubles; what he meant was that the settlement pattern was not dominated by manor house or church. Instead there were small hamlets, where people had come together as a result of work patterns, the communication network, and cultural changes. Some of the hamlets, as their names suggested, were centuries old, but others were testimony to more recent developments in religion, trade, and industry. In each parish there were perhaps one or two hundred farms, many standing apart and alone. On the rolling hills and moorland that constituted parishes like Cilycwm in northern Carmarthenshire the Welsh lived in isolated houses with a density of only one person per ten acres. Here the farmhouse was the centre of all social gatherings, and the place at which Rebecca outrages were planned. Class-consciousness might have been difficult in such a thinly populated area, but the secrecy, solidarity, and speed of communication of these people astonished military commanders.

For centuries this society had moved to a rhythm of population rise and fall, though the late seventeenth century had been a time of some stagnation. In 1708 the population was not far short of 100,000, or 43 persons per square mile. During the first half of the eighteenth century the number of persons in the three counties increased only slowly. These were years of poor harvests, disease, and riot, but the last three decades of the century saw a more favourable ratio of births to deaths. At the time of the census of 1801 the population was estimated at 166,553, and fifty years later it stood at 346,364. The population still seemed very dispersed to the urban visitor, but by the Rebecca years most of the villages, hamlets, and townships held about 200–500 people, and the parishes had grown to three or four times that amount. There was now over much of the countryside a density of at least 100 persons per square mile. In the hinterland of Cardigan, out towards the fishing and quarrying communities of Llechryd and Cilgerran, and near the southern Pembrokeshire coast the people were more closely settled, with 180:1 being a typical ratio, while in some parishes about Llanelli and Swansea as many as 300–400 persons inhabited a square mile. Only Carmarthen and Aberystwyth, amongst the older forms of community, could match this. Such figures of population growth represented unprecedented change, comparable with that of Britain generally, but made even more remarkable because it occurred in a region of limited industrial and urban development.[22]

The pace of population change is fascinating; in the forty years before the appearance of Rebecca the parishes most involved in the troubles experienced typical growth rates of between a third and two-thirds. In general, the fastest rate of increase occurred in the first quarter of the nineteenth century. With little economic activity other than agriculture the pressure on land was intense, but by the 1820s this was beginning to slow down. In some areas there was an actual decline in the number of residents within the next twenty years. Llanddewi Befi, Ciliau Aeron, Llanwrtyd, and Cilycwm, forerunners of hundreds of similar modern Welsh communities, were now places that one came from. Uninhabited houses, still haunting our upland landscape but long removed from the back streets of our country towns, were just starting to attract attention in the mid-century. One must, of course, not overestimate the number of homes left to the wind and rain, but if one sets the population history of this part of Wales against that of Britain there is a significant disjunction about the mid-century.

The variations in the statistics within the three counties are also illuminating. Some parishes reached their population peaks at the turn of the century, and others decades later. They included Henllan Amgoed in 1811 (a population of 466), Llansadwrn in 1821 (1246), Llanboidy in 1831 (1820), and Tre-lech a'r Betws in 1841 (1620). Llanboidy, like neighbouring Cilymaenllwyd, and Clydau and Cilrhedyn to the north, had a comparatively slow rate of growth, and one suggested reason for this was the constant departure of servants from these places to Tre-lech, Meidrim, Abernant, Newchurch, and St Clears, a western crescent of agricultural activity facing the bright star of Carmarthen. The increase of population in these latter parishes was truly astonishing, some almost doubling in size in the first four decades. Visitors to this district in the 1840s were delighted by the whitewashed cottages, which appeared like mushrooms overnight, and amused by the number of people crammed into stables and barns, but soon even these places were affected by the long decline in population. So many of the Rebecca parishes were at their most vigorous just before the riots. A few perceptive commentators sensed this particular shift of history, and called the troubles of the mid-century 'the last cry of a doomed society'.[23]

It is a remarkable comment on the population history of the region that such a high percentage of the parishes ended the nineteenth century with almost the same number of people as they began. Not all, however, experienced an even rise and then fall. It is evident from more erratic statistics of growth that Gwynfe hamlet, for example, and others in Narberth, Penboyr, and some of the most riotous districts were affected by the economic changes which we have described above. Periods of expansion and stagnation were interspersed as local job opportunities changed, waste land was cultivated, commons enclosed, and as the attractions of neighbouring towns and distant industries fluctuated. But it needed special circumstances for any community to break through to sustained growth. The greatest decennial increases in population were to be found in certain towns of the south-west, in parishes very close to these centres, and especially in those districts of the coalfield where mining and metal manufacture acted as a kind of overdrive or multiplier. The semi-industrial parishes of Llannon and Llanedi in Carmarthenshire, and St Issells in

Pembrokeshire, managed sound if inconsistent population growth through the mid-century, but Pembrey, Llanelli, Llangennech, and, on the Glamorgan side of the border, Pontarddulais and Llangyfelach had periods of staggering expansion, more than doubling their combined size in the forty years before the Rebecca riots began, and going on to greater things.

In 1841 Llanelli was a boom town of almost 7,000 inhabitants, comfortably the second largest town in the three counties. Here the natural increase in population was boosted both by short-distance migration and by the arrival of hundreds from the nearest English counties and from Ireland. A copper and mining town, with good ship-building and docking facilities, Llanelli gained a new civic pride after the Municipal Corporations Act of 1835, when the town's leaders decided to give the place a more respectable and cleaner face. The town was linked, in contemporary estimation, with neighbouring Swansea, which was more than twice as large but which had a slightly slower rate of growth at this time. The remaining towns fell into three categories of size: those, like Pembroke, Haverfordwest, and Aberystwyth, which housed between 5,000 and 7,000 people; those half as large, such as Cardigan, Milford, and Tenby; and a considerable number of other towns, like Llandovery, Fishguard, Llandeilo, Kidwelly, Aberaeron, and Llangadog, which were much smaller.

Eighty per cent of the region's population lived in communities with fewer then 1,500 people, a social fact which surprised all visitors, even those from the most rural of English counties. In fact, the smallest towns were hardly distinguishable from villages. Joseph Downes, author of *The Mountain Decameron*, called them 'village-towns', and, when staying at the Castle Inn, Llangadog, in September 1843, he described the long narrow street, the shallow bow-window shop fronts of watchmakers, saddlers, and shoemakers, and the corn stacks, orchards, and trees. Like other village-towns, Llangadog provided sturdy homes for clergymen and retired army officers, the professional services of land surveyors, surgeons, and lawmen, a large number of inns and public houses, and cheap accommodation for agricultural labourers.[24] By contrast, the main urban centres of the region had begun life as castle, port, and garrison towns, though a number also owed their status to the legal and administrative functions which governments had devolved on them.

Few visitors were impressed by these towns, not even by the new county halls and market-places which were built in the mid-century amidst the mud and the stench of slaughterhouses. With hindsight, the competition between these towns for the holding of Quarter Sessions, for the best institutes of learning and culture, and for the most efficient supplies of gas and water has a touch of desperation. Having sustained greater than average increases in population at the start of the nineteenth century, most of these places had begun to fall behind the national rate of growth by the 1830s. The towns which expanded most rapidly in the first half of the century were Pembroke and Milford, with their docks and shipyards, and the coastal resorts of Aberystwyth and Tenby, which enjoyed a boom for the reasons already described. Pembroke was a special case, because of the siting nearby of the royal dockyard, the arrival of several hundred workers from Somerset, Devon, and Cornwall, and the presence of a large guard of Royal Marines. The last gave panic-stricken magistrates some relief during the first stage of the Rebecca conflict.

Carmarthen was the undisputed capital of the region. In the eighteenth century it had drawn away from its main rivals, and by 1801 it was a substantial town of 5,548 inhabitants, and a major port with several hundred ships a year passing into its harbour. In the first two decades of the new century its growth was remarkable, but by the time of the census of 1831 the town's population had reached a peak of 9,526. Even so, during the Rebecca years Carmarthen was still the second town in Wales. It was situated at the bridgehead of the river Tywi, and set upon a small hill. The countryside came up to its very gates and, on its market days, through its spider web of narrow and dirty streets. Hundreds of farmers, drovers, and market women arrived twice a week to buy and sell, and on the days of the fairs the numbers were even greater. Then the town was full of the sound and the smell of the country and the sea, and the 200 public houses were overflowing.

Carmarthen's architecture revealed its importance as a centre of government, education, and religion. The shire and town hall hosted the Assizes, Quarter Sessions, and Petty

Sessions, and the other impressive stone buildings amongst the 2,000 dwelling-places were the second homes of gentry families, residences of merchants and bankers, grammar schools, Nonconformist college and academies, and assembly rooms and market halls. In the winter months elegant carriages took the wealthy to balls, concerts, and the theatre, and for the Welsh amongst them there were the added delights of the Cymreigyddion meetings and new books from the town's printers. Carmarthen was also noted for its castle, its church and Welsh chapels, its poorhouse, and its slums. Through the Priory, Lammas, Little Water, and Goose streets trailed a motley band of scavengers, hawkers, beggars, and prostitutes.

The town was a target of endless gibes. Visitors complained of the disappearing open spaces, and of the state of the quayside and of Picton's monument. Respectable inhabitants, having stopped weekly bull-running, now petitioned against the dangers of loose pigs and of horses being furiously driven. And always the stench and dust seemed to hang over this underwatered town, forcing the mayor to admit that Carmarthen was the dirtiest place in Wales.[25] Yet his town had a vitality that few could deny. Nowhere was this more true than in its politics. The corporation was a beargarden, providing the large urban and newspaper-reading electorate with a weekly treat of jobbery and financial incompetence. After the Municipal Corporations Act, which left the town with a larger debt than ever, the borough was run by a divided common council, some of whom were replaced on Charter Day every November. These elections, and even more the parliamentary elections for the town's MP, were the occasion of vigorous controversy and frequent violence. From time to time the presence of soldiers was needed to curb the excesses of the inhabitants, and it came as no surprise when Carmarthen was chosen as the military headquarters during the Rebecca riots.

The town was occupied by mainly Welsh-speaking natives, born in the town or in parishes close to it. About a tenth of the residents had come from Pembrokeshire, and about the same proportion had arrived from Cardiganshire, Glamorgan, Ireland, London, and the south-west of England. The government returns of their occupations show the predominance of trade, agriculture, and the professions over large-scale manufacture. Two-thirds of Carmarthen's working population was employed—in descending order—as farmers and agricultural labourers, domestic servants, boot- and shoemakers, general labourers, dressmakers and milliners, masons and stone-cutters, tailors, carpenters and turners, weavers, hatters, shipbuilders, publicans, seamen, blacksmiths, fishermen, and printers. The town in 1841 also had 8 resident clergymen, 11 merchants, 13 bankers and accountants, 23 solicitors and attorneys, 30 schoolmasters and mistresses, and 36 clerks. The *Morning Chronicle* reporter, who came to Carmarthen as an economic reformer, was pleasantly surprised by the inhabitants' entrepreneurial zeal and by the wide range of their incomes.[26]

The social structure of Carmarthen, and indeed of the whole south-west, was more complex than at first appeared. The most superficial analysis tended to combine the rural, urban, and industrial population into two groups, 'the upper classes', who were the gentry and leading industrialists and professional people, and 'the lower orders' of small farmers, farm servants and labourers, small shopkeepers, craftsmen, and miners. Such an inadequate analysis provided a ready explanation for the Rebecca riots and indeed for all the troubles of the period: Welsh society was too divided and inflexible, lacking the binding influence of a middle class. The royal commission on Welsh education, whose report was published in 1847, warned of the dangers of social immobility, strong class feelings, and poor relationships, made worse by linguistic and religious differences.

Of the 'lower Welsh', it stated in the report that, whether in agriculture or industry, 'they are never masters . . . his social sphere becomes one of complete isolation from all influences, save such as arise within his own order'.[27] The Reverend Griffith Thomas, vicar of Cardigan, addressing a public meeting in 1846, expressed bewilderment that Welshmen never made farm bailiffs or masters of large ships.[28] The same point was made about managers of works, land agents, police inspectors, and gamekeepers; they were, it was incorrectly claimed, invariably outsiders from Scotland and England. The *Times* correspondent, writing three years before, believed that amongst the rural population the Welsh language and a lack of ambition precluded upward movement; farmers either remained in the position to which they had been born, or, more commonly, they 'appear to be continually descending to the class of labourers'.[29]

Marriage registers provide some supporting evidence for this thesis of a rigid social structure. At least half the sons of farmers and craftsmen, and three-quarters of the sons of labourers, followed the occupations of their fathers, and the great majority of them married women from within the same social group.[30] Marriage was perhaps the most class-orientated activity in rural society. According to the very limited information on the birth, marriage, and death registers, comparatively few people experienced upward social mobility. However, there were exceptions, and it appears that education, religion, and material success were the three avenues of escape. Sons of farmers who did well at the local grammar schools and at Lampeter College sometimes became clergymen, or joined one of the other professions, and the daughters of successful businessmen, too, had a good chance of rising, by marriage, in the social scale. John Lloyd Davies, a lawyer who married well and became the landowner of Alltyrodyn, was the outstanding example of a social climber, though Colonel Love insisted that he was never comfortable in polite society.

The growth in the ranks of the middle class at this time in south-west Wales did not make as much impression on contemporary sociologists as it has done on historians. 'We have no middle class here' was a typical complaint of those seeking to explain social friction and Rebecca violence, just as it had been in the aftermath of the Chartist rebellion at Newport in 1839. This was, one could argue, an instructive comment in terms of class-consciousness, relationships, and action, but it did scant justice to the range of occupations and the limited social mobility that did exist in the region and, most importantly, it ignored the recent impact of changes in agriculture, trade, and industry.

The census returns are the most obvious source from which to re-create more accurately the society of south-west Wales during the second quarter of the nineteenth century, although the problems of using them are well known. This evidence suggests that, compared with Glamorgan and Monmouthshire, there was no major shift in the balance between agriculture, trade, manufacture, and handicrafts in the years 1801–31. The census of 1841, which provided detail of a doubtful variety, indicated that less than a quarter of the working population of Cardiganshire, Pembrokeshire, and Carmarthenshire was involved in commerce, trade, and manufacture. Of the three counties Pembrokeshire had the most even spread of occupations, for large numbers of its inhabitants were returned as domestic servants, non-agricultural labourers, seamen, and people of independent means. One of the more interesting developments, taken up by the *Welshman* in its comment on the 1841 census, was the recent growth in the number of 'capitalists', bankers, and professional men. Even so, the statistics revealed that Rebecca's country was still much more dependent on agriculture than was Britain in general.

The census of 1851, which provided better information on occupations, reinforced this point. This showed the dominance of agriculture (42 per cent of working males, and 51 per cent of females) and of related finance, commerce, and industry. It also highlighted the narrow range of female professions, with domestic service being second only in importance to agriculture, and most of the remaining women being occupied in the making of clothes and in the running of public and lodging-houses. The census further revealed the growing job opportunities in mining, the manufacture of metals, and the booming transport industry, though even in 1851 fewer than 1 in 8 of working males were thus employed.

All this, and other, information enables us to formulate a primitive social structure for the three counties. At the top were a handful of great landed dynasties like the earls of Cawdor and Lisburne, whom we shall meet in the next chapter and who were as exclusive now as they had always been. Beneath them came several hundred families who deserved the appellation 'leaders of society'. The great majority of these were landowners, though their lower ranks were leavened by people whose main income came, at least for a while, not from estate rentals. The armed forces, the Church, the legal profession, and now banking and industry were the usual sources of this external wealth. These people were represented by Captain David Edwardes, John Lloyd Davies, whom we have just met, John S. Harford, the Bristol banker, who inherited Peterwell near Lampeter in 1821, Erasmus Williams, and William Chambers. Captain Edwardes of Rhydygors, Carmarthen, was a retired army officer with a Sussex wife and seven servants, who found the whole experience of the riots most trying. The Reverend Erasmus Williams was less concerned; a man of substantial independent means, he was away from his country mansion at Myddfai for much of the year, and liked to call Marlborough home. Finally, there was the redoubtable Chambers, a Whig industrialist who succeeded his namesake at Llanelli House. In the eighteenth century it had taken three generations for newcomers to be accepted into the ruling élite, but by the early Victorian era it was much easier for such businessmen to buy or marry their way into all but titled families.[31]

There was undeniably a wide gulf between the highest and lowest members of Rebecca's society, but there was in between an important group of 'middle orders' or 'middle classes'. Regarded as small in number, they nevertheless comprised perhaps 10 per cent of the population. Their origins were varied; some had brought their money and expertise from as far as Essex and Devon or, more commonly, from Breconshire and Glamorgan, but most of these people were products of the west-Walian economy. Thus Charles Newman, who spent £100,000 on improvements to his Gwendraeth works in the early 1840s, and appointed the Englishman James Slocombe to look after the 200 workforce and his 600 acres, was exceptional rather than, as was implied, typical of the group. The parish and registrar's records reveal that many of them had begun life as the youngest sons of local gentry and yeoman families.[32]

At the top of this middle-class category were the biggest farmers and land agents, some of the more successful industrialists, merchants, and bankers, the cream of financial consultants, auctioneers, and surgeons, and men near the summit of the legal, government, and religious professions. For these people times were good; trade was booming, money was sought after, and much of the current legislation about turnpikes, tithes, public health, legal administration, and the Poor Law offered new opportunities for wealth and influence. Barristers, solicitors, and attorneys were widely believed to be the most thriving members of the community at this time.³³ 'What is to be done with all their money?', asked one radical commentator. Some, like Richard Jenkins, the mayor of Cardigan, and fellow solicitor John Williams of Carmarthen, put it into building up small country estates, while many bought houses as an alternative to land.³⁴ In the census records of 1851 and 1861 their widows can be found, comfortable in Johnstown suburb, Carmarthen, and High Street, Haverfordwest, with rents, annuities, and nursemaids.

Investment was always difficult in an age of insecurity, especially for those, like solicitors John Morgan and Henry Lloyd Harries of Llandovery, who were moving quickly up the middle-class ladder. The purchase of property had its problems, and accumulation of tithes, turnpike trust tallies, government funds, and private-pension schemes brought sound but not spectacular returns. If possible, these middle-ranking professional people hoped to make at least 8 per cent p.a., and they obtained this by lending out their capital to those above and below them in the social scale. It was a rewarding, if socially dangerous venture. The conservative press and radical balladmongers poured scorn on the financial priorities of the 'aristocracy of the breeches pocket', and, for good measure, complained of their exploitation of borough lands and charities, their cultural mimicry of the gentry, and their concern over proper titles and addresses.

These criticisms are a testimony to the subtle differences of status within the middle class. One test, by no means foolproof, was the number of servants in a household. Dr John Bowen, physician, magistrate, and Napoleonic War veteran of Carmarthen, had a footman, cook, housemaid, and groom. Morgan James, the Narberth-born attorney, had one more, and William de Winton, the Cardigan banker, one fewer. Two seems to have been the minimum requirement. Education was another standard. The wealthiest families could afford governesses and private tutors, and a few, like Martha Brown, the wine merchant of Haverfordwest, sent their sons to Oxford. Most middle-class children, however, were educated at the many grammar schools, academies, and seminaries in the region. Their formal education over, the keenest amongst them progressed to the literary, philosophical, scientific, and archaeological societies that were springing up in west Wales, and patronized the mechanic institutes, mutual improvement, and temperance societies.[35]

With wealth and education came influence. After the overhaul of corporation affairs in 1835 urban politics was increasingly dominated by bankers, merchants, and others of this social group. This was true, for instance, of Haverfordwest, where power now resided in the hands of the friends of John Walters, the mayor and banker, who acted as a financial representative for several landowners and who was appointed treasurer of his Poor Law union and a turnpike trust. Amongst others with local power were three members of formidable middle-class family dynasties: David Morris, a Carmarthen banker with strong farming contacts, who became a councillor and the borough's MP, surgeon Edmund H. Stacey, mayor of the town when the Rebecca riots were at their height, and John Hughes, the Aberystwyth solicitor, who was elected mayor in 1839 and vice-chairman of the Board of Guardians. Such people, and certain clergymen of the Church of England, had the time and qualifications to make a significant impact on politics and administration. They were sometimes mocked, usually at moments of great political and economic crises, for their lack of middle-class spirit and independence, but outside Llanelli and possibly Carmarthen it was hard to run against the landed interest.[36]

Alongside, and frequently serving, this group of people, was another, perhaps twice as large. Labelled variously as 'respectable', 'middling', and, most accurately, 'lower middle class', this social rank was full of farmers, small employers, the more prosperous tradesmen and dealers, the more respectable teachers, bailiffs, auctioneers, and medical men, many of the Nonconformist ministers, and a large number of registrars, inspectors, and surveyors. Evan Thomas, Baptist minister in Cardigan, was a good example; a native of Peterstone in Glamorgan, he had in 1851 a wife, three children, and two servants. His family moved in the same social circle as the printers and 'superior' clerks who were to be found in the High Streets of the largest Welsh towns. Clerks epitomized the problems of placing people of the nineteenth-century professions; many were on the margins of the lower middle class, but a few raised themselves to positions of considerable importance.

Each small town or substantial village contained people whose income, education, and status lifted them above the working class. These included the farmer with 40 acres, the draper with an assistant, the master shipwright, and the headmaster or headmistress in one of the permanent schools. Churchwardens, parish constables, Petty juries, and Sunday-school teachers were often chosen from this social group. They were regarded as the driving force behind Nonconformist religion, education, and politics, as well as being keen supporters of Welsh culture and Welsh periodicals, and prominent members of the Ivorites and other societies. Despite so many claims to the contrary, most of society below the ruling élite was unmistakably Welsh. Although the language of law was, and the language of commerce was said to have been, English, the majority of the middle class were bilingual, able to enjoy David Rees's *Y Diwygiwr* (The Reformer) as well as the *Welshman*. The success of the Cymreigyddion societies, founded at Carmarthen, Cardigan, and Llanfynydd in the 1820s, was testimony to this. Predictably, the exaggerated inter-class nature of Welsh culture was widely viewed across the border as one of the most dangerous aspects of life in the three counties. Edwin Chadwick declared in 1843 that those who, through associations, fostered the Welsh language amongst their social inferiors were partly responsible for the 'anti-Saxon' feeling of the time.[37] One of the first instinctive reactions in London to the mass Rebeccaite demonstrations was to portray them as a nationalist rising by the Welsh people against their English-speaking governors.

Although politics and protest always threw up interesting social alignments, there were acknowledged boundaries separating the 'lower orders', or 'working classes' (the second term gradually supplants the former) from those above them on the social scale. The 'lower orders' comprised perhaps two-thirds of the population, and contained a varied collection of men, women, and children. The most obvious internal divisions ran along occupational lines. About a seventh of these people were employed in the traditional crafts of working with stone, wood, leather, wool, silk, and rope. There were several thousand of these craftsmen and women, the biggest number (6,000) being carpenters, joiners, and boot- and shoemakers. In all these trades the differences in skill were important, and apprenticeships were passed down from father to son. These people worked together in small premises, under the watchful eye of master craftsmen like David Charles the rope manufacturer, and John Awbrey the tailor, both of Carmarthen. Certain trades, especially the smiths and masons, and to a lesser extent the carpenters, were in great demand all over south Wales in the mid-nineteenth century, and Rebecca had to compete for their services. But others, like weavers and shoemakers, faced more debilitating competition, as their strikes for better conditions and food indicated.

Craftsmen in this period had a deserved reputation for their cultural skills and political awareness. George Thomas, ship's carpenter of Carmarthen, Daniel Lewis, the Pontarddulais weaver, William Hopkins, the Llangadog shoemaker, and others led double lives as poets, harpists, and lay preachers. They participated in the local eisteddfodau, which were beginning to gain in popularity in the region, and they contributed to the Welsh periodicals. Their distinctive voice was heard at many of the popular disturbances during and after the Reform crisis, and one of the above was selected as a Rebecca leader because of his command of both Welsh and English languages.[38] Only the English press, raised on tired myths of peasant stupidity, was astonished by the fluency and content of their speeches during the great meetings of 1843.

A slightly smaller proportion of the 'lower orders' was to be found in the new industries of extracting coal, manufacturing metals, producing chemicals, and setting up modern forms of transport. In 1841 there were, for example, more than 3,000 people employed in and above the coal and iron mines of the south-western counties, and a somewhat smaller number were engaged in the lead and copper trades. Few of these people were as well paid as their industrial counterparts in Glamorgan and Monmouthshire, and they sometimes obtained secondary income from farm work, which helps to explain their interest in the Rebecca troubles. The connection between agricultural employment and transport was especially close, for a good proportion of the estimated 600 men working on the new turnpike and parish roads for a part of the year were from families of farmers, agricultural labourers, masons, and paupers. Road work was poorly paid, at 1s. 6d. a day or less, and given only monthly, but, like the railway construction just beginning, it was a welcome addition to the rural job market.

All these workers were easily outnumbered by the people we shall be meeting in the next chapter, the poor farmers, agricultural servants, and labourers, who constituted a good third of the 'lower orders'. There were, according to the census of 1841, about 30,000 of these people, though their numbers would have been larger had working women been fully classified. It is, in fact, hard to exaggerate the predominance of these people in so many of the parishes where Rebecca had her greatest support. In Llanfihangel-ar-arth, where the census of 1841 records 715 occupations (480 male and 235 female), three-quarters of the men were farmers (161), farm servants (110), and agricultural labourers (85), and 131 women were returned as servants, usually on farms. Even in the parish of Llanddarog, which had sizeable communities of colliers and craftsmen, 166 of the 280 occupations given in 1841 were of agricultural labourers (105), farmers, and farm servants.[39]

To confuse matters slightly, it has to be remembered that the census returns were not designed to record the complexities of rural employment. People often did several jobs, typically renting a smallholding and acting as a labourer on the farm of a neighbour. In the villages about Llangadog and Llanegwad workers kept a cow and a pig, and went 'on the road' and into the quarries during the slack farming season. The social investigator of 1843, bewildered by the hazy line between small farmers and labourers, dismissed them all as 'of one kind, a poor peasantry, all to a man Rebeccaites'. The women were, significantly, forgotten, but the census suggests that a higher proportion of women worked outside their home in rural than in urban society. Domestic service, usually on smallholdings, was the characteristic form of female employment, though in the parishes mentioned above a considerable number of women also found work in the weaving, stocking, and dressmaking trades.

To these rural men and women must be added the remaining members of the 'lower orders', the poorest keepers, constables, collectors of tolls and taxes, as well as the sailors and fishermen, shopworkers, general labourers, paupers, beggars, vagrants, and those who survived by criminal activities. Shop assistants were about the most admired group in this category, if only because of their conditions of work, and in the largest towns their employers were now rewarding them with shorter hours of work and reading rooms. The remainder of these workpeople were a mixture of the feared and the despised, together with that well-defined tenth of the population dependent on poor relief. These paupers, living in damp village cottages, and in the slums of Carmarthen, Llanelli, and Haverfordwest, must have been acutely aware of subtle differences of status and attitude amongst the people at the bottom of west-Walian society, but in much of the literature of the time they were all labelled and dismissed collectively as 'the poor', 'the rabble', and 'the mob'.

LIFE, DEATH, AND ESCAPE

The economic and social transformation of the early nineteenth century was founded on a delicate balance of marriage, birth, and death rates. Everywhere in this period the balance was favourable towards growth. It seems that a greater proportion of adults than in the past decided to marry, some 70 per cent of men over 20 years of age and two-thirds of such women. Marriage was supposed to be for life; hardly anyone took the vows for a second time. They married at a slightly earlier age than in the eighteenth century, encouraged, it was said, by the practice of bidding, giving gifts to the newly wed, and by the improved chances of getting parish relief.[40] The age of marriage was higher in south-west Wales than in certain other parts of Britain. Here it was common for men and women to marry at about the age 26 or 27 years, and the more rural the community the older were the families. The children of farmers were often the last to be wed; it was rare for young married couples to live with their parents on the farm, though they relied on their relatives for assistance in finding another property. In some instances, people married and then lived separately until they found a place to suit them, and poor Jane Michael of Llanfihangel Aberbythych committed suicide in 1848 when the waiting became too much.

Delayed marriages might have accounted for some of the illegitimacy of south-west Wales. Estimates of this 'evil' in both the eighteenth and nineteenth centuries were notoriously unreliable. The region, or rather its female farm and domestic servants, still had an unsavoury reputation for bastardy in the early Victorian period, but the official statistics—with all their faults—do not substantiate the worst accusations. The defenders of Wales were able to use the published information on births to support their view that the people of south-west Wales were bound by a strict code of moral and religious conduct. As soon as state registration began, in 1837, it appeared that no more than one child in ten was born to single women, confirming the impression from unsatisfactory baptismal registers of earlier decades.[41]

Of course, just a few such births a year meant friction between families, and sometimes an unwelcome addition to the poor rate burden, and costly court cases. The Poor Law authorities estimated that in 1835 almost 2,500 children, or about one in a hundred of the population, were illegitimate, that is, bastards chargeable to the parish or affiliated to, and maintained by, their fathers.[42] The figure, especially in Pembrokeshire, was higher than the national average, but it is hard to draw conclusions as the statistics can be interpreted in several ways. Other Poor Law figures have the same drawback; these suggest that in the early Victorian years there were proportionately as many single mothers in the Welsh country districts as in the towns. This is a little surprising. The rural community did exert considerable pressure on the courting couple to convert a pregnancy into marriage, but this had perhaps less effect than we realize. As we shall see later, many of the fathers were married already. Extra-marital sex was strongly condemned in this society, but it was the women who kept more faithfully to the wedding vows.[43]

None of this, as the report of the education commission of 1847 amply demonstrated, satisfied reforming clergymen. They were appalled by the courting habits of the Welsh peasantry, not least by the nightly get-togethers before the dying fires and the assignations in the lofts and barns which were home for so many servants. Courting between the same couples often took place over several years, and sexual intimacy was a precursor to marriage. The Reverend Richard Buckby of Begelly in Pembrokeshire and many other clergymen said that they could hardly recall a country girl who was not pregnant at the moment of her wedding.[44] An exceptional number of marriages took place as autumn gave way to winter, and a good proportion of the first-born appeared three or four months later. From cross-checking of selected registers, one can deduce that at least a third of the eldest children had been conceived before their parents' marriage.[45] In the three counties pregnancy, and the arrival of independent means, were the two spurs to marriage.

If it were true that the illegitimacy rate in the first decades of the century was fewer than 1 bastard per 10 births, then the phenomenon had less effect on population trends than some contemporaries claimed. These trends were well documented. The birth rate in south-west Wales had risen substantially to a crude rate of 31 births per 1,000 of the population by the 1840s, when the Registrar-General reported a pleasing ratio of 10 births for every 6 deaths. The reasons for the encouraging birth rate were, and are still, by no means clear, though contemporaries were convinced that economic growth and the introduction of the cheap potato into people's diet had beneficial effects. The death rate in the three counties had fallen to 19 or 20 per 1,000 living by the same period, well below the appalling level of the mid-eighteenth century.

In the early years of the nineteenth century there were references to a decline in smallpox and other 'malignant fevers', but no one was sure how much of the improvement was due to medical advances and the emergency Boards of Health or to natural phenomena. Hospitals were almost non-existent, dispensaries few, and, before the pioneering efforts of inoculating Poor Law doctors, most people relied on primitive remedies to ward off serious illness. Each village had its own quack doctor and bone-setter. In the 1830s and 1840s hundreds, especially of the poor, died of cholera, typhus, smallpox, scarlet fever, measles, diphtheria, and influenza epidemics. At the time of the Rebecca disturbances the 'pox' and two strains of deadly fever were making their way eastwards and northwards across the south-west, turning weekends into children's funerals.

In spite of this, life expectancy rose all the time. At the moment of their birth, boys could look forward to an average lifespan of almost 40 years, and girls to five years longer, but this improved substantially once they had survived to the age of 5 years. One out of three males and one out of four females died in those early years, and most of these fatalities were the result of premature deliveries, malnutrition, and disease. Three-quarters of the females who lived beyond 5 years could expect to reach the age of 50 years, and a greater proportion than the males lived into ripe old age. Mary Davies, who never once left Llanfynydd, was 100 years of age and had eighty-six grandchildren at the time of the Rebecca riots, and Mary Legg of Porthyrhyd and Ruth Evans of Llandeilo passed away after 102 and 108 years on earth. Country women had a reputation for being exceptionally strong and healthy. Some of them ran farms almost single-handed for years after their husbands' deaths, and very old women walked great distances every week to Carmarthen and Swansea markets. There were twice as many widows as widowers in the three counties, which was why the *Carmarthen Journal* in 1836 made such a fuss of Thomas Davies of Pantwyn, still harvesting in Cynwyl at the age of 89.

The male of the species proved less resilient and adaptable than the female. Often ill at birth, the boy grew up slowly and, especially if he were of the 'lower orders', lacked the stronger physique of his industrial cousins. Farm labourers were short and gaunt, and, if they were to become policemen or colliers, needed feeding up before they were fit for work. After about thirty-five years of hard work, often in driving wind and rain, they succumbed to the dreaded tuberculosis, croup, and rheumatism. More men, and women, died during the first cold winter months than in any other season of the year.[46] People compared agricultural employment favourably with the horrors of industrial life, but there was a growing awareness of the debilitation that struck rural men in their fifties, and of the many fatal accidents resulting from contact with horses, wagons, machines, and shotguns. At harvest, the most healthy time of the rural year, the newspapers contained accounts of men carted home with haemorrhages and fat gangrenous limbs.

The result of these population trends was an increase in family size, and the preponderance of the young and of females. Outside the upper classes, who could afford many children, the typical family of south-west Wales in the mid-century was father, mother, and three or four children, with a tendency for the urban family to be slightly larger than the rural. This was a small but significant increase on the figure a century before. The household size, which was related but not identical, was 4.6 across the three counties. Amongst those with larger households were farmers, certain industrial workers, and the inhabitants of the largest boroughs of Pembrokeshire and Carmarthenshire. Places as different as Tenby and Llanelli had, on average, more than five persons living in each house, and in parts of these towns there was serious overcrowding. Inspectors appointed during the mid-century found as many as twenty people sleeping on straw in unregistered back-street accommodation. The other large urban households belonged to members of the middle class, where the small nuclear family was supplemented by half a dozen relatives and servants.

The society of the 1840s was predominantly youthful, which gives an interesting dimension to the claim that crime, protest, and Rebeccaism in this part of Wales were the activity of juveniles. Between a quarter and a third of the population was under 10 years of age, and almost half was under 20 years. Agriculture and industry were dependent on these young people, and about half the children between 6 and 13 years were working at home or for other people. A typical middle-class farming or urban household had two young teenagers amongst their servants. Contrary to some impressions, it appears that the smallest proportion of the very young was to be found in established towns like Cardigan, Haverfordwest, and Carmarthen, whereas the reverse was true in the semi-industrial parishes of Llangendeirne, Pembrey, and Llanelli. The popular court image of hundreds of black urchins collecting coal and dust for their mothers seems to have been grounded in reality.

Over the three counties generally the sex ratio was 53 female : 47 male. In urban communities the female dominance was even more pronounced, while in the new industrial areas and the Rebecca parishes the sexual balance was almost even. Everywhere, however, the independence and vulnerability of women was apparent; amongst women over 20 years of age almost half were either spinsters or widows. Typical was Mary Ann Mathias, a Cardigan widow of middle age who ran a primary school and mothered a young Rebeccaite, and about her in St Mary's Street and High Street she had plenty of company of her own age from unmarried tradeswomen, other widowed heads of households, and the wives of mariners long at sea. The more fortunate ladies had been provided with independent means, and in places like Llandovery these women played a very prominent part in church, charity, and other society work. Most lower-middle- and working-class single women had, of course, to rely on their own skills or resources. They ran farms, schools, shops, and public and boarding houses, and they formed a large army of servants, laundry workers, and needlewomen. At the end of their days many of them fell into the census category of 'helpless pauper'.[47]

The presence of such numbers of females in west-Walian society confused some observers. They thought, wrongly, that it implied extensive migration or emigration by males. In fact, the great majority of people spent their life in the vicinity of their birth. 'They know everybody, and see everything', exclaimed an exasperated Metropolitan policeman in 1843; he had learnt a social reality so basic that it is easy to miss its influence on crime, protest, and politics. When asked by education commissioners to place Swansea on the map Rebecca's children made a reasonable guess, but when asked to identify London and Ireland their hands began to shake. The unreliable census of 1841 indicates that 9 out of 10 people had been born in the county in which they now resided, and most of the remainder had been raised in neighbouring counties. More detailed studies suggest that, in the mid-century, 6 out of every 10 inhabitants of a rural parish had been born there, and another 2 or 3 had moved there from a contiguous parish. The marriage records confirm the pattern, for the majority of men and women chose partners from within five miles of their home. Signatories of the registers were often natives of the same village, and even worked at the same farm. Only persons in the highest reaches of society proved an exception to this trend.[48]

From this evidence it is clear that the rise in population owed little to immigration. There were a few famous arrivals, like the duke of Newcastle, who bought the Hafod estate in Cardiganshire and lived there for short periods, a number of prominent English industrialists and merchants, some Scottish land agents and gamekeepers, and Jewish tradesmen, but there was no mass migration into the town and countryside of Rebecca. Each of the large port towns on the south coast, including Pembroke, Haverfordwest, and Carmarthen, had several hundred English and Irish people, but there were few strangers in the inland parishes. The only significant exceptions were about the lead mines of Cardiganshire and north Carmarthenshire, where scores of Cornishmen were employed, and in the industrial south-eastern corner of the latter county. Llanelli contained, despite William Chambers's disclaimer, families who had journeyed from the Midlands on the promise of well-paid work in the pottery works and iron foundries, but even in this area the proportion of English, Irish, and Scottish workers was small. The three counties were overwhelmingly Welsh and, outside certain south Pembrokeshire parishes, Welsh-speaking.

It was not, however, a static or immobile society. As we have seen, people had always moved from parish to parish, not least in search of jobs. For generations hundreds of Cardiganshire male and female harvesters had journeyed to neighbouring counties, and many others interrupted their agricultural work with occasional bouts of sea-fishing, mining, carrying, and droving.[49] The movement of young country girls to find work in neighbouring towns was so well established that many middle-class urban households had become identified with particular village families. A few adventurers, amongst them ministers and teachers, had always travelled further afield, and settled, at least for a time, in London and the Home Counties. This was in the nature of things, but what caused more comment after the post-war depression was the growing number of people who pulled up their roots and moved to other parishes and other jobs.

In the second quarter of the nineteenth century hundreds of country people made their way to the growing towns, and to the industrial works of Pembrokeshire and Carmarthenshire. It sometimes began as a winter break by adult males, a way of earning the rent while their wives stayed behind to look after the holding and the children. For a time they managed well, but then the poverty of Llanybyther and the temptations of Llanelli became too great, and the temporary move turned into a permanent one. The same could be said of longer-distance migration, to the ironworks and collieries of Glamorgan and Monmouthshire. In 1841 it was estimated that every seventh person on this coalfield had been born in the south-western counties, and ten years later there were over 20,000 of them working there. In the months after the Rebecca troubles hundreds of Pembrokeshire families suddenly packed their carts, and, well before the coming of the railways, this population movement had turned into a flood.

The famed insularity of the inhabitants of the three counties was not therefore complete; contacts were made with people many miles away and families were split across south Wales. Elizabeth Thomas, an 18-year-old of Abernant parish, revealed during a Poor Law enquiry in 1843 that she could not remember the ages of all her brothers and sisters, but knew that some lived locally and some at Merthyr.[50] The roads on the hills between Merthyr and Carmarthenshire were full of people crossing and re-crossing. When times were bad, as they were in the early 1840s, these migrants sought sanctuary in their native parishes, adding fuel to the Rebecca flame. The most unwelcome return in 1843 was the arrival of Welsh-speaking London policemen in the land of their fathers.

The people of the south-west also had contacts outside Britain, for this was a land from which people emigrated. The process had begun in the late eighteenth century, when hundreds of Nonconformist farmers and artisans left the country, and it gained renewed impetus in the post-war depression. Cardigan was initially the main emigration port, but by the mid-century there were several departure points. The newspapers of the time carried adverts by emigration agents like Captain Lewis of Haverfordwest and letters from happy and not-so-happy Welsh colonists in Wisconsin and

Australia. At least 1,000 people left these shores in the years 1832–40, many of them from the southern half of Cardiganshire. During the early 1840s the rate increased sharply; 80 people pulled up their roots in the parish of Clydau in Pembrokeshire in the early summer of 1840, and two years later hundreds walked the streets of Carmarthen and Aberystwyth waiting for a ship.[51] They were a mixed bag; small farmers and tradesmen, ministers of religion and schoolmasters, and craftsmen and servants. They were both the beneficiaries and the casualties of the economic social changes that were overtaking Rebecca's country.

2
Greater and Lesser Men

EDWARD Crompton Lloyd Hall, the solicitor who was to play an odd part in the Rebecca riots, spent much of the autumn of 1843 writing long letters to the Home Secretary. Most of them were not acknowledged, though they provided the government with some of the best information on rural society and its problems. Living in Newcastle Emlyn, and being heir to an estate of over a thousand acres, Lloyd Hall knew that most of the population lived on and by the land. He knew, too, that the possession of land was the most important economic, social, and political fact of existence. For him there were direct parallels between his landed society and those of the classical and imperial past; each had idyllic settings, and each shared great inequalities of property, power, and responsibility. The verbose solicitor was keen to disabuse those who portrayed south-west Wales as a land of normally happy peasants: 'the men feel that they cannot be worse off & therefore become reckless in exhibiting the real state of their feelings towards their landlords & others to whom in more prosperous times they were accustomed to crouch like the slaves of Jamaica'.[1] Others concurred; this was a society sharply divided into the gentry and the peasantry, between whom there was little 'overt sympathy as exists generally throughout England'. The picture was overdrawn and inaccurate, but not without some validity.

LAND: OWNING, OCCUPYING, AND WORKING

Such an analysis of rural inequality and oppression has to be set in the context of a distinctive pattern of landownership. There were almost a million and a half acres of land in the three counties, and a considerable portion of this, notably mountain and waste land, was still in the hands of the Crown. Much of the remainder was owned by great landowners and squires. The exact proportion is difficult to establish, but

TABLE 1. *Landownership in south-west Wales in the 1870s (major landowners)*

	Peers		Great landowners		Squires		Greater yeomen	
	No.	% of land	No.	% of land	No.	% of land	No.	% of land
Cardiganshire	1	11	8	24	48	20	110	14
Carmarthenshire	2	9	13	24	50	16	198	19
Pembrokeshire	2	7	19	30	41	19	130	18

Table 1 is an indication of the situation a generation later.[2] From this return it seems that only 45 people owned a third of the land in the mid-Victorian years, 184 owned a half, and a total of 622 people owned over two-thirds. Proprietorship was slightly different at the time of Rebecca, but the basic pattern was the same.

The other third of the land was owned by a wide collection of persons and public bodies. They were returned as recorded in Table 2. From the information given there, which omits only the small acreage of waste lands in the region, one can deduce the following averages: one in every 19 of the population was a proprietor, he or she owned 87 acres, and there were 44 such freeholders in each parish. The tables also bear out the claim that most of the owners of property in the mid-nineteenth century were small proprietors or peasants, although in fact they owned no more than 12 per cent of the total acreage.

TABLE 2. *Landownership in south-west Wales in the 1870s (lesser landowners)*

	Lesser yeomen		Small proptrs		Cottagers		Public bodies	
	No.	% of land	No.	% of land	No.	% of land	No.	% of land
Cardiganshire	304	13	1553	15	1278	0.07	14	0.5
Carmarthenshire	497	16	2093	12	5168	0.40	45	0.7
Pembrokeshire	263	12	1134	8	1492	0.08	40	3.0

The obvious explanation for this dichotomy was the size of holdings. During the Rebecca years there were a small number of great estates in the area, the most outstanding being those of the earls of Cawdor (Golden Grove and Stackpole Court) and Lisburne (Crosswood), William Powell of Nanteos, and Pryse

Pryse of Gogerddan in Cardiganshire, of Lord Dynevor (Dynevor Castle) in Carmarthenshire, and of Sir Richard Philipps of Picton Castle and Sir John Owen of Orielton in Pembrokeshire. These were the 'great barons' of Welsh rural society, with their extensive manors, huge expanses of mountain and moorland, and rich mineral deposits. All were either close to, or above, 10,000 acres in size, and the biggest, that of the Cawdors, perhaps six times that amount, the bulk of it in Carmarthenshire. Typically these estates stretched across several Welsh counties and took in whole parishes, religious livings, and town properties. They were encumbered with commitments from the past and debts for the future, and only the fortunate few, like the Cawdors, still found it easy to get cheap money. The Cawdors had large holdings in Scotland, and the Dynevors a small retreat across the Welsh border. Neither of the elderly earls played a significant role in the Rebecca riots, nor for that matter did the duke of Newcastle, who had acquired Hafod, another large Cardiganshire estate.

Beneath this first league there was almost a score of estates of 5,000–10,000 acres and twice that number in the 3,000–5,000 range. All were defined as being in the possession of 'great landowners', and their very names dominated life and government in the three counties. In Carmarthenshire the traveller moving southwards through the county passed Rhydodyn (Sir James Williams) with its new expensive façade, three-storey Glansefin (Edward P. Lloyd), mock-Tudor Glangwili (John Lloyd Price), and Maesgwynne, square and solid like its owner, Walter R. H. Powell. Crossing the border at this point, he was soon confronting the improvements at Slebech Park (Baron de Rutzen), and, moving north across Pembrokeshire, eventually reached Ffynone, redesigned for the Colbys by John Nash. In this county there were a considerable number of estates with about 3,000 acres, the size of a small parish, often bound together by close family ties, and giving the Colbys, the Bowens, the Allens, the Rochs, and the Leachs an exceptional influence on Pembrokeshire affairs. There were fewer great estates to the north of the Teifi, though Hafod and Alltyrodyn, at the opposite ends of Cardiganshire, were two exceptions.

Some of these large estates, like Pentre (David Saunders Davies), were divided geographically across counties, though the compact nature of the Mynachty estate, near Ciliau Aeron, and of Tregib, near Llandeilo, homes of two of the oldest Welsh families, represented the contemporary ideal of consolidation. Only a handful of owners in south-west Wales, such as the Puxleys of County Cork and Carmarthenshire, had large estates elsewhere in the Principality, or in England and Ireland.

Another group of estates between 1,000 and 3,000 acres was described as the property of the squires or gentry. There were some 150 of these in the three counties, dotted along the main river valleys and the coast. We shall come upon these gentry seats frequently in our study. At the top of the range were Mabws, near Aberystwyth, Llysnewydd close by Llangeler, Sealyham, next to Haverfordwest, and Glanryhydw, in the parish of Llandyfaelog. Some of these gentry properties, like Sealyham, were owned by branches of the great landowning dynasties, while others belonged to smaller families claiming direct descent from the old Welsh princes. The gentry had been badly hit by family crises in the eighteenth century, and the connections which they then established with English landowners, and the attempt to revive their fortunes with the assistance of trained agents, inspired popular and poetic discontent. In the early nineteenth century they were obliged to make up the shortfall in their landed incomes from other sources, or to sell off small parts of their inheritance. A few estates passed completely out of the hands of their owners, including Middleton Hall in 1824 to Edward Adams, a Jamaican merchant, and Taliaris in 1833 to Robert Peel, cotton magnate and cousin of the Prime Minister. There were always plenty of industrialists, bankers, and lawyers waiting to put on the mantle of esquire.

Almost a third of the land of the south-west was owned by yeomen, whose holdings ranged between 100 and 1,000 acres. Contemporaries insisted that the position of the greater yeomen had declined significantly over the previous generations at the expense of consolidating landowners. This was certainly the view of three survivors, Timothy Powell and Rice P. Beynon of St Clears, and Lewis Evans of Pantycendy in the parish of Abernant. In fact, two-thirds of the yeomen's estates in the list of 1873 averaged only 170 acres. These were owned by a varied group of people, from aspiring farmers like Howell

Greater and Lesser Men 49

Davies of Cynwyl Elfed to the urban *nouveaux riches* of Carmarthen or Cardigan. At the very bottom of the list were freeholders David Morgan, forcing a living out of his boggy Lledrod farm of 137 acres, and Thomas Davies of Abernant, with a slightly smaller holding, but able to manage comfortably with a house servant and two farm workers. Like all freeholders they owed suit and service to the lord of their manor, though by the mid-nineteenth century this debt had been reduced to only a token payment.[3]

Below these in size, but more than four times their number, were freeholders, with between 1 and 100 acres. Many of the people who owned close to the average of 32 acres were to be found in the parishes of upland Cardiganshire, where land had been bought from the Crown or won from the extensive wastes and commons, and there were also large numbers of them across the northern perimeters of Pembrokeshire and Carmarthenshire. Tre-lech a'r Betws, the most studied of parishes, had 44 native and 40 extraneous proprietors, and two-thirds of their holdings were below 100 acres. The differences between them were important; freeholders with more than 50 acres had status and a little hope, those with 30 acres had a farm, and those with less were obliged to rent extra acres, or find other occupations. If Michael Bowen of Tre-lech were singled out as a representative of the larger freehold families, John Harries, the Talog miller, stood for the smallest, and both men became Rebeccaites.[4] Finally, the largest group of owners were the cottagers with less than one acre. In the 1870s there were 7,938 of these, rather more than the combined total of all other landowners, and the proportion would not have been very different thirty years earlier.

One can understand, from all this, why contemporaries were impressed by the differences between south-west Wales and other regions. This was not, like East Anglia, a tripartite society of aristocrats, large farmers, and poor landless labourers. This was a country of 'petty princes', the greater and lesser gentry families, and beneath them were ranked the owners of estates of between 300 and 1,500 acres, more numerous here than just about anywhere else in Britain. These differences had, it was claimed during the Rebecca years, adverse effects on social relationships. Firstly, whereas in the south and east of England much of the existing conflict was contained between large farmers and labourers, here in Wales tenant farmers and their employees were more ready to act together against the landowners. More hostility was directed and deflected towards the gentry. Secondly, it was widely believed that the medium-sized proprietors of the three counties were more 'penny-pinching', and thus socially divisive, than the aristocratic landowners. Rural and urban proprietors who rented out a few hundred acres have always had a bad press, but there may be several reasons for this. Sadly the disappearance of so many of their family papers makes it difficult for historians to pass judgement on them.

Perhaps, however, the most intriguing social phenomenon in this corner of Britain was the survival of small owner-occupiers. They were more numerous here than in the rest of Wales and England. This again was given as a reason why the population of the region showed such independence in the years 1839–44. It was true, as we shall see, that freeholders played an important part in organizing and directing public opinion. Such people had a status and freedom denied to the others. Yet one can easily exaggerate the point. No more than one in nine of Dyfed's farmers were freeholders or copyholders, and they shared a difficult economic experience with their neighbours, the tenant farmers. Both were often poor, the first group undermined by the constant division of the family inheritance, and hit by the depressions and banking crises of the early nineteenth century. Many people owned land in Rebecca's country, but it was generally property that provided an inadequate economic return, and added together the very smallest holdings constituted no more than a great estate.

This unique pattern of land ownership had emerged over a long period of time, but it was not static. Land was continually coming on to the market, hastened by the carelessness of Crown officials, the accumulation of debts, landowners' improvements, the process of consolidation, and the constant sale of smaller estates and farms to meet family and financial requirements. Although the sales of estates in the 1830s and 1840s never compared with those of the Great Depression, they did catch the attention of the local and national press. Amongst the large estates which came on to the market at this time were Hafod (more than once), Abermaide in the same county, and Taliaris, Rhosmaen, Manorowen, and Ystrad in Carmarthenshire, and there were hundreds of small properties bought by successful businessmen and land-hungry miners, seamen, and craftsmen. The purchase of Hafod by the duke of Newcastle added another great landowner to the élite of the south-west, and his £20,000 worth of home improvements and his support for the development of Aberystwyth were well received in the county. The subsequent resale of the 13,500 acres, and some of the other landownership changes during the difficult years of the early 1840s, were greeted with a real sense of disappointment.[5]

So long as the new owners were known, or indifferent outsiders, the people of the region were reasonably content. Despite claims that the Rebaccaites were levellers, challenging property rights, the ownership of land was not a major issue. In their way Welsh merchants and bankers and English industrialists who took over established estates were neither better nor worse than some of the old gentry families, but when the buying up of land brought new styles of management and fundamental changes in its use, occupation, and cost, there were signs of friction. There were complaints, for example, when Adams of Middleton Hall, the Peels of Taliaris, and one or two Carmarthen solicitors sought to maximize their income from their holdings, as there had been when de Rutzen, a Russian nobleman, took over Slebech from his Jamaican father-in-law.[6] Perhaps the most detested individuals were those who ignored local circumstances and customs. The outstanding example, recalled at the time of Rebecca, was the attempt by Augustus Brackenbury, a young Lincolnshire gentleman, to build a small mansion and farm an allotment which he had bought on Mynydd Bach north of Lampeter. For a decade crowds of people, disguised in women's dresses and with blackened faces, successfully drove him and his English employees off land which had long supplied them with cheap fuel. Their attitude was expressed in one of the many anonymous notes found at this time; 'There shall not be any Farms or houses [of the rich] built on Mynydd Bach, but what they shall be pulled down, without any Poor man shall come, then he shall build a House and make a Field, and we will help him.'[7]

This note referred to other methods of increasing one's holdings than by purchase and marriage. Chief amongst these in the region was encroachment on commons, waste, and mountain land. This was done by large acquisitive landowners, by farmers intent on pushing out the boundaries of their property, and by others seeking a new life. Although the open fields had long been enclosed and divided, usually by private agreements, there was still a residue of common land in many of the towns and lowland parishes of the region. Parishioners perambulated these commons each year, and jealously guarded their use. Vestries allowed some encroachment, frequently to relatives of commoners, but the creation of new holdings was not encouraged. The sending of anonymous letters and other forms of protest greeted the gradual disappearance of open land and greens in Newport, Tenby, Kidwelly, Tre-lech, and Llandybie during the 1830s.

Those seeking easy access to land looked to the large amount of upland commons and waste, over which there were contested or indeterminate rights. Lords of the manor, certain townships, and groups of commoners and freeholders claimed special rights on this land, especially the pasture of sheep on long, unfenced, walks. For a time the laxity of the Crown and other great landowners permitted the growth of small proprietorship in such places. At the beginning of the nineteenth century, under the pressure of a rising population and with the connivance of anxious ratepayers, virgin holdings were carved out of these hills. Amongst the areas affected were the north-eastern parishes of Cardiganshire, the mountains overlooking the upper Teifi, from Llangadog to Bettws along the edges of the Black Mountain, and on the hills either side of Kidwelly.[8]

Much of this encroachment, which amounted to thousands of acres, was carried out without permission from the various manorial courts, but it did conform to familiar patterns. Sometimes, as in the case of widow Mary Rees of Llandybie, a community helped to establish a new home for one of its members, and families adjoining uncultivated mountain land did likewise for their relatives. Yet, increasingly, new freeholds were being created 'secretly' by enterprising individuals, some unknown to their neighbours. 'Within the last 10 years great numbers of cottages have been erected upon the mountain wastes and sheepwalks . . .', ran a common complaint by Cardiganshire landowners at the turn of the century, 'some of whom are parishioners having settlements there and others are strangers, and they usually enclose about 6 acres of land with each cottage, under an idea that cottages erected on the waste with each a portion of land annexed to them cannot be pulled down.'[9]

The 'strangers' were the squatters of Welsh myth and history, 'the scum of this and other counties'. They settled on land under the old custom of *ty unnos*, whereby a person was entitled to the freehold of whatever shelter he or she could build in a night and of the land within a stone's throw. Such encampments were not universally popular, for they cut across the rights of local farmers, interfered with the traditional sheep walks, and there were fears that the poorest squatters might become a burden on the rates. Their homesteads became a source of 'everlasting quarrels', and of innumerable court cases. In the latter there was an assumption that if squatters were allowed to remain on the land for more than one generation they had established a legal right to it; after that they could, as one prosecutor put it, 'snap their fingers at everyone and become a Cardiganshire squire'. Before that happened people often resorted to direct action. John Jones, who was to turn informer against the Pontarddulais Rebeccaites, began to erect a *ty unnos* on the mountain only for it to be pulled down by two families who had rights of pasturage there. The same thing happened to Mrs Lewis, a widow on Clydau common, in 1838, though in this case the violence of the commoners took place twenty years after the cottage had been erected.[10]

Significantly, the man behind this destruction was Pryse Pryse, MP and lord of the manor. Since the mid-eighteenth century those landowners most aware of the latest commercial and legal developments had begun a prolonged attempt to secure and establish all their extensive rights of property. The Crown, which owned large tracts of Cardiganshire, and lesser amounts of land in the other two counties, passed slowly to the front of this campaign. By the 1830s its officers were enforcing payment of Crown rents and dues, and its courts repeatedly condemned encroachments by freeholders and squatters. Other lords of the manor took similar action, especially where the new settlements were above mineral resources. The agents of the Cawdors, Lisburnes, and Powells used one set of favoured farmers against another, and bailiffs, beadles, homagers, constables, and keepers were drawn into the legal fight. They pulled down cottages and fences, prosecuted people for trespass, and impounded their animals. Above all, the great men of Rebecca's country pressed on with private and parliamentary enclosure to enforce their claims. In Wales enclosure was always more of an exercise in power and finance than an experiment in improved agriculture, a fact which did not escape the notice of radical critics.

At the beginning of the nineteenth century much of upland districts of northern Dyfed comprised large stretches of open commons and waste. Lords of the manor, who wanted to benefit from the wartime demand for food, and from the lead, coal, and timber resources on this land, soon discovered that parliamentary action by enclosure commissioners made more sense than the private negotiations of the past. A score of Acts were passed in the years 1807–16, under which large tracts of upland Cardiganshire and smaller areas of the other two counties were divided up. In Carmarthenshire, where landowners were most active, parliamentary enclosure in the early nineteenth century was carried out in three main areas: a wide circle of hilly land around Llandeilo, a smaller area about St Clears, Llangynog, and Llanstephan, and a long strip of land from Llanfihangel Rhos-y-corn parish in the north to Llanelli in the south. In this county many of the known Rebeccaites lived in parishes where enclosure was a phenomenon of the recent past.

Landlords welcomed the legal security, higher rentals, and industrial profits that followed enclosure, a few lawyers and agents made fortunes, and even some of the smaller freeholders and oldest squatters had reason to feel pleased. At Llansantffraid in Cardiganshire the enclosure of common in the 1830s was followed by a little boom in house-building, and elsewhere there were short-lived increases in job opportunities. The census of 1811–41 indicated that rises in population in the parishes of Llansadwrn, Llanegwad, and Llanfyrnach were stimulated by enclosures and the cultivation of waste lands.

Perhaps, it has been suggested, this was one reason for the lack of extensive violent protest against enclosure.

People in the three counties did, however, suffer loss of amenities, especially the valued grazing and turbary rights, and they did complain about enclosure more than we realize. Hill farmers in the affected parishes had to reduce their flocks of roaming sheep, and labourers found it harder to keep a cow. In Cardiganshire parliamentary enclosure undermined a variety of fishing practices and gleaning customs, not least the cutting of turf, peat, clay, and wood. In the 1820s, when people finally appreciated how closely these rights had been circumscribed, there were ugly incidents, with fences pulled down, property set on fire, and animals maimed. Amongst the loudest protesters were female fuel-sellers and rush-collecting hatters. The latter declared their determination to ignore all restrictive legislation, but Pryse Pryse, who had bought part of the great Gorsfochno bog in an enclosure award, proved a worthy opponent and the most militant hatters were imprisoned in 1843 for six months.[11]

A number of squatters were legally and physically removed from their holdings as a result of enclosure, or they left their homes rather than pay the purchase price and the increased costs of continued occupation. David Gower, a carpenter of Pembrey, who built a cottage on Craig Chapel common in 1810, discovered almost twenty years later that enclosure commissioners disputed his property rights and sold his house and market gardens for £50. Thirteen years later he was 84 years of age and, so it was claimed, totally dependent on others for help, bare-footed, hungry, and unable to persuade the Poor Law Guardians to increase his relief of 2s. 6d. per week.[12] Others, like Edward Evans of Gwnnws parish in Cardiganshire, managed to survive the trauma of an enclosure award only to find years later that bright new agents and professional policemen were determined to enact the squatter's nightmare.[13]

Such was the unpopularity of certain enclosures, especially in the later years of the war and in the post-war depression, that mobs attacked enclosure officials and property. Fences, gates, and houses were pulled down or set on fire by mobs at St Clears in 1809, Marloes in 1816, and at Maenclochog in 1820, and magistrates had to billet troops in the region. When people wrote of the ignominy of military occupation during and after the Rebecca riots they forgot that in a similar depression twenty-five years before soldiers had marched into Aberystwyth, Lampeter, Carmarthen, St Clears, and other country towns.[14] Even the comparatively small number of parliamentary enclosures that were carried out in the following decades were not without their share of trouble. There were criticisms of the urban authorities in Haverfordwest and Cardigan who followed Carmarthen's baneful eighteenth-century example and enclosed the remaining 'poorfields', common land and public areas, and of landowners about Llanelli, Kidwelly, and Pembrey who were hell-bent on securing possession of possible mineral rights. During the Rebecca years the coastal hills of south-east Carmarthenshire were the scene of a long and expensive legal battle over the Kidwelly Enclosure Act, and some violence. According to one speaker at a meeting on Pembrey mountain in September 1843 it was too late to reverse this calamitous measure, but they wanted no more enclosures.[15] In general, however, as the pace of enclosure eased, so did opposition to it. By the mid-century the ownership of property was less of an issue than its occupancy and cost.

The owners of the large estates did not work the land themselves. At most, they retained a home farm, which supplied the big house with the main necessities of life. A few landowners took a keen interest in the productivity of the home farm, but outside its boundaries the degree of involvement varied. Most consulted family lawyers and surveyors, like John Harvey of Haverfordwest, over important legal, financial, and management decisions, while everyday estate matters were left to their own agents and stewards. The latter included R. B. Williams, Lord Cawdor's astute and powerful nominee, William Pitt Currie, who had been associated with Slebech Park since 1826, and Thomas Cooke, a more recent appointee at Middleton Hall. Like a number of these men, Cooke was irritated by the prolonged absences of his master, Edward Adams. As soon as the Rebecca riots were over, Cooke was keen to move away from a country where his talents 'were wasted'. Not all these middle-men were as well qualified, but those who had some training in agriculture, law, and manage-

Greater and Lesser Men

ment were able to demand good salaries and, sometimes, substantial holdings.

From their beginnings, in the previous century, the estate agents were subject to considerable hostility. They were identified with the more unwelcome aspects of the new commercialism in agriculture and, especially if they were imported from England and Scotland, they were accused of destroying the close relationship between landlord and tenant. 'The master looked cheerfully at them,' said one speaker in September 1843, 'but the steward looked at them like a lion.'[16] In fact, as the private estate papers show, agents like Cooke and Williams were sometimes more conciliatory than their employers, but they were left to perform the unpleasant tasks, like issuing stricter leases, seeking higher rents, and initiating legal proceedings. On the largest estates they also supervised lesser employees, the under-agents, the bailiffs, the foresters, and the keepers. These, too, tended to be outsiders, appointed over the heads of Welshmen. William Flutter, the Brownslade steward and farm bailiff, George McLaren, the Nanteos gamekeeper, and Philip Mackie, the Cilgwyn woodward, only heightened the language barrier that ran across so many estates.

One of the main tasks of these landlords' representatives was to deal with the tenants, of whom there were over a hundred on every great estate. Altogether, as we have seen, there were in south-west Wales perhaps eight tenant farmers for every freeholder. The amount of land occupied by these people varied considerably, both from area to area, and within one parish, but the balance tended to favour smaller holdings. The following list, of land held by tenant farmers in Abernant in 1851, is typical:[17]

Under 25 acres	21
25–50 acres	10
50–100 acres	11
100–150 acres	7
150–200 acres	5
Over 200 acres	2

In parts of the region there were more tenant farmers than this in the 250–500-acre range, and they frequently owned a little land as well. They were thus one economic step ahead of those occupying 180 acres, whom the *Times* correspondent wrongly identified as the most common form of tenant in Rebecca's country. Like John Rees of Pantsod near Llanarth, who had 200 acres, most of these people farmed all the land themselves with the help of family and servants, but there were a few who would not, or could not, do so. James Evans, who paid twice the rent of Rees for the same acreage on squire Colby's estate, farmed just over half of it and sublet the remainder to six people.[18] No one was certain of the boundary between a large and a small tenant farmer, but a rental of £50 a year was often taken as the dividing point. It was widely assumed that 40 acres was the minimum let required to secure a decent living from the land, but there were many people who struggled on with less than that, relying heavily on the larger farms for support. Those who rented fewer than 25 acres needed other jobs; at Abernant these small farmers were also labourers and workers in stone, iron, and wood.

The method of holding land had changed a good deal in recent decades. For centuries tenants had taken land for a number of lives, usually three, thus ensuring stability of tenure and rent, and giving them the impression that the land somehow belonged to them. The farmer with a lease always had a higher status than one without. Although this tenure was later criticized as being against the landlords' interest, it had certain benefits, not least the security of having a good supply of known and cheap tenants. Farmers had the advantage, or disadvantage, of rents set for long periods of time, but they were responsible for repairs under the agreements, and for certain food and service dues. They also had to pay the local taxes, as well as the tithes and land tax, and the fines and other dues payable as each generation took up the tenancies. These were heavy financial commitments, but the wartime boom in agriculture encouraged tenants to renew long leases at high rents, something which they later regretted. About the same time, too, agents on the largest estates were beginning to draft leases which restricted tenant freedom, severely limiting access to wood, game, fish, and minerals, and penalizing bad farming methods. When Rees Goring Thomas and some of his landowning friends offered new leases in the 1830s and 1840s, it was reported that the majority of their tenants were unwilling to enter long-term contracts. If this were true, the recent fluctuations in agricultural prices must have influenced their decision.[19]

The movement away from leases for lives was uneven across the three counties. By the mid-nineteenth century the situation was confusing; the Lloyds of Glansefin, the Colbys of Ffynone, and other traditionalists preferred the old contracts, at least for the larger holdings, but a growing number of landowners, especially in Carmarthenshire, offered leases for 21 years, 14 years, and for one year only. On the Harfords' and Saunders Davies's estates in Cardiganshire, and on those of the Cawdors and Dynevors in Carmarthenshire, leases of all kinds were becoming rarer, most occupiers having been converted to 'tenants at will'. Such tenants held the land for one year only, from Michaelmas to Michaelmas, with six months' notice on either side, and an increasing proportion of these men and women had to pay the market or rack-rent for their property. For landlords this change of tenure frequently brought with it the burden of repairing and draining the land, but there were two distinct advantages: rents could be altered more easily, and more pressure could be placed upon difficult tenants. For the de Rutzens of Slebech this meant the removal of farmers who did not measure up to their agricultural standards, and for the Cawdors a chance to ensure a loyal body of voters.[20]

These changes in tenure did not produce, so far as we know, a great outcry of protest, nor a rapid turnover of tenants. A selective examination of the estate papers of the great landowners reveals that during the first half of the nineteenth century over 50 per cent of tenancies remained within the same family. For example, the rentals of the more than fifty tenements of the Cawdors in Llanarthne parish during the 1840s show that occupiers had not changed greatly since the war years, and where people had died or moved away, relatives and close neighbours had replaced them.[21] On the Tyllwyd estate of Charles A. Pritchard land passed in 'very strict successsion' from farmer to widow and eldest son, and efforts were made to honour the wishes of tenants dying without heirs.[22] In close-knit rural communities, the character of the incoming tenant was a matter of the greatest importance. Contemporaries stressed the attachment which all Welsh tenants had to the farm of their birth, and their preference for a holding on the great estates where there was more security and perhaps a better chance of rent concessions and abatements. On the smaller estates it seems that the turnover of tenants was rather greater, and this was especially true of holdings of about 30 acres and under. Tenants of these properties had troubles, whether they had leases or not, and they were the people most bothered by writs and possession orders.

Despite the outward appearance of stability on many estates, there were repeated claims that the occupation of land was becoming more insecure and expensive. The shortening and removal of leases allowed greater opportunities for revaluations, which did occur in the 1830s, and those who could not afford the higher rents knew that there were others waiting to take the land. The coming of Michaelmas each year was a time of anxiety and, as we shall see in the Rebecca movement, the likely occasion of popular protests. Land hunger was endemic in the society, and the rise in population meant that there were more people in the mid-nineteenth century seeking tenancies. Spokesmen at farmers' and labourers' meetings in 1843 cited examples of 10, 20, and even 30 chasing the same piece of land, giving the less scrupulous landowner the pretext to increase the rent. Very little land, no matter how poor in quality, remained untenanted in the three counties; those agents who reduced the amount of properties for rent or who offered several holdings to favourite farmers and outsiders soon encountered the people's wrath. It was against old Celtic law, community morality, and biblical interdict.

The exceptional fluctuations in agricultural prices in the half-century before the Rebecca riots brought mixed blessings to tenant farmers. Some of the Llanarthne tenants mentioned above, and farmers on the Gogerddan, Clynfiew, and Coedmore estates, were enjoying the late benefits in 1843 of very long leases. This can be contrasted with the experience of the most unfortunate occupiers who, having signed an agreement for 21 years in the first decade of the century, then struggled to pay the rent over the long post-war depression only to face a substantial revaluation upwards when the lease expired. On the Cawdor estates a considerable proportion of the leases came to an end during the mid- and late 1830s; one or two requests for new ones were refused, and many of the smaller tenements were re-let at several times the old valuation. In 1843, for example, Martha Dawkins of Bosherston parish, Pembrokeshire, one of Cawdor's tenants, was paying £42 for the holding which, until 1835, had been leased at £14. 10s. per year.[23] This was an understandable adjustment, but it came at a difficult time.

Tenants at will quickly found that flexibility had its problems; rents moved upwards in the good years, but never fell like prices. One persistent complaint against Welsh landowners, heard especially in 1816, 1822, and 1843, was that their refusal to make permanent reductions in rents had produced a generation of poor farmers. John Hughes and Thomas Cooke, in their surveys of the Nanteos and Middleton Hall estates in the early 1840s, stated that the demands on the tenant farmers were too great, and, in spite of the land hunger, Cooke found it hard to let property at the price which his master wanted.[24] The precise increase in rents is not easy to chronicle, nor was it uniform. At a conservative estimate, there was a general rise in the rents of three counties of at least 100 per cent between 1793 and 1843, and for some of this period the pace of the increase was exceptional by British standards. During the war years the rentals of the Crosswood, Nanteos, and Golden Grove estates doubled and trebled. The subsequent collapse of prices brought temporary rent concessions, but there was a high level of rent arrears, distraints, and bankruptcies, and in later depressions, too, many farmers struggled to meet the twice-yearly settlement.[25] The Cawdors, and James Child with his commercial interests, could sustain annual rent arrears of several hundred pounds, but lesser landowners, with substantial debts themselves, found it harder to be so generous.[26] The law gave them a preferential claim over those of other creditors on the property of tenant farmers in difficulties, and the removal of those worth less than £20 p.a. could be done at Petty Sessions. The court records suggest that they made good use of these advantages.

Compared with most areas of Britain the holding of land in south-west Wales seemed inexpensive. Estimates put the average yearly rental in Pembrokeshire at 75p–£1 an acre, in Carmarthenshire at 60p–75p, and in Cardiganshire at 50p.

Some observers claimed that this low price of land helped to explain the backwardness of its peope and its agriculture. The argument ran along these lines: people with small amounts of capital could afford such rents, and tradesmen, artisans, and servants were tempted to become either full- or part-time farmers. Sons of farmers similarly used the money received at the time of the weddings to take on fairly large tenancies. All too soon these people discovered that they had overstretched themselves; houses were allowed to fall into ruin, the land went uncropped, and only loans and sales kept the wolf from the door. When at last the rents and debts proved too much, and the tenancies were terminated, David Griffiths of Llanfyrnach, Martha Evans of Nevern, and other farmers of the 1830s and 1840s refused to quit their premises, or, on departing, turned their spite on the few improvements which they had made.

There is some truth in this popular picture of low rents and inadequate tenants, but it is not the whole story. Over great tracts of poor and rocky landscape the rents were indeed low, but in the valleys and especially close to the main towns the situation was comparable to that in England. In the vicinity of Carmarthen, Newcaste Emlyn, and Llandovery, for instance, rents were nearer £3 an acre, and more, and to these were added tithes and other rates and taxes. Sir James Graham and the *Times* reporter were adamant that 'high rents and oppressive local rates' had much to do with the Rebecca troubles; landowners were accused of keeping them artificially high in hard times, and the 'oppressive burden' was passed on, through poor employment and low wages, to the labourers. Farmers' representatives in 1843 maintained that rents were about 20–5 per cent above an acceptable level.

In the way of things, the small farmer who could least afford it paid proportionately more than the man with 100 or 200 acres. One reason for this was the system of subletting which, though not common across the three counties, was popular in certain districts, despite leasing conditions. In southern Cardiganshire, where the quality of land was mixed, tenants frequently cropped the best acres, and let out the rest to cover the rent. In some of the best farming country of the Teifi valley and southern Pembrokeshire there were examples of under-tenants paying twice the going rate for land. The *Times* correspondent, who made rather too much of the 'evils of middle-men', described how families struggled to survive in these circumstances, with the wife farming and the husband labouring. They became, to use his memorable phrase, 'a morose, vindictive and dangerous population'.[27]

Table 3, abstracted from the census of 1851, tells us a good deal about the holding and working of land. This was not, like East Anglia, a region of high farming, with large farms and a great force of proletarian agricultural labourers. Only about Carmarthen, Laugharne, and St Clears, and in districts of southern Pembrokeshire, were large farms a common sight. John Evans of Clogyfran, St Clears, had 331 acres, and in 1851, when he had a wife and six children aged from 1 month to 18 years, he employed one house servant, a nursemaid, three dairymaids, four farm servants, and, depending on the season, one or two labourers. From the books of such a farm it is possible to reconstruct the seasonal pattern of work; spring lambing, the first sales, sowing and weeding, the long summer harvesting and other field work, late autumn stock sales and root harvesting, and, at the end of the year, threshing, ploughing, hedging, and wintering the stock. From March until late October the large farm was exceptionally busy, and casual labour was required. Each group of workers had their own special tasks; the *gwas mawr* and the other male servants looked after the horses, the female servants fed the poultry, milked the cows, and helped to make the dairy products, while the day labourers carried out general duties like hedging and ditching. Occasionally additional help was sought from the estate workers, and from independent craftsmen like the molecatcher, the mason, blacksmith, and carpenter, and, where large sheep walks were attached to the farm, from the shepherds who tended flocks for several masters.

In the districts of large farms there were, according to the census of 1851, three times as many male outdoor labourers over 20 years of age as farm servants. In the parish of Llanddarog near Carmarthen there were on average two male labourers to each farm, with more being the rule where the farm was above 100 acres, where the family was very young, and where the head of the household had died. Female

TABLE 3. *People occupied on the land in 1851*

	Carmarthenshire		Pembrokeshire		Cardiganshire	
	Under 20	Over 20	Under 20	Over 20	Under 20	Over 20
Males						
Proprietor	1	89	—	78	—	69
Farmer	8	4,032	7	2,449	7	4,539
Grazier	—	2	—	8	—	5
Male relation of farmer or grazier	767	1,456	383	802	857	1,672
Farm bailiff	4	38	—	39	—	48
Agricultural labourer (outdoor)	211	2,819	247	2,956	122	2,054
Shepherd	4	12	10	6	67	52
Farm servant (indoor)	1,821	1,340	1,381	1,020	2,310	1,399
Other connected with agriculture	—	18	3	11	1	10
Other connected with woods	1	8	—	10	1	8
Other connected with gardens	7	116	16	116	11	99
Females						
Proprietor	—	68	1	53	—	63
Farmer	2	638	2	417	3	732
Wife of farmer or grazier	4	3,404	—	1,952	3	3,666
Other female relation of farmer or grazier	865	1,256	532	955	1,002	1,613
Agricultural labourer (outdoor)	41	176	37	201	7	117
Farm servant (indoor)	1,683	1,753	1,132	1,408	1,994	1,907
Other connected with agriculture	—	—	—	—	—	—
Other connected with gardens	—	6	—	10	—	15

labourers were more numerous than the census records indicate; they were often employed for two-thirds of the year, doing all the work on the farm except that with horses, and being in special demand for hoeing, barking, pulling turnips, and harvesting. They were preferred by certain employers because of their stamina and concentration, besides the other advantage of cheaper labour costs.[28] Although their male counterparts worked very long hours, estate agents and large employers complained of their inefficiency, and even imported 'foreigners' to perform skilled tasks. These Scots and Englishmen, like all non-parish workers, were resented by the village community.

Most of the region was, as we have seen, dominated by small farms, both freehold and tenanted. Those between 50 and 100 acres employed only one or two people, and those with fewer acres sometimes one. In the parish of Llanfihangel-ar-arth, above Llandysul on the Teifi, where there were more male servants than labourers, and even more female servants, as many as a third of farmers had no outside help. The nature of their work varied from district to district, but a typical small farmer on the lower land kept a few cows, a couple of pigs, and poultry, and raised fodder crops and a little barley. Each week in summer his wife took eggs and butter to market, and once or twice a year they sold much of their stock to pay the mortgage or rent, the loans, rates, and taxes. The value of pastoral farming was the lack of labour required, but even then hay was needed and harvesting this meant reciprocal help from friends and neighbours.[29] The farmers, labourers, and village craftsmen who followed Rebecca with pitchforks over their shoulders were accustomed to working together.

All the small farms relied as much as possible on family labour, especially when the male head of the household worked at other jobs to supplement their income. The women and children had to labour hard for little reward, as we can see from this interview with a lady living on poor land at the foot of Allt Cunedda, north of Kidwelly, in 1843:[30]

She told me her husband was born on the farm, and she had been on it 23 years, ever since she was married. She had two children, a son and daughter, grown up, who worked on the farm. The farm there was 24 acres, but her husband had a few acres of land elsewhere . . . The rent was so high, that though her husband was a good farmer, and anxious to lay by money, never spending a shilling, they could save nothing—they could just live . . . It grieved them that they could not give a little money for the children. 'We work as hard as labourers', said she, 'and live as hardly. You must not suppose that that bread I have handed you is what we usually live upon; we keep a little of that sort for strangers and chance callers . . . we cannot save [for old age], and it grieves us we cannot give a little to start the children'.

On many of these small farms there was insufficient work for all the children. It made good sense for the eldest son to rent, or labour on, another farm. The Morgans of Cwm Cile, near Rhydypandy on the Carmarthenshire–Glamorgan border, who will figure in our story, found a small property at Tymawr for their son Matthew. Two of his brothers and sisters continued to labour on their parents' holding, and the rest went into service. Girls in their early teens were wanted in both town and country houses as nurse-, house-, and dairymaids; the work was hard, the terms poor, but at least it was a roof over their heads. Meanwhile, the youngest children on the small farms helped their mothers with the cows and milking, chased sheep across the moors, gathered stones, and scared the crows. At harvesting the baby was often carried on his mother's back, drowsing over the land that he might one day inherit.

The children of farmers and labourers who left home usually obtained accommodation at their place of work. This suited their families, as it did the larger pastoral farmers who needed the constant help of just one or two hands and preferred to give board and food rather than regular cash wages. In contrast to the more advanced agricultural regions of Britain, there were as many indoor servants as outdoor labourers in Rebecca's country. Most of the servants were taken on when they reached the age of about 12 years, and the age gap between them and the 20-year-olds created a form of apprenticeship. Once installed, the girls slept in the farmhouses, and the boys, with the farmers' sons, in lofts, stables, and barns.[31] At meal times some servants had their meals with the farmers' families, but on the bigger farms it was customary to offer them food in the back kitchen. Partly because of their domestic arrangements, and their youth and nightly courting habits, these employees had a reputation for mischief and indecency, and their involvement in the Rebecca riots surprised no one.

Servants were hired in November of each year, at the time of the Allhallowtide fairs, when the streets of the country towns were packed with young men and women in holiday mood, seeking to bind themselves into a twelve-month contract. The most poignant sight was the 'round ruddy faces and bright eyes' of 11-, 12-, and 13-year-olds standing in line down Lammas Street, Carmarthen, and hoping for little more than a bed and food in return for a year's hard labour. When hired the servants received a very small part of their salary, the remainder being paid towards the end of the year. Although not of the written kind, the contracts entered into at the hiring fairs had the force of law, a fact increasingly publicized as alternative industrial work encouraged desertions and rebellion. All too often the promised settlement was not paid in full when the twelve months were over, but still the age-old pattern of yearly hire continued, until the men and women married. By their mid-twenties most men had ceased to be indoor servants, but, as Table 3 indicates, female servants stayed longer on the farms. There were, after all, more women than men in this society, and their choice of occupations was limited. Female servants were the dogsbodies of rural society, vulnerable and abused but never totally passive.

Some of the male servants, before they left their masters, had saved enough money to take a smallholding themselves, but most when they married joined the ranks of the agricultural labourers. Although there were still yearly contracts to be had, an increasing number of labourers worked to a month or week's notice. Cash payments were monthly, or at more frequent intervals, and in many areas these were supplemented by perks and favours. In some parishes labourers received very low wages indeed, but benefited from food at work, cheap rents in farm cottages, and assistance with fuel, corn, and potatoes.[32] As this was the land of small family farms, the demand for extra labour was not constant throughout the year, and this was recognized in the customs of the region. In mid-Cardiganshire, and several other districts, there was a kind of bond-system, whereby farmers, in return for supplying rent-free cottages and pasture for a cow, had first call on the occupants' services. There was also the very common practice of allowing people to plant and manure a hundredweight of potatoes in the farmers' fields, on condition that their families helped with the corn harvest.[33] These, and other customs, bound the labourers into the farming community, and offered ways of surviving when the economic hurricane broke. It was a precarious existence, but the most fortunate of them had employment for two-thirds of the year, a cow and a pig, a large garden, odd jobs in the late evening, and the sweet recurring dream of their own farm.

LORDS, MASTERS, AND PEASANTS

When asked in detail about the nature of their society, the inhabitants of the Welsh countryside usually replied that it was composed of three main elements: 'the great men' (the great landowners and squires), 'the masters' (the yeomen and big farmers), and 'the people' (the peasantry: small farmers, and cottagers with and without land). Every popular analysis has its simplifications. Amongst the top 200 families, for instance, there were great social differences between the ennobled landowners in their splendid mansions, and those with somewhat smaller rentals who deserved the suffix 'esquire'. In parts of southern Cardiganshire and down the Pembrokeshire–Carmarthenshire border the gentry were comparatively thin on the ground, but most parishes of the south-west were within five or six miles of a big house, and a few places, like Llangadog and Abergwili, were surrounded by them. Some of the mansions, including Dynevor Park, mock-Gothic Hafod, and Palladian Nanteos, were imposing homes, set off by splendid trees, gardens, and long drives, and housing a dozen and more servants. Building improvements, begun in the eighteenth century, continued into the nineteenth. During the second quarter a remarkable number of country homes were completely rebuilt, increasing the status of old and new families in the region, if never quite placing them in the style and comfort of their English counterparts.

The condition of the big houses was given as one of the reasons why the great men chose not to live in the Principality. Non-residence was less common than in Ireland, but it was both a feature and an issue of Welsh life. The Dynevors made their home at Barrington Park in the Cotswolds, Lord Kensington, the Pembrokeshire landowner, lived in London, John S. Harford, inheritor of Peterwell in 1821, never resided in

Cardiganshire, while the Reverends Erasmus Williams and Thomas Lewes preferred Marlborough and Oxford to Llandovery and Newcastle Emlyn. Others, too, had been brought up outside Wales, and never found anything to match the pleasures of Eton, Oxford, and the Carlton club in London. John Mirehouse of Brownslade, Angle, in Pembrokeshire, a common serjeant of the City of London, was criticized by James Mark Child, a fellow squire, for living in the capital 'like a prince'.[34] Foreign travel appealed to some; the occupant of Picton Castle left in 1833 for a seven-year stay in Italy, and Captain James R. Lewes Lloyd of Dolhaidd, near Newcastle Emlyn, a Napoleonic War veteran, and Walter Rice of Llwynybrain, another military man, liked the quality of life in France. Such landowners, from a comfortable distance, were inclined to ignore or ridicule the first accounts of trouble at home.

For these men and their wives a stay of more than six months on this side of the Welsh border proved mentally and physically exhausting. 'County business, association with odd people, (whose habits are as foreign to him as their language,) living at a public Inn in a country town, . . . this cannot but have been rather disagreeable,' mocked the *Welshman* in its appraisal of Colonel Rice Trevor, Carmarthenshire MP and son of Lord Dynevor.[35] In fact, some of the latest occupants of the big houses, like Edward Adams, the excitable eccentric of Middleton Hall, and William Chambers of Llanelli House, barely concealed their contempt for those who did not appreciate their sacrifice in living on their estates. They were aware that in the countryside where the Rebecca riots began, and in the semi-industrial areas where it ended, there were few resident gentry. Rees Goring Thomas, who moved upwards from Llannon, perhaps the most neglected spot, to Llysnewydd and temporarily to Iscoed, regretted the fact that 'many of the landlords do not live in the country'.[36] He was joined in this view by another Tory upstart, John Lloyd Davies of Allytyrodyn; both of them warned the government of the dangers of leaving large tracts of land without the guiding hand of a great family. A worried Home Secretary launched a survey of the domiciles and efficiency of magistrates in the three counties and was taken aback by the number of 'absentees'. Not all of them were, as assumed, enjoying the delights of south-east England;

some simply preferred to live off the main estate, perhaps in a town house.

In fact, most of the leaders of Rebecca's society were Welsh in spirit if not in name, and their presence, especially in Cardiganshire, amidst the native population was said to have largely contributed to the low levels of crime and protest. It was stated repeatedly, in all the major commissions of the nineteenth century, that country people respected great men who spent a good part of the year on their estates and who had long pedigrees, legitimate wealth, and respect for the land of their birth. Amongst the descendants of old Welsh gentry families were the Colbys of Ffynone, the Bowens of Llwyngwair in Pembrokeshire, the Gwynnes of Mynachty in Cardiganshire, and the Lloyds of Bronwydd in the same county and of Glansefin, in Carmarthenshire. They had fine collections of Welsh books, and gave a little help to Welsh poets, singers, and musicians. Although Anglicanism united these landowners, they were nevertheless regarded as being sympathetic to the aspirations of their Welsh Dissenting tenants.

The Pryse Pryses of Gogerddan, one of the largest dynasties, took a special pride in their Welsh ancestry. According to one of their admirers 'the Pryses live amongst them, spend all their money amongst them, farm amongst them, and bring up their children amongst them, . . . The Pryses, instead of despising the Welsh, make a point of being Welsh themselves, and take a pride in speaking the Welsh language and teaching their children to speak it . . .'[37] Few of the great men now spoke Welsh, except perhaps a half-remembered version to half-remembered tenants. Only Thomas C. Lloyd of Bronwydd, some lesser squires, and yeomen like Lewis Evans of Pantycendy were confident enough in the language to act as interpreters and intermediaries during the Rebecca troubles.[38]

By the 1840s the culture and language of the great landowners and the great churchmen were predominantly English. One of the regrets commonly heard at Whitehall was that the withdrawal of these gentlemen from Welsh cultural, religious, and social life had allowed the development of separate, and potentially dangerous, social structures. One speaker at a farmers' meeting at this time brushed aside the instinctive call for a return to gentry paternalism with the words: 'we never see the great men . . . let us rely on our own people'. The landowners' Church, like so much else, was that of England, and indeed, such was their appropriation of its wealth and purpose that here, more than anywhere else in Britain, it had become little more than a poor spiritual vehicle for the gentry and their dependants.

The wealthiest clergymen, closely related to the prominent families, occasionally joined the aristocratic train out of the Principality. From four to eight months every year the great families were away from their estates, moving to the spas and resorts during harvesting, part-wintering in the fashionable London society, and ending with perhaps a trip to Bath or Gloucester for the spring. The other indication of aristocratic status was the amount of time and money spent on leisure pursuits. Gambling, dinner-parties, the latest dance crazes, theatre-going, and racing filled the months away from home, and much of this was continued on their return to the country.

The most fashionable landowners were roundly criticized in the local press for failing to patronize the 'old sports', and were reminded of the standard set by John Jones of Ystrad MP. Such criticisms were only partly justified, for fox-hunting remained popular with those who stayed at home between mid-October and the end of March. Members of the Teifiside, Carmarthen, Pembrokeshire, Gogerddan, and Begelly hounds met twice a week, or even more frequently, and some of the gentry found time to support hare coursing, otter-hunting, and fishing. In addition there were regular shoots, with patridges and hares the main victims, and steeplechases at Aberystwyth, Haverfordwest, and other places. For those still in Wales the summer months were enlivened by endless visiting and the fun of yachting, cricket matches, race meetings, and the Assize, while the autumn brought agricultural shows, dinners, and the hunt week, and the winter was made bearable by balls, concerts, and the theatre. It was important, not least for the lesser gentry, that the great landowners should appear at a selection of these events, though, as in all good Victorian dramas, their presence was infrequent and the cause of much rejoicing.

Celebration and display were important in this social world;

birth, entering one's seniority, engagement, marriage, death, and even the going out and coming into one's estate were marked in a spectacular way. Thus on 26 September 1843, when the Rebecca riots were at their height, bells rang out and cannon boomed to announce the marriage on that day of the eldest son of the Lloyds of Glansefin to Georgiana Caroline, the youngest daughter of the late Colonel Gwynne of Glanbran. Nine carriages took the bridal party through crowds of well-wishers to Llandingat church.[39] Glansefin was said to have been held in special affection, for the Lloyds, who had been friends of the first Tudor king, had a reputation for keeping up the old customs. A month before the wedding, about a hundred harvesters had been thanked by the squire himself, and given supper and beer.[40] On other occasions the family hunted for otters with their tenants, supported local cultural festivals and schemes to benefit the Church and religious education, and handed out charity. These actions, and indeed all face-to-face contact between the rich and poor, was intended both to illustrate the unique position of the great men, and, to paraphrase the vicar of Moylgrove in Pembrokeshire, to reveal the superiority of the giver over the receiver.

The bridge between this rather exclusive social world and that of the yeomen was provided by the many squires of south-west Wales. These men had about 1,000–3,000 acres of land, and were resident on it for much of the year. They had a more lasting interest in the local community than the greatest landowners, were the captains of the yeomanry and militia, leading magistrates and chairmen of Boards of Guardians, and sponsors of village shows and festivals. When the Rebecca riots broke out, it was people like 'timid' Edward Lloyd Williams, who had returned from the Midlands to Gwernant on his father's death, Walter R. H. Powell, the master of the Maesgwynne foxhounds, John Hughes Rees of Cilymaenllwyd, Pembrey, and James Mark Child of Begelly House who chaired the meetings of reconciliation. Child, who was a man of industry as well as agriculture, shared some of Williams's liberal opinions. He told one assembly that, unlike some landowners, his family had lived in Wales for centuries, and he had 'mixed much with all classes of society', including the aristocracy. 'He had associated with farmers, and a more generous and patient class of men under their sufferings did not exist under heaven.'[41]

Child's friends included substantial yeomen, junior members of the Roch, Waters, and Carver families on either side of the Pembrokeshire–Carmarthenshire border, and Lewis Evans of Pantycendy, the largest proprietor in Abernant parish. Evans, a 'selfish overbearing bull headed Welshman' who was once accused of being in league with the Rebeccaites, owned several hundred acres and was widely regarded as the archetypal yeoman.[42] This 63-year-old man was both a magistrate and a captain of the Royal Carmarthenshire Militia, and his love of horses matched his formidable pride and independence. Together with Rice P. Beynon of St Clears, he was a Tory paternalist, deeply suspicious of political economy, factory towns, the Poor Law Amendment Act of 1834, and the Rural Police. He shared with other prominent yeomen the abiding belief that recent economic changes and an uncaring government had undermined the position of 'the masters' of rural society.

In fact, the evidence suggests that for much of the early nineteenth century the largest farmers had done rather well. In 1822 a landowner, tired of the tales of agricultural woe, described the higher standards of comfort, living, and culture that these people had come to enjoy during the previous quarter of a century.[43] The farmer with a couple of hundred acres could afford fine clothes for his wife, good food and drink for himself, and books and musical instruments for his children. They were persons of some wealth and importance, 'the masters' of popular literature, and, when they expanded their holdings, the enemies of Rebecca. In districts where the gentry house was missing they filled the social gap, especially if they belonged to well-respected families and attended the established Church. John Thomas, who had more than 200 acres at Crunwear in Pembrokeshire, was said to have been the 'ruler' or 'master' of that small parish, and so, to a lesser extent, were the Thomas family of Cwrt Derllys, Merthyr, and the Bowens of Plasparcau-uchaf in Tre-lech. Even in Llangadog, where the gentry families, Lloyds and Lewises, held sway, large farmers like James Jones of Cwrt-y-plas and William Williams, of Gwynfe hamlet had a special position and authority.[44] For this reason their response to popular movements was critical. *Yr Haul* and the English daily newspapers never forgave them for not denouncing the Rebecca movement in more strident tones, and Sir James Graham went further, implying that 'the yeomanry and the peasantry are banded together against the magistracy and the gentry'.[45]

Status required that the families of the leading farmers should have the right education, a good marriage, and a secure future. It was a costly business. Tutors were hired for the youngest children, and then they were off to learn classics, religion, and manners in the grammar schools, academies and colleges across south Wales. When they came of age, the children of John Thomas or James Jones were expected to marry well. The bride's family provided a dowry, sometimes in the form of both money and land. Even at this level a breach of promise was a serious matter. When the Reverend Francis Thomas, setting out on a promising religious career, turned his back on Caroline Williams of Penycoed, daughter of one of the oldest farming families in Pembrokeshire, it cost him £500, and there were many who paid a similar price.[46] The sons of large farmers, on marrying, received help to set up in business or on another farm, though one of them usually stayed at home and inherited the family property as a reward for looking after his parents.

Although the richest farmers were accused of aping the gentry, and of being tied economically and politically to their interest, these parish rulers were overwhelmingly Welsh and, by 1843, ever more Dissenting in their culture. Nor did independence end there, for in the growing number of farmers' clubs and shows, their sponsorship of friendly, religious, and Welsh societies, and their particular view of the Game, Fishing, and Militia Laws, there was a growing awareness of the farmers' own interests. They also had their own sphere of influence, as smaller farmers owed them favours, and so did others in the village community. 'When the master's horn blows, we run', was the motto of the countryside. Despite Child's favourable picture, the growing independence and status of the largest farmers did not always work to the advantage of their employees, for this class of masters now rarely shared a table with their servants. A few of them, in the manner of eighteenth-century Welsh gentlemen, subjected male and female employees to appalling levels of physical violence and sexual harassment. Rice P. Beynon of St Clears and Howell Davies of Cynwyl, who became two of Rebecca's main targets, were notorious for their drunkenness, bad temper, and moral lapses.

There was a clear distinction between Howell Davies and other freeholders and occupiers with a few hundred acres, and the rest of rural society. The latter were described variously as 'the people' or 'the peasantry'. The terms covered small farmers, labourers and servants, and the craftsmen and women who were a part of the rural economy. Contemporaries were keen to demonstrate the links between these people. Of the small farmer, Thomas Phillips wrote in 1849 that he 'is but little removed, either in his mode of life, his labourious occupation, his dwelling, or his habits, from the day-labourers by whom he is surrounded'.[47] Yet there were important divisions of status, both between the landed and non-landed poor, and within each group. Those with land commanded greater respect in social, political, and religious life, and were more likely to be chosen as jurymen, constables, deacons, and the like. The landless poor, unless they were related to important village families, were of lesser consequence, and their voice was seldom heard.

People tried to explain the popular movement with reference to these social divisions. The *Morning Chronicle* reporter put his slide rule between those with more and less than 30 acres, and declared that those in the second group were too weak and dispirited to become leading protesters.[48] Others put the mark higher, and made the contrary claim that those with over 50 acres and voting rights, or at least the freeholders amongst them, were always unwilling Rebeccaites. All such theories proved incorrect, for in this movement, as in some other fields of culture and action, the lines between these people were more blurred than at first appeared. Class feeling did exist, even within the Rebecca movement, but of greater importance was the fact that the peasantry shared the bonding experience of poverty and a close-knit social and religious life. This, Edward Laws of Pembroke believed, was the reason why Rebecca's activities were so unrestrained, widespread, and successful.[49]

Those contemporaries who called these people 'the peasantry' used the term to imply not simply a particular position in society, but also certain attitudes to land and to the community in which they lived. The first wish of the peasantry was to retain or secure a few acres; the possession of land was, to paraphrase the words of incredulous observers, placed above its quality and exploitation. Small farmers, butchers, innkeepers, blacksmiths with money, and ambitious servants and labourers were prepared to do almost anything to obtain small pieces of land. Land was 'stolen' from the hills, bought from dubious sellers, and taken on very unfavourable terms indeed. There was an acute interest in the imminent death of owners and occupiers, and in the wills and verbal promises which they had made. All the members of a freeholder's family cherished the hope that they would ultimately obtain a share of the property. David Williams of St Clears, along with many others in the mid-century, was taken into custody for snatching his father's will as it was about to be read in chapel, and tearing it to shreds before the astonished guests.[50] Battles over the succession to property raged continuously through the courts, houses were double-searched and bodies raised from the dead to ensure that no documents had been deliberately 'lost'. Boundary disputes were another speciality of the Welsh peasants, the costs of court cases far outweighing the value of the few disputed metres of hedgerow. No one in south-west Wales gave up land easily.

The peasantry had a special attachment to their place of birth, and it was no accident that many of them were called by the farm or cottage where their families had lived for generations. Even when the families of small farmers were large, great efforts were made to find a settlement nearby for the children. Property was bought, leased, and exchanged to accommodate their requirements. Marriage was difficult, if not impossible, without such provision. As we have seen, these people usually married partners from within the village community and the same occupational group. As in other parts of remote rural Britain, there were veiled criticisms from medical experts on the dangers of inbreeding in the more isolated village communities. Whatever the truth of this, on the occasion of weddings kin and neighbours ensured that the local couple had a good start. Thus at the bidding of William Lewis, a farmer of Myddfai, and his wife in December 1845, the considerable sum of £34 was collected in gifts, dues, and loans.[51] His spouse was expected to develop an 'animal pride' in her new home and her offspring. Unlike the detached gentlefolk, and the cruel Poor Law Commissioners, said Stephen Evans, farmer of Cilcarw, 'the poor know what it is to love their children'.[52] Such were the ties of kinship and community that migrants from the region took their place-names with them to the industrial towns, and often returned to their birthplace in search of a wife or to help with the harvest. In old age some returned for good, and were buried in the plot that they had long fancied. For all but paupers, a Welsh death, like marriage, was a celebration of history and community, a time when all the extended family came together.

The many critics of Rebecca's society were appalled by the mental world of the peasantry. 'To judge of the Welsh & their feelings & modes of thought and action by anything with which an Englishman is acquainted wo.d only lead to error,' wrote Edward Lloyd Hall from his superior cottage in Newcastle Emlyn, '. . . there being hardly any distinction betn. the farmers & the labourers, & their language isolating them from the rest of the kingdom, they have scarcely emerged from the state of comparative barbarism & poverty induced by the intestine troubles of this part of the country so graphically depicted in Rees's History of South Wales.'[53] The correspondent of the *Morning Herald*, and one or two other English papers, fell upon this image of a half-conquered and half-civilized people with all the relish of amateur anthropologists. So did education commissioners a few years later; when their assistants, David Lewis and William Morris, stepped out of the towns and into the countryside of northern Carmarthenshire they were met by dark faces and a monoglot silence. These Welsh peasants still had, it was presumed, the criminality, the morality and the irrational reactions of the old Celts. Cut off in their own mental world, they were deeply suspicious of strangers, and opposed to all change. Beneath a veneer of religious respectability they were said to have indulged their secret passion for drink, petty theft, superstitious practices, strange courtship rituals, and all manner of sensual and violent recreations.[54]

Information on the peasant's use, or misuse, of time is fascinating in itself, and also because it helps us to understand Rebecca organization and ritual. 'One of the principal characteristics of the people', said William O. Brigstocke, the landowner-magistrate of Blaenpant in the Teifi valley, 'is their having little idea of economising time; they will, without consideration, devote a whole day, which might be more profitably employed, to auctions, funerals, fairs, markets, and meetings of all descriptions, though they have no particular business which calls for their attendance; farmers and the labouring class are alike in this respect.'[55] In fact there was a difference; while farmers attended the markets, the labourer in full employment had only three or four holidays a year. Even so, the rural year was dotted with festivals, which were the occasion for drinking, dancing, merriment, the collection of alms, and other customs. The coming of spring and winter, and of Christmas and New Year's Day in the old calendar, were still observed as holidays in certain places. So were All-Hallows and All Souls' Eves, with their burning bonfires and ghostly white ladies. At Aberystwyth, Pembroke, and Carmarthen, drum and fife, flags, burning torches, fireworks, and gunfire destroyed all peace on Christmas Eve, and Easter and Whit Monday were marked by equally noisy perambulations of the town boundaries. Each parish had similar festivities, during which people dressed up in masks, women's clothing, and animal skins. Shrove Tuesday was universally celebrated by a bloody and riotous form of football kicking, a ritualized version of a pastime which had once been the popular Sunday entertainment.[56]

Some of the festivities, like the carrying of the Mari Lwyd on Twelfth Night and of the harvest mare, were a legitimized form of levying money and food, while the crowning of popular queens, mayors, and fools had an element of the mock trial. Animals played an important role in the people's entertainment, even if it were an unpleasant one like cock-fighting in the churchyard, bull-running through the towns, and the wild-riding of horses and shooting of guns at weddings. In 1844 both horses and a young rider were killed at a Llangeler wedding but, in spite of repeated attempts to stop it, the practice of chasing the bride continued, a reminder that the old customs contained the triple ingredients of licence, danger, and rebellion.[57] All the pleasures were, of course, under attack in the eighteenth and nineteenth centuries, especially from clergymen, town councils, and their new policemen, and there were a few pioneering landowners who tried to replace 'Welsh debauchery' with 'manly English sports'. By the second quarter of the nineteenth century they had achieved some success; football had been stopped at Llanwenog, Llanelli could be entered at the weekend without a bodyguard, and Sunday-school parades were replacing youthful excesses at Llandysul.

None of this brought much joy to those, like Archdeacon Beynon and Edward Lloyd Hall, who were trying to bring the Welsh peasantry into the nineteenth century. They claimed that the intellectual ambitions of the *gwerin* were as limited as their economic ones. The families of small farmers were singled out for criticism; locked in their hill farms, with only the evening darkness and a Welsh Bible for company, they experienced an even more narrow existence than those craftsmen, carriers, and fishermen whose occupations entailed social intercourse and travel. Educational opportunities were certainly sparse, and expensive, for the poorer inhabitants; there were about 500 schools in the three counties in the mid-century, many of them recently established and rather more than half of which were of the private enterprise kind.[58]

Many of the private schools were opened during the winter months only, when the part-time untrained teachers could be reasonably sure of good attendance and income. The quality of these schools varied enormously; some were held in barns and sheds, where the boys, who usually comprised almost two-thirds of the pupils, were given a basic reading and writing knowledge, and the farmers' girls a short course in home economics. 'I have no hesitation in saying that a child might pass through the generality of these schools without learning either the limits, capabilities, general history, or language of that empire in which he is born a citizen', wrote Ralph R. W. Lingen in October 1846, at the front of his section of the education report. 'The ideas of the children remain as hopelessly local as they might have done a thousand years ago.'[59] Of one typical school, a visitor commented: 'it is conducted without any system, most of the children being ranged on benches struggling by themselves with the adversities of the English spelling-book, or hammering at the Bible'.[60]

This particular institution, at Aberaeron, was one of the public schools in the region. These comprised a small number of endowed grammar schools, parish schools, workhouse schools, works schools, and a larger number of denominational schools, mainly supported by the Church of England and, to a lesser extent, by the Independents. Some of these, like Madam Bevan's Circulating Schools, founded for clean Welsh-speaking Anglicans, were subsidized or free. Church or National schools were sponsored by Ebenezer Morris at Llanelli, fellow clergymen Herbert Williams and John Pugh of Llanarthne and Llandeilo, and by a few enthusiastic landowners. In 1841, for example, Walter R. H. Powell, David Saunders Davies, and Rees Goring Thomas shared the plaudits for starting church schools close to their country homes. At first many of these schools were used indiscriminately by the unruly penny-paying children of the local population, but education soon became the focus of independence and conflict. As parish, inter-denominational, and church schools were pushed to the front line of the Anglican revival in the early 1840s, the more hostile Dissenters withdrew their children from these schools, or removed clergymen and their wives from the controlling committees.

It was estimated that up to 50 per cent of the children aged between 5 and 10 years attended the day schools, at least for a third or half of the year. Parents seemed to welcome the comparatively rare chance of free schooling and, in the case of the eldest male children, they appreciated the benefits which education could bring. In the private schools children were taught through a mixture of Welsh and English, but in Anglican schools English, like the performance of the Catechism, was often obligatory. According to some observers, admittedly fairly biased, there were many small farmers, craftsmen, and labourers in the mid-nineteenth century who wanted their children to learn English rather than Welsh. As David Davies, a farmer living near Llandeilo, put it: 'a little English is enough to carry a man through the world, but there is no doing without it.'[61] Much to the annoyance of government commissioners, the gentry and their religious friends had shown insufficient interest in this matter. In fact, for all the educational missionary work, and the accompanying tea-parties, clothing clubs, and annual examinations, there were about a hundred, mainly rural, parishes which had no schools at all for years on end.

It was easy to deduce from the published statistics that the Welsh peasantry could not have sustained a literary culture. The first Registrar-General's returns showed that over a third of the men and two-thirds of the women of the three counties signed the marriage forms with a mark. A private survey of 1,000 people in two Cardiganshire parishes in March 1846 by 'a creditable person' indicated that three-quarters of them knew little or nothing of writing and arithmetic, and 40 per cent could not read adequately.[62] Other investigations revealed that the homes of the small farmers and labourers had few books, apart from the Bible, two or three religious tracts, and perhaps a child's spelling-book and a copy of the latest monthly periodical. Yet in this region, the home of Griffith Jones of Llanddowror, reading skills had always been higher than statistics and outsiders suggested. It was estimated that *Seren Gomer*, *Y Diwygiwr*, and the score of short-lived journals published in south-west Wales between 1830 and 1850 were seen by as many as 10,000 people every month. 'They are generally read by the country people,' said Thomas Williams, magistrate's clerk of Lampeter, 'and form the staple means of information.'[63] The interest in religious affairs, local matters, and political issues, which these papers fed, was clearly shown at the public meetings of the Rebecca years. 'I saw nothing', wrote one of the educationists of 1846, 'which induced me to connect the outbreak of those disturbances with popular ignorance . . .'[64] On close inspection, it was difficult to believe that Rebecca's children were primitive unthinking rebels, easily swayed by unscrupulous leaders; the general and glowing conclusion, of all but the most hostile observers, was that these were the most literate, numerate, religious, and articulate peasants in western Europe.

The education of the young, old, and especially female villagers owed much to the religious community of which they were a part. It is difficult now to estimate the religiosity of these people, not least because of the subjectivity of contemporaries. Most were impressed by the sight of crowds of well-dressed people going to chapel on wet mornings, by their interest in religious argument and literature, and by their willingness to finance new chapel buildings and repairs. One would have expected the presidents of the Dissenting colleges at Carmarthen and Brecon to be effusive in their praise of the scriptural knowledge of the *gwerin*, but so was the head of St David's, the Anglican institution at Lampeter. Biblical allusions were a natural part of popular speech, and of threatening letters, and the evening protest meetings resounded with Old Testament warnings about God's vengeance and the overthrow of corrupt rulers. Rebecca prided herself on carrying out Jehovah's will, even on a Sunday.

'Rob not the poor, because he is poor: neither oppress the afflicted in the gate' (Proverbs 22: 22), was perhaps the most apt reference in all the speeches of the early 1840s, but it fails to do justice to the millennialist tone of the rest. The people's enemies were told to prepare themselves for 'famine, fire and the appearance of great beasts'. Even the most hardened reporter, in the worst days of 1843, was inclined to believe his strange informant when he said that they were living through one of Daniel's visions.[65] Only a few cynics queried the depth of genuine religious interest, noted a slowing down of revivalist fervour, and talked obliquely of the persistence of 'bad habits' and superstition. Church missionaries, scorning all other testimony, described how they had visited homes on remote hills where people neither had nor knew the Bible. At Haverfordwest a few intrepid characters even declared that religion was a fraud.

This should not confuse us. South-west Wales was a cradle of revivalism, and of late-flowering Nonconformity. In the late eighteenth and early nineteenth centuries one religious revelation after another inspired the people of these counties, and attendance figures, which may have fallen to about a quarter, began to double and, in places, triple that amount. 'We are God's chosen', sang a Llangeitho farmer in 1811, and the Church of England, unable to respond, receded before a swelling tide of Independent, Methodist, and Baptist chapels and Sunday schools. In some of the newest communities along the Teifi valley the Church disappeared almost completely.

Northern Cardiganshire was exceptional in the strength of its Church and Calvinist Methodist connections, but in the parishes where Rebecca rode it was a different story. From Aberaeron to Felinfoel, new Independent chapels appeared every year and the brooks were full of bobbing Baptists.[66]

Eleazar Evans, who arrived in Llangrannog from the richer pastures of England, confided: 'we have generally the richest and the poorest'.[67] By the mid-century two-thirds of the population had withdrawn their nominal support for the Church, many to form self-governing and Welsh-speaking Dissenting congregations. Their confidence and gregariousness in religious worship astonished visitors; chapel buildings rang with discussions as well as hymns, farmers' homes became the venue of evening prayer meetings, and at weekends, out in the warm fields, thousands came to hear David Rees of Llanelli and half a dozen other favourite ministers. From the admittedly limited information on individual Rebeccaites, it seems that they shared in all this. Some were Sunday-school teachers, and some were arrested in chapel.

Landowners later protested that they tolerated chapel building and the changes in the tenants' religious allegiance without fully realizing what was happening. Their mouthpiece, the *Carmarthen Journal*, had always challenged the official figures of chapel support, and found many examples of residual respect for the Church in rural parishes like Meidrim. It needed the trauma of the Rebecca riots, and the investigations of *The Times* and other correspondents, to discover the truth and to apportion blame. The responsibility for the change in religious observance, as indeed for much of the general disaffection amongst the peasantry, was placed in 1843 at the door of the Church and more especially at the door of those laymen who had reduced it to such penury. The majority of church livings in the St Davids diocese were worth no more than £150 p.a., parsonages were empty, and services infrequent. If this picture ignored the good works of some notable clergymen, and the gifts of land and money by the Cawdors, Dynevors, Lisburnes, Pryses, and Colbys, it was a fair testimony to the neglect that characterized one of the largest and poorest dioceses in Britain.[68]

Outside districts such as Llanfihangel Genau'r Glyn in

Cardiganshire, and others in northern Pembrokeshire, where the 1830s and 1840s were marked by enthusiastic clerical prayer meetings and by the opening of new churches, too few clergymen had the language, resources, and will to stop the decline in support. It was said that the established religion inspired neither respect nor affection, whereas the chapels and their ministers were, if also poor, at least 'of the people'. David Rees of Capel Als, Llanelli, for example, the Congregationalist who figured prominently in the Rebecca chronicle, was the son of a Tre-lech farmer, and his extended family kept him in touch with the movement. Although Rees despised non-resident vicars, the most unpopular churchmen were probably clerical magistrates like the Reverend James W. James of Robeston Wathen and High-Church busybodies like Ebenezer Morris of Llanelli and David Prothero of Eglwyswrw. Even Herbert Williams of Llanarthne and respected fellow curates discovered in this period that the Tithe Commutation Act and other legislation placed them on the wrong side of the rural divide.

The religious census of 1851 indicated that, during morning services, the places of worship were half full, and almost half the population attended. Many of these people also went to worship a further two times on a Sunday, and sometimes during the week. In addition, many of these worshippers were to be found in Sunday schools. Most of the chapels, and perhaps a third of the churches, had these schools. They were intended for adults and children, and they were praised for the levels of literacy and religious understanding which they reached. In those schools attached to chapels, men and women took the elder members on a long journey through the Bible, and questioned children until answers became second nature. The Sunday schools and the chapels were also at the heart of the campaign to have a more sober, thrifty, and independent population, and one which engaged in rational recreation. Expulsions from chapel membership, when they occurred, were usually for drunkenness and unchastity. By the mid-1840s the chapel and Sunday-school festivals, tea-parties, picnics, outings, and thanksgivings were an indication of the change in popular habits. Altogether, it was argued, the peasantry of this part of Wales was being transformed into a

God-fearing, industrious, and literate people; but the political consequences of the Dissenting triumph were more difficult to assess.

POWER, INFLUENCE, AND CONFLICT

Until the early nineteenth century the structures of power and influence were hardly disturbed in rural society. Power was based on extensive and legitimate landed wealth, and its force was felt at every level of society. In a sense everyone was dependent on the great landowners and squires, and their authority was mediated through the Church, courts, and almost every other institution. Most people did not come into face-to-face contact with these people, but their influence was pervasive. Even in the largest country towns of the south-west, the landowner was never far away, and on election, Assize, and Sessions days his presence was almost guaranteed. From time to time angry tenants, burgesses, and labourers made a stand against Rebecca's 'lords of the earth', and, a generation before, some of the first seeds of Welsh radicalism had been sown in this countryside, but individual rebellion was not the same as organized and articulate opposition.

The rise of Dissent, especially in its more democratic form, was the first major challenge to established authority in the region. So long as the population remained respectful, this social revolution could be accepted, and indeed landowners like the Pryses, the Cawdors, and the Lloyds of Bronwydd provided land for chapels. What caused concern in the decade and a half before Rebecca was the appearance of an aggressive religious press, which raised issues not only of Church and State, but also of the political role of landowners and their relationship with tenants and labourers. The monopoly of the *Carmarthen Journal* (1810–), the voice of High-Church Toryism, was thus broken by the reforming zeal of these new Welsh journals. In 1827 the *Greal y Beddyddwyr* (Baptists' Miscellany) appeared, hostile to Tory landlordism and dedicated to rather lukewarm Reform. Eight years later the Independents brought out the more strident *Y Diwygiwr* (The Reformer), edited by David Rees, and there were other shadowy radical ventures in the region, like *Y Freinlen Gymroaidd* (The Welsh Charter) of 1836. Although they gave limited space to politics, these magazines quickly placed themselves in the van of campaigns against slavery and for church reform. In 1832 the most successful reforming newspaper, the *Welshman*, was launched in Carmarthen, and it soon became the self-appointed voice of Liberal politics.[69]

Perhaps understandably, in the search for explanations of the Rebecca troubles, people were inclined to place too much blame on the press. The low circulation figures in the three counties meant that every paper struggled to survive, and had less influence than was thought. The *Carmarthen Journal* and the *Welshman* each sold about a thousand copies a week in 1844, or perhaps a little more, and their new rival, the *Pembrokeshire Herald*, half that figure. English morning newspapers were hardly read in this part of Wales, at least before the Rebecca troubles, and the small Nonconformist press relied heavily on the chapels for support. Those who enquired about such matters in the mid-nineteenth century were convinced that amongst the peasantry there was a demand for only Welsh religious publications. This was interpreted, rather arbitrarily, as a sign that the Welsh were an obsessively religious people, who lacked a real knowledge of the world and its politics. 'There is . . . no Public in Wales; there is no Public Opinion, and, on a necessary coincidence, no Public Press,' moaned the *Welshman*, whose stated purpose was to build a political home for the growing child of social democracy.[70]

The *Welshman*, and the elderly *Seren Gomer*, were never as unequivocal as *Y Diwygiwr* in their advocacy of church reform, the extension of the franchise, and the ballot, but the editors of these publications tried harder than anyone to reveal to the Welsh people the source and abuse of political power. It was not an easy business, for authority in the three counties changed subtly from one level to the next. At a national level virtually no one from south-west Wales had any political influence. The region in the 1840s was represented by eight MPs, and these were renowned for their lack of attendance at Westminster and more so for their reluctance to enter parliamentary debate. Sir John Owen, perpetual MP for

Pembrokeshire or its boroughs, found it hard as the years passed to make the distance across London, and even to come home for elections. David Morris, Carmarthen's MP, one of the more assiduous attenders, was silent in the Commons during the Rebecca troubles, and so were almost all his colleagues, a fact which drew unflattering comparisons with Irish and Scottish representatives in the House of Commons. The reforming press, both Welsh and English, was highly critical of their performance, and the *Welshman* was convinced that the 'non-representation' of the Principality lay 'at the bottom of all the Welsh "grievances" '.[71]

Their critics wondered why Welsh MPs bothered to enter politics and why people voted for them. The answer was straightforward: a seat in Parliament gave prestige, as well as authority and power within the counties. Much of the legislation of this time was of a local character, and the MP was in a position to influence enclosure, turnpike, railway, and other improvement Acts, and to profit by them. Of greater importance, perhaps, was the patronage which an MP always had; certain county offices and church livings were effectively in his hands, and this brought its own rewards.

To secure a seat involved the expenditure of money and careful planning and canvassing. There were four county MPs in 1841, Carmarthenshire having an extra one, and four borough MPs representing Haverfordwest, Cardigan, Pembroke, and Carmarthen districts. They were elected by almost 15,000 voters, or 5.7 per cent of the population, making the region one of the least democratic in Britain, and putting considerable influence in the hands of those landed families who could offer the support of their tenants and friends. The crucial decisions were made before the elections, for on that day the gentry and their beribboned agents just walked the block votes to the nearest polling station. When a county election was uncontested, political landowners gave these voters the required attention and promises, but when there were several candidates even the Cawdors, in their anxiety to stay in power, resorted to the time-honoured practice of creating new voters and intimidating others. Although the boroughs were less open in Wales than across the border, here too the wise landowner tended his middle-class constituency with care.

In general, the personality of the candidate was at least, if not more, important than his politics. In Cardiganshire a Peelite conservative, William E. Powell of Nanteos, was returned unopposed in county election after county election, whilst a Liberal, Pryse Pryse of Gogerddan, enjoyed a similar experience for the Cardigan boroughs. When his son was challenged in 1849 by a comparative newcomer, John S. Harford, the electorate were reminded that Pryse was the only 'true Liberal' amongst the local MPs, from a family which had supported Catholic Emancipation, the Reform Act, the abolition of the church rate, an end to capital punishment, and the introduction of the ballot. Harford, who almost won, was shouted down by the mob, and mocked for being an English candidate.[72] In Pembrokeshire the conservative Owens of Orielton dominated proceedings, giving one borough seat to Sir James Graham in 1837, though the county borough of Haverfordwest was contested on several occasions. Sir Richard Philipps of Picton Castle, who held off these challenges, was widely regarded as 'a nonentity' whose 'Whig' label was as false as his adopted name. The political terms meant little; within a generation, as radical change became unavoidable, almost all the landowning dynasties who appear in this chapter had declared their allegiance to conservative principles.

During the decade before the Rebecca riots the county and district boroughs of Carmarthen witnessed the greatest number of political battles. The borough was unusual in that it had a sizeable proportion of independent voters, many of them farmers from outside the town who demanded favours from the competing landowners. In the Reform crisis, against a background of formidable violence and intimidation, Carmarthen's electorate almost rejected the conservative politics of their 'old darling' John Jones of Ystrad. Within ten years the situation had settled down, with George Rice Trevor, son of Lord Dynevor, and John Jones now taking the county seats, and David Morris, a Whig banker, winning a succession of borough elections for Carmarthen and Llanelli. On his death, in 1842, Jones, the explosive independent, was replaced by a more relaxed country gentleman, David Saunders Davies of Pentre. With this by-election, the great landowning families had reached, in this difficult post-Reform Act era, a settlement about the apportionment of politial power.

The most important task for a parliamentary representative was to act as a bridge between national and local government. In this he was on a par with the lord-lieutenant. Lord Dynevor, Sir John Owen, and William Powell were the three appointments at the time of the Rebecca riots, all of them Tories. Only George Rice Trevor, who acted for his father in Carmarthenshire, impressed. Colonel Love believed that such 'milk and water Lieutenants' prolonged the Rebecca riots by their lethargy and indecision.[73] The lords-lieutenant had important roles in the administration of justice and the military defence of their counties. Most of their authority derived from the influence which they had over the appointment of magistrates; the Dynevors, for example, worked hard to get fellow conservative David Pugh of Greenhill, Llandeilo, appointed as chairman of the Carmarthenshire Quarter Sessions in place of the deceased John Jones. They had firm views on the composition of the magistracy; outside the largest towns they insisted that the *custos rotulorum* and the Grand jury reflected the landed wealth. In 1843 there were over 200 county magistrates in the region, a roll-call of the important families.

An examination of the records of the Quarter Sessions reveals that in each county the proceedings were dominated by up to a score of well-known figures. In Carmarthenshire, for instance, the Cawdor and Dynevor families were supported regularly in the shire hall by people like Sir John Mansel of Llanstephan, John E. Saunders of Glanrhydw, Rees Goring Thomas of Llysnewydd, John Lloyd Price of Glangwili, William Peel of Taliaris, and John Lloyd Davies of Alltyrodyn. These people administered the county, setting rates, sanctioning appointments, overseeing the upkeep of highways and bridges, and making decisions on police and criminal matters. It was demanding work, and not surprisingly landowning justices relied heavily on people such as David Pugh, with his legal training, Richard J. Nevill of Llangennech, with his business acumen, and the Reverend Ll. Llewellin of St David's College, Lampeter, one of the few clergymen who had the status and influence to make a mark at county meetings.

At the district level one or two of these leading county figures, though more often the lesser squires and clergymen, came into their own. The three counties in 1843 were divided into 29 units of about a dozen parishes, administered by Petty Sessions. The work of these Sessions was divided fairly evenly between criminal business, and matters connected with the Poor Law, roads, and the police. As so many magistrates lived far from the centre of a district, or even outside, and others were too old or ill to be bothered, power and responsibility once again passed into the hands of a few persons. In the district of St Clears the dominant characters were Rice Price Beynon, Timothy Powell, John Waters, and the Reverend John Evans of Nantyreglwys. In the urban courts the situation was similar; at Llanelli, R. J. Nevill, William Chambers, and John Hughes Rees controlled things, and at Carmarthen Edmund H. Stacey, the mayor, and William Morris, of the banking dynasty, were rarely absent from the Bench. On some occasions, not least if the weather was poor, two or even one magistrate only appeared; at Tre-lech a'r Betws these were likely to be Lewis Evans of Pantycendy and the Reverend Benjamin Lewes of Dyffryn.

Many of these people who officiated at the Petty Sessions were also prominent figures in two newer institutions of district authority, the turnpike trusts and the Boards of Guardians. The management of the trusts had to be sanctioned by Act of Parliament. Although the number of trustees was large, very few of them attended the meetings where decisions were made. Frequently the keenest administrators were those who had provided the largest loans. A good example was John Lloyd Davies of Alltyrodyn. He was a major controlling voice on three trusts between Cardigan and Carmarthen, and had invested over £600 in them. In Pembrokeshire, Davies's counterparts were John Henry Philipps of Williamston, William Edwardes of Sealyham, Hugh Owen, vice-lieutenant, Lancelot B. Allen of Cilrhew, and James M. Child. Other prominent trustees and creditors whom we have met already were Rees Goring Thomas and John Hughes Rees. Like William G. Hughes of Abercothi Cottage, Lewis Lewis of Gwynfe, and David Lloyd Harries of Llandingat House, men of influence in the trusts north-east of Carmarthen, all the

Greater and Lesser Men

above were important property owners and almost all were magistrates.

The Poor Law Boards, set up under the Amendment Act of 1834, provided another tier of influence. The Boards, which administered the Poor Law over a considerable number of parishes, were elected annually from a small constituency, though there was an additional number of ex-officio appointments for prominent local people. The powers of these Guardians were limited by Somerset House, and it was not always a popular position, but the chairman and his deputy had an exceptional degree of influence on proceedings. Henry Leach of Corston House was chairman of the Haverfordwest Board, as well as of Pembrokeshire Quarter Sessions, for many years, and James R. Lewes Lloyd of Dolhaidd, William Chambers of Llanelli, David Davies of Carmarthen, and the Reverend Ll. Llewellin at Lampeter also presided at Board meetings. Most Poor Law unions covered a wide geographical area, and their Boards were full of gentlemen, yeomen, and large and medium-size farmers. At Merthyr John Thomas of Derllys became, almost inevitably, his parish's first representative on the Carmarthen Board, as did William Bevan Gwynn for Llangain. In the large Carmarthen union Gwynn, David Evans, a Newchurch farmer and Guardian, and Lewis Evans of Pantycendy directly confronted two urban reformers, William Morris, banker, and Dr John Bowen, Edwin Chadwick's *alter ego*.[74]

The confidence of these farming Guardians is a reminder that there were other levels of authority in Rebecca's country. Not every parish, after all, had either a resident squire or a permanent clergyman. Over areas of south Cardiganshire, and along the Carmarthenshire–Pembrokeshire border, communities were more open, and some of the status of the non-resident landowners was taken by the largest and oldest farming families. Many of them were freeholders. Thus at Tre-lech a'r Betws John Bowen of Plasparcau-uchaf and Thomas Lewis of Blaendewi Fawr, who owned and rented over 1,000 acres between them, were important local figures, men who were used to political and judicial service. There were lesser characters too, distinguished by their political influence, family reputation, religiosity, legal knowledge, and general education. They included tenant farmers William Evans of Llanboidy, Daniel Thomas of Bremenda-issa, Llanarthne, John Thomas of Llwyncelin, Llanginning, Thomas Davies and William Rees of Llanfihangel-ar-arth, and Walter Griffiths, the Abernant butcher. All were keenly interested in the affairs of their parish and all were chosen as Poor Law Guardians.[75] Like two suspected Rebeccaites, William Davies of Pantyfen, Llanegwad, and John Lewis of the Llangadog area, they were widely recognized as community leaders. Of John Lewis, the 100-acre farmer who represented his friends before Charles Bishop, landowners' agent, at one large meeting in 1843, it was said that he was 'a man of considerable influence in that part of the country, and who is indebted for that influence, not so much to his property, which I have heard is considerable, as to his rough-hewn talents, strong sense, and independent spirit'.[76]

These people, mockingly called the 'parish gentry', attended parish or vestry meetings. The typical vestry meeting was chaired by the local clergyman and discussed matters relating to church property, local rates, common land, poor relief, and charity. About a dozen people usually attended the meetings, and a few, like William Williams of Llangadog, Thomas Tobias of Llanarthne and John Davies of Plasparcau-issaf, Tre-lech, hardly missed a session. Some were clearly 'the vicar's men', but others, like David Griffiths of Lletty and Phillip Phillip of Pound, Llangunnor, regarded themselves as spokesmen 'of the Inhabitants'.[77] All were used to making speeches and taking decisions. Until the major changes of the 1830s and 1840s, the government of the parish was firmly in their hands. During the Rebecca troubles it was often these people who called the protest meetings, and conducted the inter-parish gatherings. Evan Davies of Cwmann near Lampeter, and David Lewis the 200-acre freeholder of Tyddynddu, who chaired the huge Pen Das Eithin and Llanarth demonstrations in 1843, did not spring upon the world unseen, nor, for that matter, did those farmers of lesser standing, radicals David Gravelle of Cwmfelin, and Stephen Evans with his 90 acres at Cilcarw, in the parish of Llangendeirne.

Such persons were important in church and, more commonly, chapel matters. Increasingly they were Dissenters, and their influence, and that of their ministers, was a source of growing annoyance. John Lloyd Price of Glangwili declared in 1844 that 'a wish to put down the power & authority of the Gentlemen of the County . . . has been fostered by the greater part of the dissenters for some years'.[78] The chapel was the centre of a new type of authority and influence, and, to some extent, a different set of values in the rural community, though the claim that the ministers were 'the new tyrants' of south-west Wales was much exaggerated. David Rees, the Llanelli Independent, always reminded his critics that, in the voluntary religious system, the people chose their leaders. Many Methodists and leaders of other religious persuasions were conservative and deferential, and so, no doubt, were some of their congregations. The polarization of society into church conservatism and Dissenting Liberalism was a long distance away. What is certain, however, is that scores of Independent, Baptist, and Unitarian ministers became more critical of the established authorities in Church and State in the second quarter of the nineteenth century. They were moving spirits behind the parish and chapel petitions against negro slavery in 1825–33, the pleas for the repeal of the Test and Corporation Acts in 1827–8, and the frenzied clamour against Catholic Emancipation in south-west Wales a year later.

In the Reform crisis, and more especially in its immediate aftermath, ministers and their congregations were again vociferous in their calls for greater religious and political freedom. The Reverend Henry Lewis Davies, curate of Troedyraur, on his return after twelve years in England, said that he found a change for the worse in the population; people had become more narrow in their religious education and outlook, and preoccupied with issues like the relationship between Church and State.[79] Dissenters had even begun to debate the rights and wrongs of Disestablishment in public. As we shall see in the next chapter, the 1830s and early 1840s witnessed growing tension over matters such as education, church rates, and tithes. When Sir James Graham brought forward his controversial Factory Education Bill in 1843, parish after parish rose in protest. At Treodyraur curate Davies, who supported the measure, was called a Catholic, lost some of his congregation, and had to close his school.

These incidents were by no means always the responsibility

of Dissenting ministers, but a few of them were becoming men of considerable local influence, spokesmen for farmers and their servants, and ever more critical of the big house, the vicarage, and their politics. Some, like Joseph Williams of Bethlehem chapel, St Clears, and the Baptist Joseph Evans of Llannon, Carmarthenshire, acted as spokesmen and intermediaries during the Rebecca troubles. William Day, assistant Poor Law Commissioner, believed that it was an Irish situation, with most of the tenantry now in the hands of their preachers.[80] The *Carmarthen Journal* never went as far as this, but it did suggest that such men, and their press, were creating the feelings of independence and even conflict necessary for a greater popular uprising against the aristocracy.

It is difficult to characterize the relationships between the leaders of this society and the rest. Contemporaries were inclined to compare the great men unfavourably with those that had gone before, if only because of their responsibility for the recent economic changes in rural society, but by the late nineteenth century the view was different. Needing a standard by which to judge the landlords of his day, Dr Enoch Davies chose in 1894 to praise the old Lloyds of Bronwydd, Alltyrodyn, and Dolhaidd, and Saunders Davies of Pentre. He claimed that the kindness and courtesy of these families in the mid-century had bred 'a homely feeling' between landlords and tenants.[81] People like Pryse Pryse of Gogerddan were singled out for having non-commercial attitudes, particularly for moderating the impact of tithe commutation and for their willingness to make temporary concessions on rents. But there was more criticism of landowners at the time of Rebecca than Dr Davies realized, and even Pryse Pryse was attacked for not doing enough for his tenants at the height of the crisis of 1842–3.

Some of the praise of landlords in the conservative press of the 1840s was a reminder to others of their duties under the social contract. Landlords were expected to be 'the protector, the benefactor . . . and hospitable'. The good landlord maintained friendly contacts with his tenants, and made some provision for the religion and education of the poor. It was agreed that not enough had been done in this direction, but after all this was not Ireland. Efforts were made on some of the Welsh estates to keep honourable if distressed tenants on holdings which had, for generations, been in the hands of their families, and every winter the gentry offered cheap or free food, clothes, and fuel. There were other occasions when their wives or, more commonly, their clergymen dispensed aid. Edward P. Lloyd of Glansefin, who, as we have seen, kept up the old traditions, was cherished, even during the Rebecca riots, as a leader who was 'favourable to them'.[82]

For their part, the people of the three counties were widely praised for having the natural instincts of loyalty and deference. It was said that they greatly respected the ancient Welsh landowning families, and distrusted the easy English words of the political agitator. The following comment, made in September 1843 by a Scottish gentleman who had lived in the region for twenty-eight years, exemplifies this view: 'there is no peasantry that I know of more faithful than the Welch, or more attached to their masters and their landlords; they will obey you faithfully, and labour for you cheerfully; you may lead them with a silken thread, but you cannot drive them; and depend upon it, neither the bayonet nor the bullet will coerce them.'[83]

This was the ideal: a dutiful landowner and a deferential population. But the reality fell short of this, for even the most unbiased observer admitted that some of the leaders had shown a remarkable indifference to the wants and ambitions of the less fortunate members of society. The fiercest critics of the gentry argued that social contact between classes had declined, that charity had become standardized and depersonalized, and that in a predominantly Welsh and Dissenting society the ruling élite gave neither time nor money to recapture 'the good spirit of the past'. At the height of the riots anonymous newspaper correspondents, who had close contacts with landowning families, condemned the latter for being too distant and uncaring in their economic relationships. John Lloyd Davies of Alltyrodyn believed that the whole Rebecca affair had grown out of indifference towards, or ignorance of, the growing social and economic problems of the rural population.[84] He never quite forgave those great men who returned to regal prominence after the troubles, having left their agents, and magistrates like him, to become the targets of popular anger.

'I know many landlords, during many years, who never see their farms,' said Thomas Williams, blacksmith at a Llandyfaelog meeting in September 1843.[85] To reinforce his point, there were frequent declarations during and after the Rebecca riots that disaffection was confined to those districts where landowners and clergymen were non-resident. By contrast, in Walter R. H. Powell's Llanboidy, Edward P. Lloyd's Llangadog, and Lewis Lewis's Gwynfe, all was apparently well. When such arguments were shown to be at variance with the geographical truth, rural conservatives retreated, as some historians have done, to the second line of defence: most of the trouble in the countryside was due, not to the old landed families, but to the *nouveaux riches* with their agents and capitalist mentalities undermining the established pattern of paternalism, patronage, and deference.

The *Times* correspondent, another upholder of traditional values, offered a different perspective on the relationship between the greater and lesser men. After three months of interviews in rural Wales, Thomas Campbell Foster concluded that there were only two classes: 'the oppressed, the abject and the cringing; and the oppressive, the tyrannical and the haughty . . . It cannot be denied that the people look upon the landlords and the gentry and magistrates, as a class, with hatred and suspicion . . . this arises from no Chartist or political feeling, but solely from oppression, and insulting, haughty, offensive demeanour.'[86] He argued that the language of the threatening letters and the attacks outlined in Chapter 6 simply confirmed this opinion. As his other comments show, Foster hoped that this was just a transitory stage, and that Welsh rural society would soon return to the 'good old days'. He preferred not to think too much about the structural economic and social changes described in this chapter, which lay behind much of the feeling he described, and he never satisfactorily defined 'oppression' nor explained why it was tolerated less now than in the past.

The charge of oppression and of exploitation in Rebecca's society was often heard. Landowners protested that they were not responsible for the 'evils' described in Chapters 3 and 6, and it was true that the Lady Redresser found many of her enemies in other reaches of society. Yet, time and again the debate over grievances returned to the central role of the great men. Things will not change for the better in Wales, Thomas Cooke told his mother in March 1844, until 'its landlords and rulers . . . become more meek and christianlike, and less arbitrary and tyrannical'.[87] 'Landlord tyranny' was a popular theme of Edward Lloyd Hall's tirades, David Rees's articles, and government correspondence. Perhaps nowhere else in Britain was the concentration of power so obvious, and the opportunities for abuse so tempting. Those who seriously investigated the turnpike accounts, county finance, and the administration of justice were convinced that some of the gentry, and their appointees, had benefited unduly from their positions of trust, and that it was virtually impossible for anyone to seek redress against them through the courts. There were claims, too, as we shall see in the next chapter, that in matters of the Poor Law, tithes, rents, and rates, landlords and clergymen 'have not assisted us as they should have done'. Often talk of 'tyranny' meant only that.

A few landowners were prepared to use their economic power in a ruthless fashion, as tenants, cottagers, and squatters testified. The de Rutzens of Slebech Park, James Lewes Lloyd of Dolhaidd, and Thomas Lloyd of Coedmore could be vindictive towards tenants who offended them, and there were several accusations during the 1830s and 1840s that farmers had lost their holdings in the Teifi valley for voting independently on church rates and in parliamentary elections. In 1838 Parliament actually took up the complaints of blatant intimidation against Thomas Tobias and others on the Cawdor estate during the last county election, and there were similar accusations elsewhere in 1832, 1839, 1841, 1849, and 1852. Phillip Phillip, whom we have met before, said that people were afraid to speak out because of the whip upon their backs.[88] Such 'oppression' was probably fairly rare, and no worse than it had been before, or as it now was in other parts of rural Britain. But the character and sensitivity of this society was changing.

This can be seen in the multiplication of complaints about the personal behaviour of landowners and magistrates in the mid-nineteenth century. Those who detected hostility in Rebecca's society felt that it was largely caused by the 'selfish greed, the petty peculation, dense stupidity, and stiff necked obstinacy of a few Welsh squires'.[89] Their arrogance was said to have turned deferential tenants and labourers into angry protesters. Edward Lloyd Hall, who was not above abusing his own tenants and workers, singled out James Lewes Lloyd of Dolhaidd as a man who by his manner had lost the respect of the community, while Liberal Edward Adams of Middleton Hall, a natural ally of the people in the politics of protest, also allowed his fiery temper to destroy a promising relationship with his tenantry. The language became especially loud in the courtroom, when magistrates cursed clever lawyers and their Welsh-speaking clients. A Cynwyl farmer told the *Times* correspondent that the county magistrates 'look upon the people as if they were beasts and not human beings, and treat us with the greatest indignity.'[90] Hugh Williams, whose job brought him into close contact with these justices, called them 'kind and courteous', and then led a campaign to replace them with stipendiary magistrates. 'It is quite clear,' wrote Sir James Graham, 'that the people have been bullied; [and] that they have turned . . .'[91]

The role of the magistrates in mediating between the greater and lesser men of Rebecca's society was endlessly debated in Parliament, press, and public meetings. This extract, from an adjourned Quarter Sessions in September 1843, conveys the range of feeling:[92]

Sir James Williams: 'God knew there was not much reason to talk of the dignity of the Carmarthenshire magistracy; for my part, I firmly believe . . . that the magistrates have completely lost the confidence of the people . . .'

Mr. J. Lloyd Davies: 'I believe that the magistrates of this county have acted with an impartiality and integrity never equalled by the magistracy of any other county.'

Colonel Love, the military commander of the south Wales district, when asked to evaluate the responsibility of the leaders of society for the riots, replied that they had displayed a dangerous mix of insensitivity and inactivity. 'I grieve to say', wrote Love's political superior, Sir James Graham, the Home Secretary, to his Prime Minister, 'that S. Wales bids fair to rival Ireland. Poverty and the misconduct of landowners are at the root of crime and of discontent in both countries.'[93]

3
Poverty, Despair, and Crisis

THE people, the masses to a man throughout the three counties of Carmarthen, Cardigan, and Pembroke, are with me [wrote Rebecca]. Oh yes, they are all my children—when I meet the lime-men on the road covered with sweat and dust, I know they are Rebeccaites—when I see the coalmen coming to town cloathed in rags, hard worked and hard fed, I know these are mine—these are Rebecca's children—when I see the farmers' wives carrying loaded baskets to market, bending under the weight, I know well that these are my daughters. If I turn into a farmer's house and see them eating barley bread and drinking whey, surely I say, these are members of my family—these are the oppressed sons and daughters of Rebecca.[1]

'The main cause of the mischief is beyond doubt the general poverty of the farmers', declared the *Times* correspondent. 'They have become thereby discontented with every tax and burden . . .' 'The root of the evil', he continued, 'is not the toll-bars . . .'[2] Rebecca, added Michael Thomas, the Gwynfe farmer, is 'nothing but poverty', and if they wished to bury her they must first bury poverty.[3]

POVERTY AND DESPAIR

South-west Wales had long been synonymous with poverty, and was to remain so. The government commissioners and the newspaper reporters of the Rebecca troubles claimed that, for mile upon mile, the country 'assumed a poverty-stricken aspect'. According to the two most pessimistic members of this society, Lewis Evans of Pantycendy and LEX of Haverfordwest, an anonymous letter writer and farmers' spokesman, almost everyone from the squires downwards existed in a state of permanent indebtedness, and this brought a cascade of oppression and misery down upon the people at the bottom. It was, said LEX, 'a grinding poverty, which renders it impossible for landowners and titheowners, whether ecclesiastical or lay, to forgo one pound per cent of the amount of their annual income . . . a grinding poverty which makes it impossible for tenants to pay either rents, tithes, rates, taxes, or tolls . . . [There is] no just reason why the poverty of Wales should not be placed on the same footing as the poverty of Scotland and of Ireland.'[4]

Some of the people did comparatively well, confirming the *Morning Chronicle*'s first favourable impressions of the region, and there were the usual cautionary tales of farmers whose dress and life-style belied their hidden wealth, but the majority of rural inhabitants lived from hand to mouth. People were discovered in terrible conditions, emaciated, naked, fainting for lack of food, and diseased from head to toe.[5] When Anne Davies of Cardiganshire, Mary Anne Aubrey of Llangadog, and other bedraggled characters crawled before the Poor Law and police authorities, even hardened magistrates were moved to protest. Dead infants were so thin that coroners sometimes started murder enquiries, and no one visited Carmarthen in winter time without being struck by the hollow faces of the unemployed and the slow gait of hungry mothers. The poorest in this region still gave birth in stables, and died hardly noticed on cold and frosty nights. In this Nonconformist land the only miracles were Sarah Jacob of Llanfihangel-ar-arth, and a succession of other strange girls, who lived for months without food.[6] All the government enquiries, including the Poor Law commission of the early 1830s, concluded that hundreds of families in this region had insufficient income for even basic requirements.

The evidence of poverty was unmistakable. It could be seen in the people's bodies, their diet, their dress, and their living conditions. 'Welsh workmen are ye slowest animals I know', commented Edward Lloyd Hall. 'The labourer is necessarily so badly fed', wrote Thomas Campbell Foster, 'that he cannot get through his work.'[7] Only, it was said, the custom of giving food with wages ensured that people had the energy to complete a day's labour. During the extreme poverty of the post-war years there was talk of general 'feebleness', 'unfitness for work', and the 'blank stares' of the weak and old. Physical strength returned in better times, but the peasant of south-west Wales was never able to match the vigour of the Scottish and English immigrants, nor that of the native industrial worker. Coal-owners in Pembrokeshire told government inspectors that it was necessary to feed up the small farmer and labourer before letting them go down the pits.[8]

Confirmation of this difference in stature was provided in 1865, when Dr Hunter, medical officer, made a special report on the population of the area. He found the people, and especially the males, decidedly lacking in strength and energy. In Hunter's opinion the decline in their physical condition had begun about eighty years before, and was partly associated with changes in diet and clothing.

The farmer in Wales [he informed the government] as well as the labourer must be taken to mean a person generally badly lodged, and insufficiently fed and clothed. The poverty of the soil and the damp climate have rendered agriculture a trade which offers little attraction to persons of capital, and in which nothing but the sternest frugality can hope to gain. The evil effect of poverty upon the health is rather increased by the frugal habits of the Welsh farmers ... it may be truly said of this class that in many districts the farmer himself does not eat fresh meat once a month ... The farmer's labourers who board in his house live proportionately worse. A morsel of the salt meat or bacon is used to flavour a large quantity of broth, or gruel, of meal and leeks, and day after day this is the labourer's dinner ... In Cardigan district a medical practitioner described the children as 'pining for want of food as soon as weaned', and thought that if the climate were colder the whole race would perish ... These things ... in the main are due to insufficient earnings, and the effect is seen in the nearly universal prevalence of tuberculosis or scrofula.[9]

Twenty or thirty years before this report, in the pre-railway era, the character of people's diet, clothing, and housing was at least as bad as that described by Dr Hunter. Barley bread, oat porridge, potatoes, and long-lasting cawl were the staple foods, the mix varying from area to area. The more fortunate, particularly those in parishes close to town and industry, also had a little of their own cheese, butter, bacon, and tea. Meat was a rare seasonal treat, and the farms near the sea had a taste of fish. Alcohol was consumed, but all were agreed that there was insufficient income for this to be taken in large amounts. An examination of the agricultural labourer's diet, based on idealized family budgets in 1841, suggests that the daily

1. A Cardiganshire cottage in the late nineteenth century (Welsh Folk Museum, St Fagans)

calorific intake in a large family was about 1,500–2,000 per person. As for the many in a less favourable position, 'how they managed to obtain the other absolute necessaries of life and keep body and soul together God only knows'.[10] Wives and children were the chief sufferers. It came as no surprise when the first medical officers appointed under the Poor Law Amendment Act of 1834 reported that babies were exceptionally small and families generally took a long time to recover from illness. For only a little extra food, said the next generation of medical men, these people would be much more efficient employees. In Rebecca's letters the masters were asked how they would like to live on a starvation diet.[11]

As for clothing this was also basic; men lived in coarse cloth shirts, and rough woollen jackets and trousers, while women wore old coats, thick petticoats, long dresses, and shawls. Reporters for the national newspapers were struck by the sight of poorly clad children and by the bare legs and feet of so many of the women. All their garments gave poor protection in wet weather, a contributory factor to the early onset of rheumatism and bad chests. Families tried to keep a decent set of clothes for the hiring fair and Nonconformist Sunday, but there were many descriptions of people unable to attend school and chapel because of the lack of suitable attire. Although the wives of the gentry and clergymen encouraged the poor to contribute their pennies to clothing clubs, there were other, more pressing, priorities. Footware was a costly item; clogs were the traditional cheap alternative to leather boots although only the oldest children had these. 'They look like a tribe of nomads', commented one visitor, 'on every day except Sunday their bodies are covered in dirt and sweat.'[12]

This was hardly surprising, for some of the farms and cottages were built of and in mud. Landowners took little interest in the quality of accommodation, and enterprising tenants and squatters built their own homes. A typical labourer's cottage was a one-storey building with about 20 square yards of space. It had low clay or stone walls, straw, or—in some districts—slate roof, and a couple of openings for windows and a chimney. Inside the building the floor was of hard earth, and the furnishings were minimal. The farmer needed more accommodation for his family, workers, and animals; his house was a little larger, with kitchen and parlour on the ground floor, a second storey or large attic, and adjoining outbuildings. Wills of even the better class of farmers reveal that a few chairs, a table, and a sideboard were their only furniture, with, of course, bedding and rugs. Sleeping arrangements were primitive, although many of the bigger farmhouses now had partitions for privacy.

Whitewash and female care gave the cottages of rural Wales a cosy image, but because of the Rebecca riots the inquisitive stepped inside the door. This is Thomas Campbell Foster, travelling to a meeting near Newcastle Emlyn:[13]

On my way I entered several farm labourers' cottages by the roadside, out of curiosity to see the actual condition of the people, and found them mud hovels, the floors of mud and full of holes, without chairs or tables, generally half filled with peat packed up in every corner, the only articles of furniture being a wretched sort of bedstead and a kettle. Beds there were none; nothing but loose straw and filthy rugs upon them. Peat fires on the floors in a corner, filling the cottages with smoke, and three or four children huddled around them. Nearly all the cottages were the same. In the most miserable parts of St Giles, in no part of England, did I ever witness such abject and wretched poverty. Yet, according to some opinions, this state of misery ought to be one of happiness and content. Were it so the people would deserve their fate. Content to live like swine they should be fitly treated as such. But the people, to their honour be it said, are not content with this.

Many of these cottages were in desperate need of repair, but both rich and poor were reluctant to spend the necessary capital. As a result people like 70-year-old William Thomas of Abernant slept, without blankets, under open roofs and against collapsing walls. There were others who fared worse; male farm servants throughout the nineteenth century lived in barns and outhouses. On a Cardiganshire hillside a man in 1843 came across a family living in a rude shelter under a rock, and there were other horror stories. This was not Ireland, but it was not Arcadia either.[14]

These conditions of life left physical and psychological marks on the population. The farming population was renowned for its frugality and resilience in the face of grinding poverty; to use a contemporary phrase, people 'pinched themselves' and

Poverty, Despair, and Crisis

kept going. Land held, money made, and possessions acquired had a special value, and were not to be squandered. Their extraordinary sacrifices, their meanness, and their fascination with wills and heirlooms have passed into our literature, as have the more hidden feelings of anxiety and despair. The latter occasionally surface in the records of the Rebecca period. There are references to people depressed by the hard life, and to numerous attempts at ending it all. Some of the 'morose' inhabitants whom English nationalists seemed to have met at every turn refused to speak, eat, and work. Eventually they were found walking on the mountains, or lying in sheds, with neck burns, throats half-cut, and protruding bowels. Such incidents hardly deserved a comment, but when a young man of Johnstown near Carmarthen took his own life in 1845 it was time to remind people of their duty. The man had lost his job and his sick club had refused to assist his family. His suicide was heavily criticized, being, it was said, an unwelcome exception to that resignation which God required of the poorer classes.[15]

At this time, when the registration of births, deaths, and marriages was in its infancy, it is impossible to give accurate statistics of suicide cases. The Registrar-General's return of nineteen for Wales and Monmouthshire in 1840 cannot be taken seriously; the level of self-inflicted death was much higher than this.[16] It is a difficult area. Were, as the newspapers assumed, Mary Thomas of Talley, and the other women who swallowed too much hemlock, simply intent on securing an abortion? 'Found drowned' was another of the coroners' verdicts which hardly covered family embarrassment. The present writer has examined fifty of the more obvious cases of suicide in the mid-century. There was a similar number of men and women in the sample, but their method of operation was different; most males hanged themselves, while females used razors, drank poison, and walked into deep water. Typical victims were rejected sweethearts, pregnant girls, young and inadequate married men, people who had lost employment, servants who had been chastised or were about to be charged with a crime, rebellious and fearful soldiers, alcoholic tradesmen, over-burdened mothers, guilty Dissenting ministers, and aged paupers, tired, confused, and dreading the workhouse.

The end came quietly, the decision made because hope had gone. When Samuel Jones, a 'respectable farmer' of Penboyr parish, could not find the money for a debt due on the morrow, he left his wife and twelve children to pay it.[17] Here was the classic suicide in rural Wales, the depressed middle-aged farmer, who, having struggled half a life against bankruptcy and the elements, made the walk to the barn for the last time.

There were other manifestations of poverty, exhaustion, and despair, and some of these will be discussed in the next chapter. As we shall see, in spite of Stephen Evans's largely justified comment on their love of children, this was a land of mysterious child deaths, desertion, and infanticide. The papers are full of accounts of village children accidentally smothered, burnt, bitten, and drowned. Editors bemoaned the neglect, or malevolence, of mothers whose place was the home, or rather inside the bolted door. 'If women did remain in their own homes, and not trouble themselves about their neighbours' concerns', all would be well.[18] The very same women were required to help with the farm and perform other outside work. No one will ever know how many babies died because of tired or uncaring parents. In an age when people were reluctant to report infant deaths because of the possibility of a coroner's inquest and the cost of a proper funeral, families kept small skeletons in the cupboard. The first bursts of unwelcome crying were stifled forever, and little bodies were slipped between the furrows. In nightmares, which happened in south-west Wales, neighbours opened the door to find young children hanging like rabbits inside.[19] Many of the other assaults and rapes on young children, both within the family and at work, were kept in the dark. Of incest, and inbreeding, nothing can be said, for the silence surrounding it was deafening. When one disturbing case came to the public's attention, at Cynwyl Elfed in the winter of 1843–4, the family sold up and moved away.[20]

'How little we know', admitted the *Welshman*, in an unusual mood of reverie, 'of the darkness in which so many fellow creatures here abide.' The Reverend Benjamin Lewes, magistrate and vicar of Cilrhedyn, suffered from delusions throughout the Rebecca years, and eventually became so worried about his income that he stopped wearing clothes and refused to pay the harvesters. A year later, at Llandysul in the autumn of 1848, Elizabeth Evans was also declared insane; all her adult life this spinster had carried an imaginary child in her arms.[21] Many years elapsed before people like Evans and Lewes were confined to a home; in the early nineteenth century most 'idiots' and 'lunatics' remained in their own communities. Each village tolerated its fools, and, if the newspapers are to be believed, its female religious fanatics, but less sympathy was shown for those who suffered serious mental breakdowns and were regarded as a danger to society. Some were locked away in outbuildings like animals, or tied to chairs and rings in the wall.

The Principality had, as Lord Ashley told Parliament at this time, the fewest and most inadequate asylums in Britain; for years the efforts of Lord Cawdor and his colleagues to establish a central institution for south Wales floundered because of costs and inter-county rivalry. Instead the wealthy scoured Britain for the best of the licensed homes and hospitals, and for the rest there were only an overcrowded building in Haverfordwest, the darkness of a room in Amroth Castle, a few cells in local workhouses and gaols, and the distant terror of Briton Ferry and English asylums. Families preferred to remain quiet about members like old John Howell of St Clears, who fished with a Bible attached to each wrist and with a rod of brambles in his hands, at least until their wills came to be read.[22]

It is impossible to estimate the numbers of people who suffered, like Howell, with temporary or permanent mental illness. Dr John Bowen, surgeon of Carmarthen, believed that there were more of them in the three counties, especially in the countryside, than in comparable regions elsewhere. At the end of 1868, when the new lunatic asylum at Carmarthen was housing 217 patients, its medical superintendent claimed that there were 626 of 'unsound mind' chargeable to the Poor Law unions of the three counties.[23] There were, of course, other deranged people than just 'lunatic paupers' at large in the region, and a few of these, as the criminal records show, found catharsis in horrific physical and sexual violence against their immediate families and the animals that lived with them. It was astonishing, said minister John Pugh at a Llangyfelach meeting during the riots, that there were not more like them;

in his opinion whole communities had been driven to the point of madness by the strain of making a living.²⁴

MAKING A LIVING

The notion of a people grown 'mad and careless' by hard times was a popular one, but this was a land used to poverty. To assess the precise contribution of economic tension towards the Rebecca riots is a difficult exercise, partly because the government was anxious that matters such as rents and tithes should not be 'specifically mentioned in the terms of reference' of the commission of enquiry of 1843.²⁵ According to the *Times* correspondent, and some of the farmers' spokesmen, these were the very issues that prompted them to become 'children of the night'. Of course, the pressures on the small owner-occupier, tenant farmer, and labourer were somewhat different, and it was the balance between them which largely determined the nature of rural protest in the nineteenth century.

The freeholder with perhaps forty acres was concerned about the cost of borrowing money, the level of rates, and especially the price of farm stock and produce. It was estimated that well over a third of these farms were burdened by mortgages and debts, for each generation of owner-occupiers had to make provisions for the women and children of the household. As a consequence the small freeholders were never able to give their farming the capital it deserved. The need for money was insatiable. Fields and buildings were sold when all else failed, and cash borrowed from bankers, private financiers, family, and friends. In the commercial crises of 1825–6 and 1832, as one bank collapsed after another in the south-west, scores of small proprietors went bankrupt. Ten years later the mortgage interest rate stood at 5 per cent, while bank loans were at least half that again and, for the reckless, short-term advances from solicitors and auctioneers were as high as 15, 20, and even 30 per cent. Six months' credit, given at the sales of livestock, eased the immediate problem of cash-starved farmers, but the prospect of repossession soon replaced it. William Goode of St Clears and Thomas Williams of Llanwrda were in perhaps the most advantageous position in the rural economy; as auctioneers, valuers, and farmers themselves they knew which owner-occupiers were in serious financial trouble, and which securities for further loans were worth having.

The living standards of freehold farmers depended ultimately on the level of agricultural prices and the burdens placed upon land. So long as the demand for lamb, beef, butter, and cheese was high, life was bearable, and families could buy more property and take on outside labour. Although these farmers had no rent to pay, there were debts, dues, and rates which had to be met. One of the most important of these was the tithe charge, which amounted on average to about 1s. 6d.—2s. 6d. in the pound, and which increased in most areas once the Commutation Act of 1836 had been put into operation. In addition, there were the poor, county, and highway rates, the land, income, and malt taxes, the church rate, and the road and market tolls. Altogether, the burdens meant that in a normal year up to 25 per cent of a freeholder's income trickled away like a hill stream.

The tenant farmers, of course, had another and even greater problem, the twice-yearly rent. On, or soon after, Lady Day and Michaelmas they had to find the cash equivalent of perhaps a third of the expected annual income from their property. It was their first priority, and taken as seriously as their religion. To pay the rent in bad years money had to be raised on extortionate terms, and produce and stock sold when they were most needed on the farm. Even then, some tenants could not find the full amount for their landlords, and had to seek deferred payments, 'Dr Lawrence is a good master,' said Mary Thomas of Cilferi, the mother of a Rebeccaite, who rented land worth £40 p.a. in Llanelli parish, 'and will take what rent I can get together by rent-day, and trust me for any little matter I cannot make up for a while, and I never failed him yet.'[26] In the mean time there were tithes and other rates demanding farmers' attention, and non-payment of these was also a serious matter. The pressure was unrelenting: they were, said a Llannon farmer, 'daily afraid of the bailiff coming to distrain them both for rent and rates'.[27]

The following examples show how hard was the struggle and how desperately they fought. George Smith of Knock in Clarbeston, Pembrokeshire, took a farm at Lady Day 1833, for

21 years at £140 p.a. rent. He had £100 to stock the farm, and sublet some of the land. Over the next two years he had to look after both his father, with his accumulated debts, and his brothers, and then pay the landlord, titheowner, and banker. As so often in the farming story the initial enthusiasm quickly evaporated. Within two years Smith was at the insolvent debtors' court in Haverfordwest, the declaration of his bankruptcy being opposed by two of his principal creditors.[28] More common was the plight of a smaller farmer, William Williams of the parish of Lampeter Velfrey. His decline was a long one; indeed, he had so many sales on his property to pay for his rent that neighbours had to lend him furniture. When bailiffs came for the last time, to find £7 worth of tithe for the Reverend William Seaton, he and his wife took a hatchet to them.[29] Another way of losing a farm was experienced by John Davies, who rented a small piece of mountain land in Cardiganshire. As his debts increased the old tenant was obliged to offer Lampeter banker David Evans, one of the owners of the land, a share of his possessions instead of rent, and some, and then all, of his sheep. Before giving everything to Evans, Davies had the sense to hand over the last forty-seven of his sheep to his son on a neighbouring farm, saying they were payment for past labour services, and then, with a wicked flourish, sued the apoplectic banker for a slanderous remark about the transfer.[30]

The position of the small farmers was related, in a variety of ways, to that of farm servants and labourers. All were anxious that employment opportunities should not be curtailed. So long as agricultural produce was in strong demand, work prospects were reasonable and wages moved ahead, but after 1815 the return of men from the war dramatized a long-term labour problem in the countryside. 'I declare to God,' wrote David Williams from Cardiganshire during the post-war depression, 'from what I see and hear, that I fear that half the labouring poor will perish as things are, before next harvest in this neighbourhood—nor did I ever conceive before that human nature could bear up amidst such Privations . . .'[31] Labour-cuts became a fact of life, especially in the winter months, and more casual workers took their place alongside the regular labourers who lived in the farmers' cottages.

Poverty, Despair, and Crisis

One newly arrived landowner in Cardiganshire, anticipating the political economy of Poor Law reformers, said in 1821 that 'the labouring class of Society is too numerous in proportion to the land now under cultivation'.[32] In fact, the situation was not as serious as he suggested; most men and many of the women could find work at this time, and even in Cardiganshire, the most agricultural of the three counties, there were new jobs being created in the lead mines of the north, and in the quarries and fishing fleets. Women were still employed in large numbers by some of the gentlemen farmers of that county, as as well as on the arable farms of north-eastern Carmarthenshire. Thirty miles to the south, where the population was fairly mobile and attracted by urban and industrial work, few of the farm labourers now had land, but they were just beginning to feel the benefits of improved bargaining power. If they had steady work, and children of an employable age, they could afford a smile. Yet the picture for these labourers was by no means uniformly promising; in a land of small farms labour was often part-time and the employers' first response to hardship was to rely totally on assistance from the family.

Although there was nothing here to compare with the figures of short-term unemployment in industrial south Wales, there was nevertheless a great deal of hidden poverty. As the men appointed to enquire into the state of Welsh education crossed half a dozen parishes in northern Carmarthenshire in 1846, they were impressed by the irregularity of work, and the 'extreme misery' of the 'labouring classes'.[33] When the people's respresentatives addressed public meetings in 1843 they criticized the gentry for taking money and work out of the area, and the masters for reducing labour bills. 'Let them give food and employment to the people,' said one small farmer, 'and Becca would soon disappear from them.'[34] Families who had for generations worked on a particular farm wanted a return for their loyalty, and were incensed when outsiders came into the parish to harvest or make road repairs. Yet their hostility was muted, now and throughout the nineteenth century. People were surprised by the reluctance of labourers to organize openly for better conditions. Instead, they chose to move longer distances in search of a few months' work, and, inceasingly in the 1830s and 1840s, they packed their bags and took their anger and their diseases to Carmarthen, Swansea, and Merthyr Tydfil.

Prospects of employment in the biggest towns of the region waxed and waned during the mid-century, and their inhabitants were sometimes in as much poverty as those in the rural parishes. In anticipation of the Rebecca troubles, A FRIEND OF HUMANITY warned the urban authorities in 1840 of the trouble to come if the poor of both town and village were simultaneously affected by unemployment and price changes.[35] Five, and seven, years later, when Carmarthen's distress was again acute, the *Journal* of the town called on the respectable to give more charity in the winter months when there was less outdoor work and more unemployment generally across the three counties. Winter was synonymous with hardship and the workhouse, and gentlemen like Thomas Johnes of Hafod and Abel L. Gower of Castle Malgwyn, who kept up a good complement of workers during this season, were widely praised.

With people chasing jobs, it was hardly surprising that the level of wages was about the lowest in Britain. By the early 1840s the average wage for male agricultural labourers was some 6–9*d*. a day with food, or 10*d*.–1*s*. 2*d*. without. Earnings were reduced in the winter months, and were highest in the harvest period and in parishes such as Llannon and Pembrey where dissatisfied workers could find alternative employment. Female labourers obtained two-thirds of the wages of their male counterparts. Indoor male servants received board and lodgings and an annual income which varied from £3 to £9, according to type and district, while the women received a sum at the bottom of or just below that range. Children under 14 years of age could expect little more than food for farm work, and sometimes lodgings, and it was another three years before they received the full adult wage. To put this in perspective, a miner in Pembrokeshire earned about 10*s*. a week at this time, a Llanelli collier and lime-burner 13–15*s*., a Swansea copperman 17*s*., an ironworker at Merthyr 20*s*., and a railway builder near Cardiff 24*s*. For the young the comparisons in income and independence were especially odious; 19-year-old Charlotte Childs earned more in a month cutting coal near Merthyr than in half a year as a maid in the three counties, and her young male friends from Pembrokeshire, working alongside her, already had enough money to marry and set up home.[36]

In justification of low wages contemporaries argued that the farmworkers of south-west Wales preferred to take part of their remuneration in kind, and that items of expenditure were lower here than in many other areas. Cheap food and milk were obtained from generous farmers, and rents and fuel were often subsidized.[37] Other perks have been mentioned in an earlier chapter. Over the years, however, payments in kind had become less common, and, according to labourers in 1843, they had not been replaced by higher money wages. There were other complaints, too. Farmers, who hardly saw cash themselves, were notorious for ignoring labourers' pay-days, and miscalculating the settlements due to indoor servants. In one of the many court cases over these matters, at the Lampeter Petty Sessions in 1841, an employer denied ever hiring the female servant who now successfully claimed half a year's wages of £1. 6s. 6d. On another occasion John Williams, servant of Sarah Evans, was awarded £75. 17s. back pay for his services over many years. If the lowest wages mentioned above were paid regularly to the farm labourers, acknowledged William D. Phillips, clerk of the vast Carmarthen Poor Law union, they would be content and out of his hair.[38]

SELF-HELP, CHARITY, AND RELIEF.

'How do they survive?', was the question most asked about the Welsh peasantry. In their battle for existence they relied on their own initiative, on the support of kin, and on the private and official aid of the wider community. When times were relatively prosperous, families tried to provide for the inevitable downturn in their fortunes. 'Thrift' was one of the virtues that the people of the region were said to possess, though it was never on the scale recommended by Samuel Smiles. The early nineteenth century witnessed the establishment of many friendly or benefit societies, first in towns and then in the villages. In some places, including the ports of Aberystwyth and Pembroke, a considerable proportion of the population was enrolled. Initially these friendly societies were local attempts to provide for sickness and death, and they took their names from the village public house in which they met, but by the mid-century the Oddfellows and True Ivorites had begun to capture the market. The latter were a Welsh phenomenon, expanding from their Wrexham and Carmarthen bases; their appeal was not only to those who wanted 'Unity, Love and Charity' but also to those interested in 'Morality and the Cultivation of the Welsh Language'.

In the years 1839 and 1840, when agricultural prices were good, and even for a while after that, new lodges of the Ivorites and Oddfellows were established in quick succession. Carmarthen, Haverfordwest, and the largest towns had several lodges of each, and these in turn encouraged similar societies in fishing and farming villages. In June 1839 the officers of the Carmarthen district opened the Loyal Farmer's Refuge Lodge of Oddfellows at Talley, and this example was followed at Nantgaredig, Llanarthne, Llangadog, Tregaron, Llanboidy, and other places.[39] The Ivorites did likewise. Both societies were noted for the colour and respectability of their funerals, processions, festivals, and excursions. They were outwardly loyal, devoted to the Crown and Church. At St Clears, Timothy Powell, Rice P. Beynon, and the Reverend John Evans were three of the leading Oddfellow 'Brothers', and they led the anniversary processions to the local church.

A typical lodge had about fifty people who could afford the weekly fees, the silk regalia, and the hospitality of membership. Those tradesmen, small farmers, and craftsmen who made the effort to support this form of self-help accepted its disadvantages; arrears of dues meant expulsion, as did activities deemed to be criminal, immoral, and seditious. As the Rebecca riots came to an end the Oddfellows and Ivorites in Carmarthenshire expressed their pride that no member had been prosecuted for the outrages, and promised a hostile reception for those known to be sympathetic to the cause.[40] Payments to members in cases of accident and illness were closely checked, and there was, especially in difficult times, an aversion to helping people who seemed to be always 'on the box'. Officers of the smaller societies appeared many times in the courts, charged with stopping or reducing hand-outs. Like the savings banks and other institutional forms of self-help that were so applauded in the press and so rare in the region, the benefit clubs were in fact of marginal importance to the poorest in Rebecca's society.

In the last resort, the peasantry had to rely on themselves, their relatives, and their friends. For those who were in ownership or occupation of land, self-help included selling pieces of the family holding, subletting and re-mortgaging, and off-loading most of their stock. When times were very bad the farmers' youngest children were taken from school and put to service, and heirlooms and wedding rings were quietly pawned. People bought time, hoping that the extended family and the community would rally round until the next rise in agricultural prices. A meeting of farmers at Tre-lech in September 1843 cited two such survival cases, and actually produced the first of these to convince a newspaper reporter. David and Mary Owen of Abernant, who had eight young children, farmed land valued at £25 p.a., and sold everything to pay the rent, rates, and taxes 'regularly'. At this time they were still clinging to the land, but they were in rags and starving. A farmer who gave them food said that 'they subsist entirely upon the charity of their neighbours'.

In the second case, also from the parish of Abernant, the assistance was more practical. David Jones, who had a wife and seven children in great poverty, had been a tenant farmer but he was now glad to find any work as a labourer. At a large vestry meeting it was decided that everyone should help him to build a house on the common. When it was completed they provided him with a little furniture and coal. A few nights later, under the direction of a county magistrate and two Carmarthen bailiffs, a gang of men destroyed the lot, and thereafter the family had to rely on the charity of friends. 'This was the first nocturnal outrage', said an angry spokesman at the Tre-lech meeting, 'and was what set an example to Rebecca.'[41]

These people had a reputation for helping one another. Food, clothing, furniture, stock, and equipment were offered by kin and neighbours to families in distress, and there was the more organized help of biddings, auctions, and chapel collections. Here, as elsewhere in rural Wales, custom dictated that no reasonable and proper request for relief should be ignored. One of the functions of the farmers' wives was to know who needed a little buttermilk and flour. Visitors were given food, as grateful social investigators discovered in 1843, and this tradition alone had, it was said, saved many lives in the years of famine. Orphaned, illegitimate, and deserted children, poor widows, and aged parents owed their existence to near relatives and neighbours. Elizabeth Thomas, the 18-year-old daughter of an old and ill Abernant pauper, admitted that she 'often went out to beg and went around the neighbourhood to try to get work, and when I failed they gave me a meal. I took part of what I got by begging home to my father.'[42] Not everybody was lucky; poor John Griffiths, a diseased farm labourer of Verwig, Cardigan, and Mary Lewis, a famished and frozen 6-year-old bastard of Llansaint, fell between the nets of family concern and Poor Law responsibility, but such deaths were comparatively rare.[43]

The least fortunate inhabitants of the three counties were those who lived on the very margins of society. John Thomas, an 80-year-old man of Llanstephan, Catherine Davies of Llandysul, who was ten years younger, David Jones of Abernant, and many others picked up scraps of work where possible, and begged and turned to crime when necessary. Some travelled the same few miles of countryside all their lives, exchanging labour in return for food and temporary accommodation, and selling cheap household items and fuel. It was a short, but irretrievable step, for these familiar visitors to throw in their lot with the vagrants who passed by on bigger circuits. Tramps received less sympathy from the resident population, especially in the worst agricultural years, when their numbers alone were intimidating and when they were blamed for the increased incidence of crime and disorder.[44]

The vagrants, like all those at the bottom of society, relied heavily on private charity and public relief. Charity meant many things; a piece of bread and a drink of milk, subscriptions for the distressed poor begun by village clergymen and town corporations, and the elaborate gifts of resident landowners. The gentry were, as we have seen, judged by the assistance which they gave to Bible societies, missionary work, and the poor. The big house provided linen for expectant mothers, coal and blankets for the aged poor, and perhaps a little meat and corn in due season. The Cawdors and the Dynevors also gave money for imprisoned debtors and workhouse inmates to celebrate Christmas Day, and, at the moment of election victories, contributed a little to the places and people who had served them well. The dead were also helpful. Haverfordwest and Pembroke had over twenty charitable trusts, Laugharne had its annual custom of giving free loaves on 21 December, Carmarthen had charity schooling, and Tre-lech and other villages their endowments of property and money for the poor. One complaint was that these bequests reached the poor in emasculated form, a point conceded by the charity commissioners of the 1830s, and there was further annoyance at the parish lands and cottages sold to build workhouses. Rebecca did not make too much of these grievances, but she did criticize the new generation of employers and agents for moving away from traditional forms of charity.[45]

David Williams of Bronmeurig, who organized help for the Cardiganshire poor in 1817, said at the time that 'promiscuous doles neither bless the giver nor the receiver', and that attitude became more influential in the next three decades.[46] Charity became scrutinized, standardized, and, where possible, the means of improving the morals and habits of the deserving poor. Lady Williams of Rhydodyn, the Reverend Benjamin Evans of Llanstephan, and others began clothing and saving clubs, and supported free medical and educational schemes to improve the quality of life of those around them. Two wealthy Neath ladies adopted the girls and boys of a Llannon school, and on the occasion of their visits the children's families were summoned to tea and cake by the church bells.[47] Some of this giving, or rather the accompanying sermons on thrift and the Thirty-nine Articles, caused resentment amongst the Dissenting peasantry, and anyway, it made little impression on their standard of living. When things were at their worst, the press reminded the rich of this fact, and once again the distressed put their trust in exceptional acts of private generosity and public subscriptions. At a Tre-lech parish meeting in April 1847, for example, it was reported that over £32 had been collected towards relieving the destitute poor, and, in time-honoured fashion, corn and bread were offered to them at a reduced price.[48]

The relationship between these private acts of kindness and poor relief grew closer as people became more concerned about both types of assistance. Sympathetic friends claimed that the Welsh peasantry were too proud to ask for help from the parish, thus making a mockery of the official government statistics. In the mid-eighteenth century the number of people receiving relief was very small, but during the years of the Napoleonic Wars, and immediately afterwards, hundreds were added to the overseers' lists. In the year ending March 1819, a peak year, the amount raised per head of the population towards the poor rate was almost 10s., with Carmarthenshire at the top of the county table, and this high level continued for some time. When money was scarce the overseers took hens and butter for cash. In the early 1830s the amount spent on the poor again rose, but not to the same degree as before. During 1834, a fairly bad year, the money raised per head of population was only about half the figure of 1819. Across the three counties about one person in eleven received parochial assistance.[49]

At the heart of the old machinery of poor relief was the parish vestry. This set the rate, and checked the overseers' accounts. In this poor region of Britain there was a fine tension between the interests of the ratepayers and the requirements of the paupers. This worked to both the advantage and disadvantage of the poor. People were given help by the vestries to settle in other districts, and money was sent to them; so long as the iron trade prospered Merthyr Tydfil saved these rural parishes a small fortune. At the same time non-parishioners who promised to be long-term burdens were removed. Hundreds of removal cases were settled in and out of court, and no one suffered more as a result than pregnant women and widows with children. The historian can still follow Mary Morris and her two children around Pembrokeshire in the 1820s and 1830s; from Dinas, her home parish, to Llanychare, to Dinas, to Newport, to Dinas, and so on. In one famous case, which involved a string of expensive appeals, the ringing of church bells at Haverfordwest announced that one wandering family of paupers had at last been successfully dumped on a neighbouring parish.[50] Employers who hired non-parishioners as servants offered contracts of just under one year's duration, thus ensuring that settlement did not become an issue.

Those who received regular help from the parish were usually children, women, and the old. Every effort was made by the vestries to force men to look after these expensive burdens; in the case of illegitimate children the reluctant fathers were threatened with gaol on the unsupported word of the mother.[51] Even so, every parish had its bastards, and a smaller number of orphans and deserted children. Money was given for nursing these children and in time they were found work. Parish boys were placed with the larger farmers, and the girls became servants or were apprenticed to seamstresses. To make them attractive to employers, the parish offered a small sum of money, clothes, and illness insurance with each child. Many women also needed poor relief, especially single mothers and widows. Because men died early there were several hundred widows with young children in the three counties. Married men appeared less frequently on the lists; those who requested help had large families, low incomes, and perhaps an ill or dead wife. When they finished their working life, all these men and women turned to the parish for some security, and they received a small weekly payment and perhaps a place in a parish cottage.[52]

Besides these regular paupers there were individuals who received temporary relief. Physical and mental illness brought some families to the vestry, and there were single payments to help with migration and emigration expenses. In addition, there was the recurring problem of casual employment. Carmarthen fishermen, weavers, and shoemakers asked for help when their trade was slack, and families of agricultural labourers needed hand-outs in the winter months. Small farmers, too, had difficult periods. 'We have a great number of reduced farmers receiving relief;' moaned John Morgan Howell of Abergwili, a master of 300 acres who contributed £60 to the poor rate in 1833, 'their money generally goes to the beerhouse.' 'There is no sense of shame left', he added, 'except among a few old people.' 'All look to the parish,' exclaimed another critic of the old Poor Law, David Thomas of Llanpumpsaint, 'and they are upon it as soon as they have a child.'[53] As he implied, single men were rarely on the overseer's round; those who appealed for assistance were encouraged to visit all the farmers in the district and offer their labour in return for board and lodging.

Parish relief took many shapes, and one of the most popular was the payment of rents. Cottages were rented by the parish on behalf of paupers, and other infirm and destitute people were boarded at the houses of small farmers. For those poor tenants and labourers who lived at home, there were yearly payments to enable them to meet the landlord's demands. Much to the annoyance of the better-off, farmers sometimes made common cause at vestry meetings, exempting their weakest brethren from rate assessments, and turning a blind eye to rate arrears. Vestries also authorized special gifts of clothing, shoes, food, fuel, and medicine, as well as money for house-building, repairs to water- and fire-damaged property, and contributions to funeral expenses and a trip to the seaside for the dying. Weekly doles were the most common form of relief, especially in Carmarthenshire, and in the first decades of the nineteenth century these amounted to one, two, and even three shillings a week. Hannah Morris, a Merthyr (Carmarthenshire) widow, who obtained 1s. 6d. a week, and £1 for her annual rent in 1836, stands perfectly for them all.[54] The actual figures, however, were not always what they seemed; in a society where money was a scarce commodity, these weekly payments were sometimes given in the form of tickets, redeemable at the local shop. As intended, everyone benefited from the old Poor Law system.

Each rise in poor rates stimulated questions about the purpose and organization of relief. David Williams in 1817 was one of the first to seek practical ways of benefiting both paupers and the community. Indirectly, this had always been done; parishes like Tre-lech and Cynwyl Elfed had connived at the encroachment on commons and waste, knowing that it would reduce pauperism. One project tried in many places was setting able-bodied paupers to work on the roads, but this proved less successful than had been anticipated. In fact, the system of relief in the three counties was increasingly criticized for being administratively chaotic and demoralizing. There was none of the professionalism which had been introduced into some English districts. The poorhouses erected after the Napoleonic Wars did not have the disciplined regimes of their successors. Carmarthen was exceptional; exasperated by the number of resident and vagrant poor, the town in the early 1830s considered a work test, and established a method of monitoring the amount of stones which the paupers broke for the roads. It caused a howl of protest and violence against the Poor Law officials.[55] In a society where so many people were close to poverty, the nature of parish assistance was a matter of great importance. The recipients argued that they had a right to relief, and to relief in a form that was sensitive enough to meet the requirements and crises of a peasant economy.

'UNNATURAL IMPOSITIONS'

'No peasantry likes paying taxes,' said one contemporary. Poverty was bearable, but 'unnatural impositions' at such a time were a different matter. Colonel Love told the government that over much of the region tithes and rents were the principal grievances, and there was growing resentment against the poor and church rates. By the early 1840s it was estimated that the combined amount of tithes, poor rate, highway rate, and church rate was the equivalent of a third of the rent, and to these were added the tolls at turnpike gates, mills, markets, and fairs, as well as other indirect forms of taxation. Some of these taxes were now imposed more frequently, and, perhaps of greater significance, payment was increasingly demanded in cash and on time. The strain of this proved too much. Replying to a claim that only small amounts of money were needed for each of these rates, farmer David Davies of Llanrhystud said: 'Everything is not much; but all these things come together.' 'When everything comes from the same purse', added a friend, 'it comes to a good deal.'[56] On this both freehold and tenant farmers were united.

Of all the financial burdens, the church rate was the least oppressive, and yet it was a perpetual source of annoyance to many ratepayers and Dissenters. It was usually set in the mid-nineteenth century at $1d.-4d.$ in the pound by the parish vestries, and amounted in total over the three counties to only a few thousand pounds. The money was mainly used for church repairs, vestments, and communion services. In some years no money at all was collected for these purposes, and amongst those to benefit were the ratepayers of St Clears, Narberth, and Carmarthen. This was a reflection of clerical apathy and, more commonly, of the growing power of Dissent.

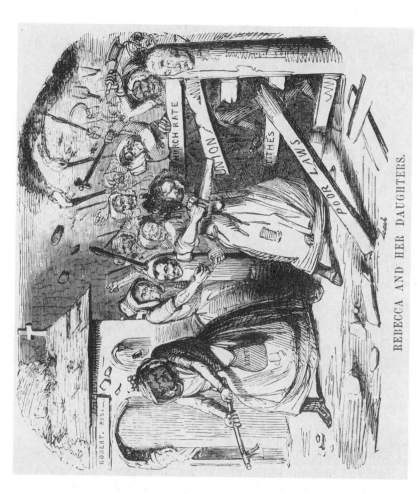

2. *Rebecca and Her Daughters* (*Punch*, 1843)

Poverty, Despair, and Crisis

A spokesman for the Llechryd Dissenters in 1836 said that they had no wish to pay for someone else's church and its sacraments. In the elections of the early 1830s church reform and taxation became an important issue, to which even George Rice Trevor had to give his attention. Lord John Russell and other prominent politicians promised reform of the church rates at this time, but their opponents won the votes and much of the argument. The meeting of mainly Independent ministers near Carmarthen in June 1834, which demanded the separation of Church and State as well as an end to church rates, confirmed the suspicions of the conservatives. The Reverends Ebenezer Morris of Llanelli and Eleazar Evans of Llangrannog were convinced that the Church itself was in danger.

Although parliamentary action over church rates was deferred for a generation, much could be done at the local level. There were anti-church rate demonstrations in various parts of England and Wales when the annual assessments were made. The latter had to be sanctioned by vestry meetings, and there were parishes where determined inhabitants were able to prevent or sabotage decisions. Between 1835 and 1844 at least 45 of the 332 parishes in the three counties refused to agree to a church rate, or obtained a reduction and postponement.[57] In Cilgerran and Tre-lech the church party was defeated year after year, and there were similar victories in a number of Pembrokeshire parishes. In Carmarthenshire, the county most affected by the protest, the ratepayers had their greatest successes across the northern countryside, in the semi-industrial parishes between Carmarthen and Llanelli, and in the largest towns. To the annoyance of the *Carmarthen Journal*, even the Adams family of Middleton Hall, and Sir James Williams of Rhydodyn, meddled in this church-rate rebellion.

The conflict owed something to the aggressive tactics of clergymen such as Archdeacon Bevan of St Peters, Carmarthen, and Ebenezer Morris. They were determined to improve the condition of churches which were crumbling indictments of the poorest diocese in Britain, and they wanted to put a little money aside for church schooling. This aspect of the Anglican revival nourished the growing resentment of poor ratepayers and local Dissent; people complained that the church-rate money was being used for illegal purposes. At Llangunnor, where the parish church was rebuilt, vestry meetings in the early 1840s denounced 'some of the leading men of the Parish' for incurring such a large debt, and resolved that contributions to its liquidation should be related to people's wealth.[58]

Opposition to church finance was directed by *Y Diwygiwr* and *Tarian Rhyddid*, and was often organized by leading Independents and Baptists, men like Henry Davies of Narberth, fellow minister David Rees of Llanelli, and David Gravelle, the farmer of Cwmfelin. Vestries were flooded with their supporters, friends were elected to key positions, and resolutions passed querying accounts and salaries. The outcome of the conflict depended greatly on the way in which the leading farmers of the parish were split; some were Anglicans themselves, or saw the upkeep of the church as something which deserved general, if voluntary, support. At Llanfihangel-ar-arth the Reverend Enoch James and John Lloyd Davies usually triumphed, but at Llangadog the battle which ministers and farmers fought with John W. Lloyd of Danyrallt, Lewis Lewis of Gwynfe, and other powerful landowners was long and hard. When they lost the vote, these rebels challenged the legality of the decision and, in the last resort, refused to pay the churchwardens. Because of this, and reasons of poverty, arrears of church rate rose alarmingly, and Phillip Phillip of Llangunnor was amongst several future Rebeccaites who figured in the lists of debtors.

Scores of parishioners were summoned to appear at the Carmarthen, Pembroke, Narberth, and Llangadog Petty Sessions during the years after the Reform crisis, and orders were made against them. Although there were doubts about the legality of the exercise, cows and furniture were removed from the property of the most obstinate protesters, and sold in the streets. In several instances, as when Messrs T. Griffiths, S. Bowen, and D. Davies and Mrs A. Thomas were summoned for refusing to pay church rates at Cilgerran, petitions were presented to magistrates on their behalf, and there were even appeals to the Court of Arches in London.[59] If these failed, packed vestries agreed to pay their fines. Perhaps the most unfortunate victims in the struggle were David Jones and John James, two Dissenters who were appointed churchwardens of Llannon and Llanelli in 1837–8, and subjected to the anger of

Ebenezer Morris, who held the livings of Llannon and Llanddarog as well as Llanelli. Both were taken before the local ecclesiastical court, prosecuted for neglecting their religious duties, and imprisoned. A public subscription was raised for Jones, and his case was mentioned in Parliament, but, on his premature release from gaol in the summer of 1839, the 'Llannon martyr' quietly passed away, depriving Morris of his planned legal revenge.

There were obvious connections between church rates and tithes, and both had been regarded in this Nonconformist society as in need of reform. Yet there were significant differences between the two taxes; the tithe was a much greater burden, extending way beyond the church boundaries, and it was more difficult to resist. Its origins were endlessly debated at farmers' meetings in 1843; according to David Gravelle of Cwmfelin the tithe had been established under Holy Writ to benefit the Church, the community and the poor, but over the centuries this laudable intention had been largely forgotten. Clergymen in south-west Wales received less from the tithe than their colleagues in any other part of Britain; six-tenths of the income had been appropriated by landowners like the Cawdors, David Saunders Davies, Baron de Rutzen, and Sir Richard Philipps, and, to a lesser extent, by the bishop of St Davids, Jesus College, Oxford, and other religious and educational bodies. Each year almost a fifth of the tithe crossed the national boundary, thus impoverishing an already poor diocese. For Rees Goring Thomas, with his annual income of £1,000 from tithes, they were an investment to be bought and sold, like any other property.[60]

Before the 1830s tithe payments were often made in kind, and, where compositions had been agreed, the amount owing was set below the full tenth of the produce of the land. Tithes, as everyone knew, were lower in the three counties than in most regions, though they had been a source of dispute for years. When the Whig reformers claimed that a commutation Act was necessary to tidy up the whole business, most Welsh politicians accepted it as such. The Act of 1836 authorized a change to cash payments, the level of which was either to be privately agreed between titheowners and landlords of the parish, or imposed on them by arbitration. In the event Welsh farmers, who paid the imposition rather than the landlords, were angry with themselves for allowing sweeping changes in the tithe payment to occur without properly voicing their opposition. When Charles Bishop of Llandovery told a public meeting in 1843 that he had been responsible for commutation in several parishes, members of the audience protested that no doubt he, and the landowners, had done well out of it.[61] The tithe commissioners reported to the government that the process of commutation in Britain had caused very little unrest, except in south Wales.[62]

By January 1841 the commutation of tithes had been confirmed in the majority of Welsh parishes, but in the next three years there were a number of prolonged and acrimonious cases. At Penbryn in Cardiganshire, where the apportionment was sanctioned in 1842, John Rees of Llanarth said that the sows on his farm could have done the work better. The men responsible had miscalculated the statistics, raised the tithe by a third, and apportioned it unequally across the farms. Furious parishioners refused to pay the new assessment, and offered Charles A. Pritchard and John Walter Phillips, trustees of impropriator Walter Rice of Llwynybrain, and the vicar John Hughes 2s. on the rack-rent and no more. After three meetings, the intervention of Edward Lloyd Williams, and threatening letters, concessions were made. At Abernant, where the value of the tithe payments was increased by an even greater proportion, commutation also caused acrimony, and took a heavy toll on the health of David Lewis, the elderly vicar, and on the temper of Lewis Evans of Pantycendy.[63]

The impact of the changes was impossible to calculate, because of the complex arrangements of the old tithe, and the casual manner in which it had been paid. After commutation the tithe rent-charge was typically 2s.–3s. in the pound, which represented an average increase of perhaps 7 per cent on voluntary agreements and half that again on compulsory awards. This itself was a shock, but there was the accompanying realization that the rent-charge had to be paid twice a year, promptly and in cash, with the owner given summary powers of distress and entry. Finally, the precise yearly amount of the commuted tithe depended a little on a moving average of seven years' corn prices. Gone was the old flexibility; tenants in

1842–4 had to pay tithe not fully adjusted to the collapse of the value of farm produce.

The degree of discontent over tithes was not uniform across the region. In north Cardiganshire, where the pace of commutation was slowest, there was comparative peace over tithes, and therefore, it was suggested, less need of Rebecca. In the Teifi valley, where the increase in tithe payments was especially high, and in parishes about Haverfordwest and Llanelli, the opposite was true. The last example deserves a brief examination. Despite reassurances from the agent of Rees Goring Thomas, the farmers of the Llanelli district had to pay the titheowner substantially more after commutation. For Mary Thomas, whom we have met before, the increase was the final straw. In the eleven years after the local commutation Act of 1831 her payments rose from £2. 10s. on her £40 rent to £6. 10s. 10d. In January 1843 she and her husband paid another increase, to £8. 16s. 5d., but six months later could not offer more than 7 guineas. Edwards, the agent, refused to take the money, even as a deposit.

> I thought that very hard, [Mary told the commission of enquiry in December 1843] for it is very hard times with us, for our stock is getting very low. I have nursed 16 children, and never owed a farthing that I did not pay in my life, and we have struggled on as we thought through the hardest times of a family, for some of the children can work now, but we are worse off a deal than ever. Yet my husband has not spent sixpence on beer these 20 years, and we can scarcely hold our heads, nor can I or the children go to church or chapel for want of decent clothing. We perhaps might have gone on, but now this tithe comes so heavy; six or seven pounds is a great deal of money; a difference, indeed it is, more than perhaps gentlefolk think, and I do not know how it is to be got.[64]

Although a few titheowners, including the Reverend Augustus Brigstocke of Newcastle Emlyn, and Llandeilo landowners David Pugh and David Thomas, voluntarily offered cuts in the rent-charge in these difficult months, most of them were wary of making concessions which might be interpreted as a precedent for colleagues and successors. The keenest clergymen saw commutation as a chance for the Anglican Church to recapture former glories. The Reverend Francis G. Leach of St Petrose, Pembrokeshire, argued that the tithe was not a matter that concerned his poor parishioners. The sum was a fixed charge on land, which had been set for all time. He told one meeting that the landlord was bound to pay it and, if that meant increases on the rent, the tenant should remember that he was only his lord's agent and had 'little stake in the country'.[65]

James M. Child, Edward and Robert Waters, and Lewis Evans, who took a very different view, helped to arrange many of the meetings held at the time of Rebecca to petition against the 'deceptive and insidious' Commutation Act. They demanded a maximum tithe rent-charge of 2s. in the pound, 'God's tenth'. Requests to titheowners from assemblies at Penbryn, Abernant, Llanarthne, and half a dozen other parishes achieved some reductions, but it was not an easy cause to fight. Ranged against them were the agents of lay impropriators, and a formidable array of unbending clerics: Sir Erasmus Williams, William Phillips of Crunwear, Eleazar Evans of Llangrannog, Joseph Evans of Llandeilo Abercowin and St Clears, and many more. These titheowners, or their representatives, distrained for tithe arrears, and pursued debtors even after they had given up their farms under the weight of rent and taxes. Phillip Phillip was determined that it would not happen to him. On his 61-acre Llangunnor farm the tithe had advanced from £3. 10s. to £8, and when he attended an appeal meeting he was not allowed to speak against it. In the spring of 1843, as he was sowing, a bailiff walked across the field and demanded twelve months' tithe upon the grain yet to appear. 'Could persons put up with impositions of this kind?', he asked a packed meeting.[66] His practical answer formed part of the Rebecca riots. In some areas people were hardly affected by the tollgates, but most knew 'the iniquities of rents, tithes and the New Poor Law'.

The roots of the disillusionment with the Poor Law Amendment Act reached deep into peasant society. It was partly a question of money; many were convinced that, in spite of the promises, the new Poor Law was more of a burden than the old. At first glance the statistics do not support this argument; the amount of money levied per head of population was a little less in 1843 than it had been nine years before, and Pembrokeshire was the main beneficiary. At about 6s. a head, the level of the poor rate was, in the British context, comparatively modest. Yet there were problems; there had been general rises in 1839 and 1840, and district ones in, for example, the Carmarthen and Aberystwyth unions in 1842. It was no longer possible, as it had been in the past, for parishes to make their own financial arrangements; small farmers in the large Cardigan, Newcastle Emlyn, and Carmarthen unions were angry that they now had to bear the cost of urban poverty as well as their own. To make matters worse, after the implementation of the Poor Law Amendment Act parishes had to be prompt in rendering their accounts, and officers of the unions were liable to fines for failing to collect all the dues. Nor were the ratepayers able any longer to delay payments and to give them in kind. As predicted, the new Act proved unsuitable for such a poor and thinly populated region; when the depressions came, hundreds could not pay the rate, and reluctantly the unions had to consider loans and legal action. Cardiganshire Guardians threatened to resign rather than place the overseers, and whole parishes, in the courts. Edwin Chadwick in London, and assistant Poor Law Commissioner William Day, who was sent down to check on developments, were not amused.[67]

Such was the annoyance at the rules and bureaucracy of the new Poor Law that parishes in south-west Wales were soon petitioning for its removal. Tre-lech, Llanddeusant, and several other vestries tried at first to carry on 'as usual', and were disappointed by the failure of the national campaign against the Act. Rees Goring Thomas forgot all this when he claimed that people in the region had shown no opposition to the Act of 1834 until the *Times* correspondent brought it to their attention nine years later. People like Michael Thomas of Noyadd, Gwynfe, had for years insisted that it was 'at the bottom of all the oppression', and of greater importance than turnpike gates or tithes. Indeed, for all the Tory and Bible radicals on the fringes of the Rebecca movement the new Poor Law was 'the climax of all their local grievances'.[68]

Few things so undermined people's faith in the social and political leaders of west Wales as their acquiescence in the Act of 1834, and nothing was more fascinating during the riots of 1843 than the manner in which James Lewes Lloyd of Dolhaidd, D. Saunders Davies, and other gentlemen denied ever giving it enthusiastic support. History tells otherwise. Lloyd, for example, along with the Reverend Benjamin Lewes and two or three other landed gentlemen about Newcastle Emlyn, was mainly responsible for the building of the workhouse there in 1838. The owner of Dolhaidd had been an early chairman of the union, but only, he declared in 1843, to 'moderate' the worst aspects of the legislation. In truth, he had always found the Welsh farming Guardians obstructive, resisting all change, and when an able-bodied pauper threatened to shoot Lloyd for stopping outdoor relief to him and his friends, he had one more reason to leave for France.[69]

Like the campaign against the Tithe Commutation Act, the campaign against the Poor Law Amendment Act reached a climax about the time of the Rebecca riots. One reason for this was inevitable delay in implementing all the provisions of the Act. The parishes of the three counties were slowly gathered into unions, and some of these, like Haverfordwest and Narberth, with over a hundred parishes between them, were very big indeed. By the summer of 1837 Boards of Guardians had been elected to run these districts. Henry Leach, the Reverend Thomas Martin, and George Roch of Haverfordwest union, and Dr Bowen, Edmund H. Stacey, and David Davies of Carmarthen, enthused over the merits of the new administration. They argued that it was more economical than the old, particularly during times of hardship, and they believed that the workhouse system improved the cleanliness, health, and morals of the 'lower orders'. Not all the wealthier Guardians shared their devotion to the new regime, but Stephen Evans of Cilcarw, like so many others, made the simple distinction between the farmers' representatives on the Boards, and the 'unfeeling and brutal great men and magistrates'.[70]

Four aspects of the Poor Law administration caused Evans and his friends concern: the change in responsibility, the expenditure of the poor rate, the attempt to prohibit outdoor relief, and the treatment of mothers with illegitimate children. According to William Day these complaints lacked authority, coming as they did from people with a vested interest in the old system.[71] The transfer of power was, however, more than simply a matter of politics. People who had different geographical horizons from Day genuinely felt that the Boards of Guardians were too remote from the applicants for relief, and not as accountable as the parish vestry. The new officers appointed were criticized for their lack of local knowledge, even of the Welsh language, and for their arrogance and their indifference. Coroners' juries condemned relieving officers for refusing to give temporary help to families in great distress. The dying poor were occasionally wheeled for miles on carts in an effort to locate the nearest person who had the responsibility to make a decision. Walter Griffiths, Guardian, finally got a Poor Law enquiry into such apparent heartlessness.[72]

Under the new Poor Law there were fewer removals of unwanted paupers from parish to parish, and the very young and old in the workhouses had perhaps a better diet, medical care, and education than their predecessors, but that apart there was less economy and humanity than had been promised. Ratepayers in rural parishes complained that they paid more in cash per year towards the upkeep of the poor, and less of it was actually reaching them. In 1843 one in twelve of the population was a pauper. There was a strong and justified suspicion that, under the new system, a larger share of the poor rates went into the pockets of non-paupers. A quarter of the money obtained was either put towards the county and police rate or used for legal fees, the costs of workhouses and medical relief, and the salaries and expenses of Poor Law officers.

In 1843 there were in the three counties a hundred relieving officers, clerks, and workhouse and medical men on incomes totalling well over £5,000. This was a significant increase in the number of salaried officials since the days of the old Poor Law. In one meeting after another farmers complained that union men were riding about the country like contented barons, while the giver and receiver of poor relief did not have a blanket between them. In an effort to hold down expenses, new appointments were delayed and salaries set lower than London requested. The requirement, outlined in a circular from headquarters in March 1841, that every district had to have its quota of permanent medical officers, with English qualifications, caused much controversy, and when prices collapsed in the Rebecca era there were furious attempts to reduce the incomes of such people. This was stoutly resisted by the Poor Law

Commissioners, on the grounds that in such a poor region the officials had more work to do.

The workhouse became a symbol of the cost, bureaucracy, and hard face of the new Poor Law. Most of the workhouses had to be built in a hurry, though Cardigan and Aberystwyth were late with theirs, and Tregaron and Lampeter were still without one in 1843. They were expensive to build, and often staffed by non-Welsh masters and Anglican chaplains and teachers. People resented paying rates to distant workhouses. 'Let not them send their poor to be confined in large towns,' said Hugh Williams at the great Mynydd Sylen meeting in the autumn of 1843, 'but let each parish retain its own poor, and let the money which the poor receive be spent in the parish.' Although the number of indoor paupers rose dramatically between 1838 and 1844, workhouses still had accommodation for fewer than 2,000 inmates and these were outnumbered more than ten to one by the persons receiving outdoor relief. Much to the annoyance of the Poor Law Commissioners about one in five of these were able-bodied, that is men who were either unemployed or living on insufficient earnings, widows and single women with children, and people who had temporarily lost their livelihood by sickness or accident.[73]

This balance between the two types of relief reflected the poverty of the region, the wishes of farming Guardians, and hostility to the workhouse. The Commissioners quickly found that it was impossible to deny outdoor help to all paupers, especially during the winter months. Even so, efforts were made to remove widows, with children, from the outdoor lists, cuts were made in the amount of relief given to the able-bodied, and meaningless work was occasionally demanded of them. Daniel Thomas, the counterpart of Dr Bowen in the Pembroke union, successfully proposed in April 1843 that no relief in clothing or culm be given in future 'to encourage forethought and provident habits in the pauper population'.[74]

In spite of everything, the poor displayed no wish to enter the workhouse. John Rees of Llanarth said that it was regarded as a disgrace, to both the pauper and his extended family, when people entered the building.[75] The notion of the forced separation of man, wife, and children was repugnant to the Welsh and against their religion. There was little, said Rees, echoing *Seren Gomer*, to choose between gaol and the workhouse; rumours of poor food, work tests, removal of privileges, physical punishment, and a high death rate abounded. Hundreds preferred to starve outside. Aberaeron workhouse never had more than about ten or twenty inmates at any one time, and everywhere the places were occupied by a few old people, single mothers, and especially children, mainly of the illegitimate kind. At the height of the depression in 1843 the workhouses were half empty.

One of the functions of the union workhouses was to deal with the problem of the unmarried mother and her children. Under the terms of the Act of 1834 it was made more difficult for women to establish paternity. Corroborative evidence was now required, and actions against the father had to be started within three months. In 1837, for example, a case was made out at the Quarter Sessions against Evan James, a 55-year-old farmer of Llanfihangel-ar-arth, by Mary David, and he was ordered to contribute 1s. 3d. a week until her child reached the age of 7 years. He could count himself a little unfortunate, for at this time scores of similar paternity cases in southern Cardiganshire and northern Carmarthenshire were being dismissed because they were 'out of time', 'without supporting evidence', and for a variety of other technical reasons. Hugh Williams then earned his nickname of 'the farmers' attorney'.[76]

Very soon there was a wealth of documentation on the difficulties facing unmarried women and the reduced incidence of proceedings. When men no longer feared imprisonment, said James M. Child, women had less prospect of marriage or assistance. Although the legislation was intended to end bastardy by making women 'more careful', its critics claimed that it was a licence for male irresponsibility, especially if the man were 'pretty rich—a half gentleman'. At meetings in 1843, angry fathers claimed that their daughters had been seduced by 'respectable young' masters, and abandoned. Charles Brigstocke, a Carmarthen vestryman, said, with his customary exaggeration, 'no rich person is ever forced to pay'.[77]

The women in this situation knew that they could not, according to a strict interpretation of the Poor Law Amendment Act, any longer claim outdoor relief for themselves and their children. The result, as we shall see in the next chapter, was a fall in the numbers of bastards dependent on the parish, but a rise in infanticide and the desertion of children. 'Oh! it is an iniquitous law,' said one minister, 'and by its severe enactments has induced many women to murder their own children. Women are "the weaker vessels", and should be protected.'[78] There were descriptions, even by magistrates, of mothers and children wandering about the countryside, tired and half-naked. Some of them eventually entered the workhouses, and walked straight through, finding the regime and the separation unbearable. Babies were left inside and outside the gates. The master of Pembroke union admitted in January 1843 that he was embarrassed by the number of abandoned babies on the premises.[79] In that year there were at least 250 unmarried mothers and over 1,000 illegitimate children receiving relief in the region.[80] In Haverfordwest, Carmarthen, Llandovery, Newcastle Emlyn, and other workhouses they comprised the great majority of inmates. When Rebecca promised to rescue 'her children' from the Carmarthen 'bastille' she meant precisely that. As soon as troops arrived in southwest Wales in 1843 they were sent immediately to defend the workhouses. One contemporary was convinced that the Rebeccaites 'took the gates in hand, because they could not do anything to the Union'.[81]

COUNTY RATES, BRIDGES, AND ROADS

One of the arguments over the poor rate was its relationship to the county rate. The latter was a small tax, valued at almost £3,000 in 1832, and varying between $1d.$ and $3d.$ in the pound on the rental of the counties. It was collected as part of the poor rate, the overseers transferring about a tenth of their monies to the county treasurer. For many years the county rate remained virtually stationary, but in the 1830s and 1840s new valuations were made, and greater demands were placed on the county stock. By 1841 the Pembrokeshire figure had doubled over the previous ten years, while in Cardiganshire a similar increase had been achieved in only half the time. In addition there were grants in aid of the county rates from the government. The situation in Carmarthenshire was more difficult to establish, significantly because of missing information, but there was a substantial rise after 1840–1, and by the beginning of 1843 it amounted to some 15 per cent of the poor rate in the Carmarthen and Llandovery unions.

Complaints about the county rate increased with the cost. Appeals against valuations were common, but the most important grievance was about the nature of, and information on, expenditure of the rate. Of all the abuses in Carmarthenshire, said Sir James Williams of Rhydodyn, who received fourteen petitions on the subject one morning in October 1843, 'the greatest . . . was the manner in which the county stock had been disposed of'.[82] It figured in all the lists of grievances adopted at the large public meetings of that year. Amongst the resolutions passed were those for the separation of county and poor rate, for ratepayers' or parish control over expenditure, and for detailed quarterly accounts.

The main items of expenditure were not in dispute; money was needed for bridges, gaols, county buildings, prosecutions, the fees of magistrates' clerks and others, and salaries. What did trouble the inhabitants was the manner in which the money was given and accounted for. Sums were agreed in advance for repairs, attendance allowances, and the like at Quarter Sessions, and magistrates, clerks, and surveyors sometimes dispensed the cash as they thought proper. There were suggestions that not all this money was wisely spent. As Henry Leach, chairman of the Pembrokeshire Quarter Sessions admitted, in his county and in Carmarthenshire the auditing and publication of the accounts left much to be desired. Several hundred pounds a year were lost in 'incidental' and 'other expenses'. In the middle of the riots the Carmarthenshire Quarter Sessions appointed a finance committee of Rees Goring Thomas, J. Lloyd Davies, John E. Saunders, Timothy Powell, and six other colleagues to oversee the larger payments.

Not everyone was satisfied. Cynwyl farmers claimed in 1843 that some of the doubling of their county rate in the previous five or six years had been spent on the private property of country gentlemen and on unnecessary jobs for their friends. Others attacked justices of the peace for dining sumptuously on public money, but Timothy Powell of Pencoed, St Clears, one of the accused, denied it strongly. Corruption was never proved; at its worst it seems that county administrators were receiving payments for work which had once been done gratuitously. Close examination reveals that the increases in the county rate were inevitable given the rising costs of prosecution, policing, and punishment, the additional work of clerks of the peace, and the obligatory appointments of coroners, registrars, inspectors, and surveyors. After the mid-1830s an attempt was made to answer the many complaints over the county rate; some uniformity was brought to the fees and salaries granted at Quarter Sessions, and ambitious plans to build new shire halls, gaols, and bridges were shelved in the early 1840s until better times returned.[83]

One of the most controversial aspects of county expenditure was the large amount of money allocated for the erection and repair of bridges. The increased volume of traffic, especially of the wheeled variety, highlighted the need for sturdy well-built bridges over the many rivers and estuaries of the region. Most of the existing bridges, like the unpopular Llandeilo-yr-ynys one over the Tywi, had been financed by private and public subscriptions, and these became in time the responsibility of the parish, and eventually of the county. The counties found themselves weighed down with the burden of bridge repairs and replacements, and one response was to contract with the turnpike trusts for their upkeep. In the second quarter of the nineteenth century the turnpike trusts and the Quarter Sessions also pressed ahead with new developments, like the suspension bridge at Llandovery and the spectacular Loughor bridge. The Kidwelly turnpike trust spent over £10,000 on the latter in 1836, and was anxious to recoup the money from tolls. It was a lucrative business, and there were protests about the quality of the work and the bulging pockets of the builders and surveyors.

The large sums expended on bridge building were part of a general increase in the costs of road transport. In 1839 just under £5 a mile was spent on the highways of the three counties. There were three types of roadways; private paths, parish roads financed since 1835 by a highway rate and repaired by statute, casual, and pauper labour, and other routes established and largely maintained by turnpike trusts. Amongst the highways built or adopted by these trusts were main routes between towns, and access roads to quarrying and mining districts. The purpose of the turnpike trusts was to finance major road improvements by obtaining private and Exchequer loans. Most of the money was advanced against security provided by local tallyholders, who were promised a rate of interest at 5 per cent. Eventually, it was hoped that everyone would be paid off by the system of turnpike tolls.[84]

At the beginning of the century, turnpike gates were erected about every eight miles or so, and the tolls varied according to type of vehicle, width of wheel, number of horses, and the load carried. A farmer in 1825, with a cart and two horses, expected to pay 2*d*. each time he passed the Maesoland gate at St Clears, or the Croesllwyd gate on the Llanddarog side of Carmarthen, and for this he received a ticket which cleared the neighbouring gates on the same trust. If the farmer used the route regularly, he agreed with the toll collector a sum that covered a season's use, and it was common to give a reduction when the load was manure. Some barriers were only manned for twenty-four hours on market and fair days. All this suited the collectors, who were part-time shoemakers, weavers, and tailors.

Soon, however, these golden days passed, and the turnpike trusts were overtaken by pride, zeal, and excessive ambition. The appearance in south Wales of pioneer roadmakers Telford and Macadam acted as the spur, and the later 1820s and the 1830s saw an extraordinary competition between trusts to have the biggest and the best. In the south-eastern corner of Rebecca's country the Three Commotts and the Kidwelly trusts criss-crossed Carmarthenshire like scuttling spiders. Some of the enterprises launched at this time were roads from Hobbs' Point eastwards and shorter ones north and south of Cynwyl Elfed. John Lloyd Davies of Alltyrodyn and John Lloyd Price of Glangwili, who were associated with the last projects, as well as other ambitious landowners, sponsored new schemes and persuaded the counties and the Exchequer to provide the extra capital which the trusts needed. Cynics suggested that these road and bridge improvements followed the contours of the great houses, and that private walls were built and views improved with trust money. 'Generally', said John James Stacey, an expert on these matters, 'we find them [the roads] pretty good about gentlemen's houses.'[85] Those accused of using public funds for the benefit of themselves and their tenants included John E. Saunders, Lewis Lewis of Gwynfe, David Davies of Carmarthen and Trawsmawr, Sir Richard Philipps, Rees Goring Thomas, Sir James Williams, and the earl of Cawdor.

The costs of building and maintaining these turnpike roads were always higher than predicted, not all the projects were finished, and the tolls never brought in quite the amount required. Where they were permitted by Act of Parliament, tolls moved slowly upwards, to 2½d., 3d., and even 4d. for a horse and cart by the beginning of the 1830s. How much of this reached the trust is unknown; the men and women of the tollhouses received very low salaries, and the accounts were in an appalling mess. Certain gates brought in very little money indeed, while other barriers were rich sources of income. The best were probably those on the perimeter of the main towns; Lampeter, Newcastle Emlyn, and Llanelli each had at least five within close range. The Verwig gate at Cardigan, the Plaindealings at Narberth, the Prendergast at Haverfordwest, and the Royal Oak of Carmarthen were the source of much annoyance; they cut off the farmers of the rural outskirts from the rest of the towns' population, and caught the carrier passing through with lime and culm. The Water Street gate, on the Cynwyl road out of Carmarthen, was, for much the same reasons, another successful venture, despite doubts about the legality of its charges. The Water Street, the Glangwili, the Whitehouse, the Croesllwyd, the Pensarn, and the Royal Oak gates, which encircled Carmarthen, had a combined annual income of almost £2,000 by the 1830s.

The other obvious places to catch travellers were at important road junctions, like Porthyrhyd in Llanddarog parish, Ffairfach south of Llandeilo, and Bwlchyclawdd above Cynwyl, or where there was heavy short-distance traffic. It was estimated that up to a half of the Cardiganshire and Carmarthenshire tolls were obtained from people carrying lime, coal, and stone. Thus bars were set up at the Mansel's Arms, between Porthyrhyd and Drefach, to tax carts bringing coal from the pits, and at Pontarllechau, to intercept the traveller on his way from the Black Mountain lime kilns. The roads about Llandybie, Llannon, Kidwelly, and Llanelli were dotted with gates and bars. No one in 1843 could enter or leave the last town without paying dues at bars at the entrance to the parish roads, or at Sandy, Furnace, and three other gates on the turnpike roads. Newspaper reporters were astonished by the scene; one counted eleven barriers in his fifteen-mile journey between Carmarthen and Pontarddulais, and another saw seven in the eighteen miles between Lampeter and Llandovery. Had they all belonged to the same trust, this would not have mattered greatly, but south-west Wales was exceptional for the number of such bodies. In 1843 there were twenty trusts in the three counties, responsible for just under a thousand miles of road, and with, on average, a gate or bar every four miles to catch the traveller.

In spite of the proliferation of gates almost all the turnpike trusts were heavily in debt. Those in Carmarthenshire were well over £100,000 in the red; the Three Commotts had a bonded debt of £9,464 and were in arrears of £15,000 on interest payments, and even the little Carmarthen and Brechfa trust, with one road only to oversee, had a debt of £2,574. This meant cutting back on the £10-a-mile repairs of turnpike roads, seeking greater help from the parish authorities, and obtaining better returns from the renting of the gates. In the mid-1830s John Lloyd Davies's Carmarthen and Newcastle trust got parliamentary sanction for a 50 per cent increase in their toll charges, thus making the holding of their gates more attractive. Letting the gates annually to the highest bidder was a well-established practice, but in the 1830s and 1840s a small number of professional toll farmers like the Englishman Thomas Bullin and William Lewis of Swansea won large, longer-term contracts. In 1839, for example, Bullin raised enough capital to hold the gates of the Carmarthen and Newcastle, the Whitland, the Main, and four other important trusts.

As a reward for paying much higher rents, Bullin and his colleagues demanded more side-gates and bars at road junctions and were more thorough about collection of tolls. At St Clears, where the Whitland and Main trusts met, new gates and side-bars appeared, and drovers on their way from Pembroke to Carmarthen now had to pay heavily on their cattle. In the Three Commotts and Kidwelly trusts especially side-bars were introduced in great numbers, travellers were asked for the maximum 6d. charge on loads of lime, coal, and peat, as well as on other goods, and cash was demanded on the spot. This, in some cases, effectively doubled the cost of the farmer's 2s. 6d. load of lime. In the parish of Llanarthne, where there were three gates and eleven bars, seven of which were on parish roads, evasion or concessionary payments were out of the question, and even the Reverend Herbert Williams and some of the biggest farmers voiced their objections. Phillip Phillip, in a neighbouring village, said that the arrogant William Lewis, and the new race of toll farmers, did 'tyrannize over the poor, and kindle in the country a spirit of disaffection and revenge'.[86]

The 'evils' of the turnpikes figured prominently in the meetings and the petitions of the Rebecca years. The following comprehensive list of grievances and reforms was drawn up by a parish committee in Llandyfaelog in October 1843:[87]

1st. Unlawfully [erected gates] to be removed.
2nd. Letting each separately and annually.
3rd. Sums collected to be expended on roads which it clears.
4th. People employed, being parishioners, which road passes.
5th. All manure free of toll.
6th. Cart, one horse, 4d.; horse, 1d.; ass, ½d.; carriages as before.
7th. Toll once a-day.
8th. Free for religion.
9th. Magistrates not to be trustees of road.
10th. Quarterly accounts of money collected and expended in each Trust.

As this, and other statements, indicate, the peasantry of south-west Wales were aggrieved by the administration and costs of the turnpike trusts. There were doubts about the legality of the gates, the number of tolls that could be demanded over a piece of road within twenty-four hours, and the changes in toll charges. Amongst the loudest complaints were those over the toll paid on every third visit to a south Cardiganshire gate, and the sudden inexplicable increase in tolls at the Llandeilo-yr-ynys bridge. Those gates within, or on, the boundaries of Newcastle Emlyn, Fishguard, and Haverfordwest were erected in defiance of local Acts, and were eventually moved. Other barriers of questionable legality were temporary affairs; collectors appeared as if by magic, with chains trailing behind. Perhaps the most unpopular barriers were the sidegates and bars erected at the mouth of intersecting parish roads. People complained that they were being forced to pay for using only a few yards of turnpike. Certain Acts also gave trusts the right to appropriate sections of old roads, which had been built by statute labour and were still repaired by the parish. This was a particular grievance in the area of the Whitland trust, where parishioners were fined for the nonrepair of both parish and turnpike roads.

From the Llandyfaelog and the Llanboidy petitions it was clear that parishioners wished to end the secrecy that surrounded the affairs of the trusts. Their administration was in the hands of the trustees, but they were actually run by just a few individuals, like John Lloyd Davies and John E. Saunders. People were naturally suspicious when they pushed their pet projects and when interest on loans was paid to some and not to others. It was hard to understand why the first loans had not been paid off, giving people the promised free transport for fuel, forage, and manure. Despite the conditions in governing Acts of Parliament, accounts were poorly kept, and rarely published. George Ellis, who examined trust finances for the government in 1843, could not find any books or accounts for the Carmarthen and Lampeter trust. The Whitland and Fishguard trusts also had very poor management. Small wonder that people were loth to accept an extension of the turnpike system, and any merger with the cheaper administration of parish roads. Petitioners wanted the trusts to confine their attentions to major and long-distance roads, though a few radicals called for the abolition of all private agencies.

One of the counter-claims in the discussions over the Rebecca riots was that magistrates had received no prior warning that the population was unhappy about the turnpike trusts. In fact, so the argument ran, most of the gates had been established years before, their charges were more reasonable than elsewhere in south Wales, and they had been accepted because of the benefits of improved travel. This claim, pressed by George Rice Trevor and others, was not strictly accurate; several meetings had been called before 1839 to protest over county and road taxes, and, as we shall see in Chapter 5, a few people had extended this to direct action.[88] But Trevor's defence cannot easily be dismissed, and it prompts two interesting questions: did the peasantry see little advantage in polite protest, and when, and why, did the burden of road taxation become unbearable? Everyone knew that the justices of the peace, to whom complaints were referred, were frequently both important trustees and creditors. Those who claimed that certain bars and charges were illegal hardly expected justice from the Reverend Thomas Martin of Withybush or John E. Saunders of Glanrhydw; these were the very trustee-magistrates who, month after month, insisted on handing out the maximum fines for non-payment of tolls.

It is difficult to quantify the burden of road taxes and tolls on the ordinary people. They were amongst the smallest forms of taxation, and many were not directly affected by them. A small farmer in 1843 expected to pay a parish highway rate of about £1 a year, but perhaps as much as that again disappeared in tolls, especially if, like James and David Evans, he lived in close proximity to the Pensarn or Bwlchyclawdd gates.[89] Phillip Phillip, farmer and carrier, felt the pressures keenly; because of the tolls it was more expensive to lime his land, and most of the profits of his second source of income disappeared on the road. Case-studies of such farmers show that transport was costing them not much less a year than the poor rate or tithes. If the rate of road taxation was lower in the three counties than in certain other parts of Britain, nowhere at this time was it more acutely felt and more stringently imposed. Everyone, including the labouring families, was affected by farmers' problems and by rising transport costs. Although there were people, in authority and on the fringes of the popular movement, who were anxious to play down the grievance, turnpike gates brought Rebecca to life and provided the rioters with their main and most accessible targets.

CRISIS

The anger over tolls and other 'unnatural impositions' was not of itself sufficient to bring about a great popular movement. Not every district experienced the iniquities of the tollgates or the burdens of high tithes, rents, and poor rates; each area and each social group had its own list of grievances. The first outbreak of the Rebecca riots in 1839 passed largely unnoticed outside the Pembrokeshire–Carmarthenshire border, and ended once the particular cause of discontent had been removed. In normal times, even with taxation rising, it seems unlikely that any of the family of David Evans of Cynwyl or of Mary Thomas of Llanelli would have joined others in a campaign of direct action. What united so many of them, farmers, craftsmen, and servants all, was a growing sense of injustice, suspicion of those in authority, and economic disaster. In 1843, when the riots at last swept across much of the region, the population were experiencing the rigours of bureaucratic government, commercial mentalities, and a cash economy when suddenly the bottom fell out of the agricultural market. 'Oppression' and 'poverty' were the popular explanations of Rebecca's appeal, but it was the crisis of 1842–3 which sharpened the focus.

Poverty was, after all, not a new phenomenon. There had been bad times before and the peasantry had survived them, and there were to be bad times again, before the decade was out. The nature of their agriculture, the reliance on family labour, and cautious and thrifty habits kept these people afloat. Every substantial fall in prices spelt ruin for some, and stimulated a change in occupations, migration, and emigration, but most people carried on as before. Only the great slumps had a major effect on the fortunes and psychology of the west-Wales peasantry, and brought violence, social criticism, and political awareness in their wake. These were the times, as one Llannon farmer put it, when flesh was torn from the bone, and the people grew reckless and mad.[90]

Compared to that of the post-war years, the depression of 1842–3 was perhaps less serious, but its impact was greater because the people had a wider range of grievances, a heightened sense of independence, and a formidable array of spokesmen. For a while at least, large sections of the population, from farmers to labourers, and from ratepayers to receivers, were united by a common bond and sense of purpose. 'I hope you can sympathise with us', ran a Llanedi tenant farmer's address to the freeholders, 'for you also must feel the burdens of the times.'[91] No one, Stephen Evans told a large audience in the autumn of 1843, could now say that he was safe from the bailiffs or the workhouse. 'You come to me at last', echoed the Lady whom he once impersonated.[92]

The outstanding aspect of the crisis of the early 1840s was its comprehensive nature; commerce, industry, and finance, as well as agriculture, experienced a sharp downturn in their fortunes, and this in turn raised the political temperature, and stimulated the contemporary debate over social policy and administration. The agricultural decline went further than anyone had predicted. After the fall in prices about the time of the Reform crisis there had been a recovery, so that by 1839–40 most commodities were selling at good prices.[93] Not everyone benefited, however, as the poor harvests, which were partly behind the rise in prices, meant that the least fortunate of the small stock farmers had to buy in corn and hay. As the demand for cattle, sheep, and pigs in south Wales also fluctuated considerably at this time, it was becoming harder to find the cash for rents.

After the harvest of 1840 the price of wheat, barley, and oats at the Carmarthen market fell sharply, and although there was some recovery a year later, the decline continued for another two years. Initially this drop in corn prices was regarded as part of the natural ebb and flow of the rural economy, and there were some rent reductions to accommodate distressed farmers. At a dinner in December 1842 David Saunders Davies, the Carmarthenshire MP, stated that the agricultural collapse had been the result of panic, and he was confident that with an excellent harvest behind them the worst was over.[94] In fact, the depression had begun to deepen, with stock prices falling as well as others, and for most observers the next stage in the crisis was the vital ingredient in the Rebecca saga.

In January 1843 Carmarthenshire farmers were getting only 60 per cent of the price which their corn had fetched three years before. Worse still, those selling meat, cheese, and salted butter at Carmarthen and Swansea markets discovered that these commodities had fallen even more in the last two years, while the demand for timber had all but disappeared. In the spring and summer of 1843 cattle were driven to the markets of south-west Wales, and home again; drovers were not to be seen and desperate farmers accepted ludicrous bids for their produce in order to pay rents, rates, and taxes. As things grew worse in the summer of 1843 even farmers living close to large urban and industrial markets protested about their plight. In August Llannon farmers said that 'the horse we had a little time back which would fetch £18 is now sold at £8; the bullock that would sell at £10 is now reduced to £5 or less, and all the livestock in proportion. Butter, which was sold for 1s. per pound is now 6d. Cheese that was 3½d. per pound is now 1½d, and the last named articles were our chief produce to pay our rents.'[95] There was so little trade at Carmarthen, Newcastle Emlyn, and Tregaron markets that money almost ceased to circulate, and shopkeepers and craftsmen found it difficult to survive.

As part of the farmers' response to the crisis, the labourers of the countryside were subjected to cuts in employment, wages, and perks. Workers about Cardigan, Newcastle Emlyn, Whitland, Llandovery, and Carmarthen were increasingly put to part-time labour, and women and children were not taken on as usual. Even in Abernant, where wage rates had always been reasonably good, it was said in 1843 that 'many of [the able-bodied labourers] do not get constant work. They have no other means that I am aware of to support their families . . .'[96] When pleading and intimidation failed to do their work, wage rates fell, and cash payments to workers were delayed. Some of the larger farmers slashed the customary harvest increments of extra money, food, and drink. In the worst parishes farm labourers were living on about 3s. 6d. a week, and prospects of alternative employment were bleak. Although some reporters felt that, because of the low food prices, farm servants and labourers were not hit as badly by the crisis as the small farmers, no one escaped its impact.

From the parish meetings held to discuss grievances, it is evident that people were perplexed by the pace of the economic collapse and ready to blame almost anyone and anything for it. The temptation was to search for local targets and tyrants, and seek immediate solutions, but a few speakers heeded the words of the English- and Welsh-language press, and looked instead to Westminster. David Gravelle and like-minded farmers' spokesmen put the blame on Robert Peel and two years of Toryism. The Prime Minister was attacked for refusing to make changes to the income and malt taxes, and especially for damaging agriculture by his tariff and sliding scale. He was accused of being too protectionist over timber and coal, and too liberal over imported cattle and meat. The Carmarthenshire Quarter Sessions, the borough of Cardigan, and a number of the rural parishes protested to Parliament about the damaging effects of the Canadian Corn Bill and cheap imports of flour. 'Look only to the last Session', advised the *Welshman* in August 1843. 'In one short Session [Peel has] contrived either to impair or destroy every branch of industry in Wales. The Coal Trade—the Copper trade—the Iron trade—the Tariff, and the Timber trade, with that ever shifting and most destructive Sliding-scale, all attest the merits of that legislation which at one and the same time has paralysed the commercial interest and impoverished the agricultural resources of Wales to a pitch without a precedent in the annals of its history.'[97]

There were many, from David Morris, Carmarthen's Whig MP at Westminster, to Michael Thomas of Gwynfe and other local Peelites, who believed that the above analysis was too simplistic, but they readily accepted that the crisis in the three counties owed much to developments elsewhere. In recent years the farms of the south-west had provided more of the food supplies of the growing towns and industrial districts of south Wales, and their prosperity or otherwise had a marked influence on the agriculture of the region. 'Land can no longer prosper in this country,' observed Sir James Graham, 'if trade and commerce be stunted.'[98] 'All look to Merthyr' was a common saying of this period, and some people argued that its emergence as the greatest town in Wales had a baleful influence on the whole Principality. Fortunately, the mining, copper, and tin industries had, to a large extent, escaped the depression which had begun to spread to other manufactures after 1837, but in 1840 there were reductions in wage rates at the south Wales ironworks, and more in 1841. After the autumn of the latter year, the whole of British industry fell into a deep and long trough. Only a few enterprises, like the Cardiganshire lead mines, were comparatively unaffected.

In the small mining districts of Pembrokeshire and Carmarthenshire, at the copper and tin works around Llanelli and Swansea, and right across the south Wales coalfield hundreds of men and women were laid off, wages substantially reduced, and works closed down. At Merthyr Tydfil, where iron prices dropped by 300 per cent in a short time, furnaces were shut down in 1841 and 1842, and strikes and political agitation were the order of the day. A few miles to the east, at Ebbw Vale, the huge iron company of the Harfords almost went out of business in 1843, and a worried government moved the troops a little closer. The first half of 1843 was characterized by colliers' strikes over much of the coalfield and by the violent outrages of the Scotch Cattle, a form of workers' action which had a resemblance to the Rebecca riots.[99] In the copper industry along the coast between Llanelli and Aberavon it was a similar story; in the late summer and autumn of 1843 the threat of another cut in wages produced an extended strike, mass demonstrations, hungry families, and daily prayer meetings. Magistrates, police, and soldiers did their best, but it was months before the commercial clouds lifted.

Dr Picton, speaking at a Llandyfaelog meeting in September 1843, claimed that the great men of west Wales bore little responsibility for the troubles; Rebecca had entered people's lives because of the sluggish nature of the iron trade.[100] It was an over-simplification, but a good debating point. The contraction of industry in Glamorgan and Monmouthshire meant that people did not have the money to buy Carmarthen foodstuffs, and people struggling on the farms of rural Wales were denied traditional escape routes and job opportunities. In the spring and summer of 1843 the reverse process was at work, for hundreds of people from the ironworks and collieries returned to their home parishes in the south-west, renewing old acquaintances and seeking temporary assistance.

From the autumn of 1842 onwards reports tell of roads full of people moving to and fro between the rural, urban, and industrial parishes. Some were farmers with carts of unsold butter, some were labourers and travelling craftsmen with a little money in their pockets, others were half-way to becoming vagrants, and the lucky ones were making their way to the emigrant ships. John Jones and David Davies, two of Rebecca's most violent children, were about to journey westwards across south Wales, and so was Mary Phillips. Her Merthyr husband had just died, and she was coming home to Pembrokeshire for assistance. She and her young children were just another burden that nobody wanted. 'God help us all', shouted a Cardiganshire beggar outside the workhouse on a cold January night in 1843; it seemed an appropriate comment.

On the smallholdings the owners and occupiers could not remember worse times. With no market for their animals and other produce, rents, tithes, and rates went unpaid. In August John Hughes, in his survey of William Powell's Nanteos estate, pleaded with the landowning MP to make a permanent reduction of a seventh of the rental, which would give the farmers confidence and improve their spirit and their agriculture.[101] At Rhydodyn the arrears and allowances amounted to a fifth of the rental, and other estates were in a similar position.[102] Much depended on the area, or whether Rebecca had influenced the farmers; in certain parishes on the Slebech, Derwydd, Coedmore, and Cawdor estates, for example, arrears were nearer a third or more, and 'will not pay' as well as 'promised to pay' was entered in the ledgers. Landlords accepted loads of hay, promises of work, and shares in property instead of cash.[103] Rentals of the Cawdor estate show that it was often the smallest tenants who were least able to pay, though even larger ones sometimes defaulted on part of the rents. R. B. Williams, the Cawdor agent, claimed in the summer of 1843 that tenants had managed remarkably well, but a year later his return revealed that arrears on the Carmarthenshire and Pembrokeshire properties had risen by over 50 per cent in two years.[104]

As always at such times, there were requests for concessions on the rents, perhaps by as much as 4s. in the pound to take account of the low prices. 'Every landlord must know that the tithes and poor man's rate have very much advanced, and our rents remain the same,' ran a typical statement of aggrieved farmers, '. . . while the reduction in every article we have for sale is very great . . . As soon as the time will mend, and we have advanced a little from our distressed state, we will be ready and willing to take the burden back again on our shoulders.'[105] Led by the Colbys of Ffynone, Lewis of Clynfiew, and Pugh of Greenhill, a small group of landowners made the required reduction in the winter and spring of 1842–3, and more offered cuts of only 10–15 per cent or simply allowed arrears to mount.

During the worst months of 1843 people seemed genuinely surprised by the number of landlords, big and especially small, who were 'unwilling to listen to anything in the rent'.[106] Another round of promised reductions followed in the late summer, under pressure from mass meetings and threats of arson, and thanks were duly paid to Cawdor, William Brigstocke, J. S. Harford and others. Other concessions were regarded as inadequate, and criticized for the strict conditions attached to them. Sir James Graham, John Lloyd Davies of Alltyrodyn, Thomas Cooke, and Thomas Campbell Foster all concluded that it was the inflexibility of landowners which turned the crisis into confrontation. Lewis Evans, the Tory radical of Pantycendy, anxious to keep the social contract between great and lesser men, explained that the former were not in a financial position to meet the people's wishes, but most were unconvinced. 'Had our masters been generous in these bad times', said the farmers, 'we should not have turned to Rebecca.'[107]

A few trouble-makers insisted that the most 'tyrannical landlords' used the economic crisis to remove bad and unpopular tenants. It was hard to prove, and unlikely, for this was not the moment to make new agreements, nor to press for distraints. Even so, as Michaelmas 1843 approached the mood of the rural population was fearful, tense, and threatening. As we shall see in the later chapters, legal battles, distraints, and ejectments were a common aspect of rural life at this period, and an exceptional number of farms were left vacant. Sales of stock, timber, furniture, and land filled the advertisement columns of the local journals, and even fairly large estates passed under the auctioneer's hammer. To complete the picture of woe, scores of farmers, cattle-dealers, and agricultural merchants went bankrupt, as well as innkeepers, lime-burners, and tradesmen of the small country towns. Some of the debtors, owing sums between £7 and £500, entered the counties' gaols.[108] There they met an exceptionally large number of inmates, for a feature of this early Victorian crisis was the high level of recorded crime and deviancy.

4
Crime and Deviance

ALTHOUGH the Rebecca movement shocked many people, its appearance was not a total surprise. As in many other parts of Britain the late eighteenth and early nineteenth centuries witnessed rising crime rates and violent forms of protest. Food riots were a feature of south-west Wales at the turn of the century, and these were followed by a number of disturbances over enclosures and the removal of squatter dwellings. The post-war depression was a time of widespread discontent, with mass begging, attacks on sheriff's officers, and malicious damage to property being amongst the most obvious signs of poverty and tension. From the records of the Assizes and Quarter Sessions, summarized in Fig. 1, it seems that criminal activity increased at a record pace in these few difficult years. The subsequent swings in the graph were less sharp, though there were upturns in the early and late 1830s, and especially during the Rebecca era. In the three counties, even allowing for population change, the figures of committals for serious criminal behaviour peaked in the years 1842–5. This was partly a reflection of the growing willingness to use the Crown courts, but it was also an indication that times were out of joint. All the statistical and literary evidence supports the claim that this was 'our most shameful' period; prisons in the area were full to overflowing, and not just with Rebecca suspects.[1]

In retrospect it is clear that views of the Welsh and crime have affected our picture of the Rebecca riots. Defenders of Wales, county leaders, and Nonconformist ministers were anxious to portray the peasantry as exceptionally loyal, peace-loving, and non-criminal. 'For the past fifty years this has been the least criminal part of the country', was a typical comment at the time of the Rebecca troubles. There were minority voices. T. S. Biddulph of Amroth Castle told the Home Secretary that 'petty thefts and depredation' in his district

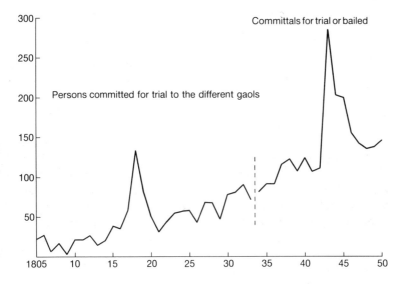

FIG. 1. Indictable crime in Cardiganshire, Carmarthenshire, and Pembrokeshire, 1805–1850

'vastly exceeds the average in a similar amount of population in the English counties'. But when, at Board of Guardians meetings, Dr Bowen of Carmarthen and Henry Leach said very much the same thing about the Welsh working class generally, they were loudly condemned by outraged farmers. The people were, said Reverend Thomas Martin and George Roch in 1839, when rejecting the notion of an expensive Rural Police in their county of Pembroke, well conducted and law-abiding, and effectively kept in order by the parochial constables.[2] This confidence, as an aggrieved landowner noted, proved awkward when the troubles broke out. The Rebecca saga had to be seen either as an inexplicable aberration from typical peasant behaviour or perhaps the result of outside agitation.

Superficially there was much truth in this flattering picture. Judges were used to finding half-empty gaols and thin calendars of prisoners for trial, and winter Assizes were almost unknown. In 1830 Judge Goulburn, speaking at Carmarthen, delivered a eulogy on the character of the native population, and, like others, hoped that the passing of the Great Sessions courts would not change this. In fact, the merging of the Welsh and English judicatures made little difference; three years later Judge Bosanquet commented on the paucity of crime on this side of the border and the comparatively minor nature of the offences. There was a feeling in legal circles that south-west Wales, particularly Cardiganshire, was not sharing in the general increase in crime, and could be excused the government circulars on punishment and transportation. In the spring of 1838 it was reported that there were no prisoners, though some bailed offenders, on the calendar for trial at the Carmarthenshire Assizes, a situation that was to be repeated a year later at Pembrokeshire Quarter Sessions.[3] Indeed, the figures of cases at the higher courts throughout the second quarter of the nineteenth century confirmed the superficial impression that the crime rate in this region was no more than a third of that in England.[4]

Judges attributed this divergence to the absence of large towns and industry in the three counties, to the special care of Welsh magistrates, and to the religious self-education of the peasantry. In Cardiganshire, too, there was praise for the landowners who stayed at home and attended to the wants of their people and the duties of judicial administration. Yet there were other and perhaps more pertinent reasons for the low judicial statistics, and the most obvious of these became a matter of deep concern during the Rebecca troubles. The machinery of law and order in the three counties neither encouraged nor facilitated that many cases. 'In Pembrokeshire and Cardiganshire', if not in Carmarthenshire, thought the Home Secretary in 1843, '. . . the constituted authorities have fallen into disrepute . . .'[5]

The administration of law in the region was, he had been informed, slow, expensive, and unpopular. There were complaints from judges about cases that had been badly prepared, of witnesses and depositions arriving too late, and of the 50 per cent acquittal record of Welsh Petty juries, which was considerably above the English equivalent. It was said that jurymen were influenced by the harsh punishment likely to be inflicted on defendants, and certainly they often recommended mercy for the men and women whom they convicted. Only about 15 per cent of the guilty offenders at the higher courts suffered death and transportation; most of the remainder received short gaol sentences.⁶ From time to time legal officials grumbled about this trend, and also about the poor attendance of magistrates and jurymen, especially in the harvesting season. In two of the three counties the Sessions alternated between the largest towns, and some magistrates made no effort to attend the furthest venue; this sparked off a long wrangle about the wisdom of having one central meeting place.

The Petty Sessions were on a different scale, for they administered justice in some twenty-nine divisions, which were comparatively small in size. In fact, one of the problems, highlighted by all popular disturbances, was the lack of active and resident justices in divisions such as Penarth in mountainous mid-Cardiganshire and the large Hundred of Kemes in north Pembrokeshire. Those reporting a crime between Cardigan and Lampeter sometimes had to walk a dozen miles and more to find magistrates at home, and Petty Sessions had to be postponed. At Fishguard it was common practice for people to ignore petty offences, rather than bother the neighbouring gentry. Nor were the latter all that interested; some of the people on the commission of the peace admitted that they lacked the legal knowledge to deal with difficult cases, and were only too willing, especially as they grew older, to pass on the responsibility to others. In the Hundred of Kemes John T. W. James of Pantsaison and the Reverend John Pugh of Ffynone covered for the absent George Davies Griffiths of Berllan and gouty James Bowen of Troedyraur.⁷ About a third of them, though inserted on the commission, never took out the dedimuses to execute the functions of a JP, and of those that did, in Carmarthenshire, another third were absent or unable to act. Altogether it was hardly surprising that the changing incidence of criminal cases across the three counties owed much to the zeal or otherwise of resident magistrates.

'There is no Law, Justice or Safety here', wrote industrialist John Vivian, in a comment on Aberystwyth magistrates during a violent strike in 1844.⁸ His words echoed those of Augustus Brackenbury, the Lincolnshire gentleman, twenty years before, and reflected the frustration in dealing with magistrates who preferred a quiet life to the problems of illegality. Edward Lloyd

Hall, the lawyer of Newcastle Emlyn, accused them of ignorance, incompetence, and bias, and although he went too far, there were cases in this period of wrongful arrest, and judicial decisions changed on appeal. 'Very few of these county gentlemen', said Colonel Love, military commander of south Wales in 1843, 'can be trusted out of your sight.' 'There is not half a dozen on the whole quorum who holds an opinion worth having, or who are of the same mind three days together.'[9] It was a little harsh; in Petty Sessions they were quite decisive, giving a conviction rate which was well above that managed by juries at the higher courts. But, as they gracefully admitted, the magistrates at their monthly meetings deferred on many legal matters to their clerks, whose power, arrogance, and charges figure prominently in the lists of Welsh grievances. For some contemporaries, including the Home Secretary, the apathy, ignorance, and fears of the gentlemen on the commission of the peace combined to turn disaffection into open rebellion, though ironically it was the most neglected districts which were sometimes the least affected by the Rebecca troubles.

Magistrates were heavily dependent on the high constables, who in turn supervised the part-time parish constables. Edward Lloyd Hall, the most damning critic of the legal system, claimed that in one emergency during the riots of 1843 no one could tell him the names of these local officials.[10] They were appointed and sanctioned yearly at court leet, parish vestry, and Petty Sessions. These craftsmen, shopkeepers, and farmers proved fairly satisfactory in identifying possible executors of crimes, especially if they were well paid for the task, but they were less good at taking them into custody. Rescues and escapes were common, as constables led people across rural parishes to the lock-ups and prisons in the neighbouring towns. On one occasion, so we are led to believe, a woman arrested for infanticide twice managed to slip the clutches of three constables in Penbryn. Even Llandysul and St Clears, communities of several hundred people with bad reputations for disorder, had no lock-up in 1842; local victims of crime apparently preferred not to seek an arrest because of the difficulties of securing a prisoner. One result of the Rebecca troubles was a sharp rise in the number of police stations in the region, as well as a general improvement in the gaol accommodation. The county gaol at Cardigan was repaired in 1839, but the borough gaol at Carmarthen remained notorious for its open character. After one of its prisoners became pregnant, and others escaped with embarrassing ease, the town authorities sacked the dejected governor, David Morris. He protested that people did not realize the problems which confronted the under-funded upholders of law and order.

By the time of the Reform crisis it became obvious that, in the growing towns at least, changes in policing were necessary. Carmarthen and Haverfordwest, which witnessed the greatest amount of popular unrest, appointed London policemen as superintendents and gave them the support of a few full- or part-time watchmen and a dozen parochial constables. Such a force set higher standards of public behaviour, limited the scale of popular recreations, and harassed beggars and vagrants. Later Aberystwyth, Milford, Pembroke, and a few other places moved hesitantly in the same direction, although the numbers of men appointed were never enough to make a decisive impact. Carmarthen, which had the biggest professional force, increased this from 4 to 6 in September 1843, but it had three different inspectors within a year, and their men were regularly dismissed for drunkenness, insolence, and cowardice. The watch committee, formed as a result of the Municipal Corporations Act, claimed that the policemen were invisible when most needed.

Like the parish constables, these new men were unable to cope with serious outbursts of disorder or execute magistrates' warrants at such times. It was, argued Colonel Love, the commander in south Wales, the 'total absence of any civil force whatever in the three counties' which allowed the Rebecca disaffection to grow to such 'an alarming height'.[11] During disturbances the parochial authorities required the assistance of special constables, the militia, the yeomanry, and even the regular soldiers. At various times in the years 1795–1832 troops were stationed at Aberystwyth, Lampeter, Llandeilo, Haverfordwest, Carmarthen, and other towns. By the late spring of 1839, when the Rebecca riots began, the Castlemartin alone of the yeomanry corps was still embodied and in training, and the only other military help close to hand was the Royal Marines at Pembroke Dock.

Most of the people apprehended by parish and police constables were dealt with at the local courts, and thus did not appear in the statistics mentioned above. Each Petty Sessional division had its weekly, fortnightly, or monthly meeting, where two or three magistrates dispensed justice. By the beginning of Victoria's reign, the courts at Carmarthen and Haverfordwest met in the town hall, but at Mathry, Broad Oak, and Llangadog sittings were still held in mansions and public houses. Inevitably, the records of these courts have not survived, though the Quarter Session papers give us an idea of the number of people proceeded against. Thus, in Carmarthenshire during the year October 1844–October 1845 some 415 Petty Session convictions were included in the financial accounts, and there would have been more had Tre-lech and Llanboidy made returns.[12] Many of these people paid fines because of their crimes; the numbers of people committed to gaol, at least for long terms, was small. To obtain numbers of prison inmates we have to turn to the inadequate gaol returns and surviving prison records. Detailed information exists for the gaol at Haverfordwest. This reveals that during the second quarter of the nineteenth century an average of 200 people were committed there each year, mostly for short periods, the number rising to exceptionally high peaks in the early 1830s and the Rebecca years. There seems to have been a similar pattern in the commitments to the county gaols at Carmarthen and Cardigan.[13]

Altogether, it is likely that at least 1,500 people were brought before the courts of the three counties in 1844 for criminal offences, and even more than that were arrested. As a comparison over 2,000 summonses were served annually by the Rural Police of Carmarthenshire and Cardiganshire when they began to publish statistics in the late 1840s.[14] Each year during this decade about one in a hundred of the population was summoned, bailed, or committed for trial. In fact, as has been argued elsewhere, the Welsh crime rate was probably closer to the English one than Thomas Phillips and Henry Richard allowed.[15] This could explain the later, and remarkable, claims that before the introduction of the Rural Police in the 1840s places like Llanrhystud, Cynwyl Elfed, and Llanboidy were 'infested with crime'.

There was, of course, a popular myth that the Welsh were particularly litigious. One anonymous and well-informed letter writer stated that the Welsh both respected and used the law.[16] Compared with the previous centuries there was now a greater readiness to proceed in civil as well as criminal matters at the Crown courts. Cases at the shire and guildhalls were often about highway affairs, personal honour, wills, bills, and leases. It was not uncommon for the *nisi prius* proceedings at Assizes to last twice as long as the criminal. How much the English language of the courts inhibited people from bringing even more cases was a moot point. William Williams MP, in his parliamentary motion on the subject, was certain that clash of languages was the weakest link in the judicial system. Some prosecutors, and defendants, undoubtedly suffered because of their ignorance of English, while others were said to have benefited from the skill of lawyers and interpreters. Edward Lloyd Hall, no friend to the native language, fed the common English assumption that Welsh witnesses and juries committed perjury under the cloak of ignorance.[17]

More daunting to the peasantry than the language of justice was its cost. Typical court fees for an ordinary Assize case were £20–30, and those at Quarter Sessions about half that amount. Although the State now gave considerable help with fees, notably for poor and successful private prosecutors, people claimed that justice was still too expensive, and the inhabitants of Llanfynydd, who felt particularly strongly on the matter, included the manorial courts in their indictment. At their protest meetings in 1843 people resolved on extensive reforms of the legal system, their intention being to bring expenses down by at least 50 per cent, so that 'the poor man may have fair play to defend his rights'. 'It is better for us', a farmer told the *Times* correspondent at a Cynwyl meeting, 'to go to a lawyer to settle our differences between us than to go before a magistrate. The lawyers cost us a great deal for obtaining justice, but the magistrates cost us a deal more.'[18] The most difficult decision facing many people in the economic crisis of the early 1840s was how to recover small debts; until the popular alternative of the county courts was established it made financial sense to rely on Rebecca.

For one reason and another many criminals in the second quarter of the nineteenth century did not pass through the Crown courts. Some matters, connected with anti-social behaviour, defamation of character, marital problems, wills, tithes, debts, rates, fines, grazing rights, trespass, enclosures, and the like were dealt with at the surviving manorial and ecclesiastical courts.[19] It had also been the practice for generations to seek arbitration in minor cases from a respected member of the village community, or to approach local justices and solicitors for their private verdicts. Even in the mid-nineteenth century, rural magistrates liked to save time and money by impressing on both sides the value of compromise. One can follow the last stages of one such case at the Castlemartin Petty Sessions in November 1844. The two parties in the dispute were James Hird the younger, a farmer of the parish of Stackpole Elidir, and his servant girl Margaret Edwards, whom he was accused of fondling. The court gave them a last chance to settle matters, and, when this was done, and Hird agreed to pay any costs involved, the charge was dismissed.[20] Police diaries of a later date confirm that constables adopted a similar approach, occasionally choosing not to proceed on information given to them, especially if the offence involved juveniles and married couples. Even at the last minute it was common in south-west Wales for prosecutors and witnesses not to support their cases in court, and many complaints were withdrawn. Associations for the prosecution of felons, which were intended to place pressure on such individuals, were few and far between in this part of Britain.

From various pieces of evidence there can be little doubt that people in a close-knit community were unwilling to press charges which might cause unpleasantness amongst family and neighbours. Thus while vagrancy offences and crimes by outsiders were well reported, it seems that family and sexual offences, petty theft and disorderly behaviour, and even fairly serious assaults and cases of damage to property were under-represented in the official crime figures. According to Jane Walters at the new country house of Glanmedeni, the local population, including parish constables, frequently withheld vital information on crimes, even murders.[21] Informers who approached the authorities in secret were unpopular. At Haverfordwest 'common informers' were driven out of town by mobs, and Benjamin Davies, who gave evidence on behalf of Augustus Brackenbury, lost his house and had to settle in England. People in the region were fined for refusing to give evidence on oath when the guilt of the accused was apparently obvious.

'The people on this occasion as on every other,' wrote Colonel Love about a brutal murder of a female farm servant near Llandeilo, 'evince a dislike to bring the criminal to justice.'[22] 'It is almost impossible to get any evidence,' agreed Colonel Hankey, the travelling stipendiary magistrate who was supposed to inject life into the commissions of all three counties during the Rebecca riots. 'The people are so combined and lie so confoundedly that it is next to impossible to get anything out of them.'[23] Even when people were brought to justice there was a feeling that the Welsh were ready to sabotage the proceedings. John Jones, MP and chairman of the Carmarthenshire Quarter Sessions for some years, said that when Grand juries presented a watertight true bill against a prisoner, wilful Petty juries took a contrary and prejudicial view of the case. The *Quarterly Review* condemned them for their 'shameless disregard of the evidence', not as bad as Irish juries but serious nevertheless.[24] As we have seen, they acquitted as many as they convicted; the proportion rose when they passed judgement on those accused of infanticide or rape and fell when robbers and burglars were in the dock. In a number of notorious instances during these years judges asked juries to reconsider verdicts that seemed to fly in the face of the evidence. When William Thomas was found not guilty of stealing hay from employer Edward Adams of Middleton Hall in the spring of 1840, the whole court broke into laughter, and an angry John Lloyd Davies insisted that in future the sheriff pick more intelligent jurymen.[25] Augustus Brackenbury, staggered by the acquittal of the ringleader of a mob by a Cardiganshire jury in 1826, pleaded with the government to remove this case to the safe haven of the English courts.

The Welsh, both individuals and communities, had shown

for centuries a preference for dealing with certain forms of illegal or deviant behaviour in their own fashion. Direct retribution and public ridicule were two popular methods. Duelling was still practised amongst the higher orders; in one instance, at Tenby in April 1839, it arose out of a violent altercation in the Petty Sessions between the mayor and another magistrate.[26] Even the most respectable took matters into their own hands. Gentlemen farmers fought each other over many things, and pulled down tombs, fences, and houses in dispute. The very term 'Welsh ejectment' referred to forcible eviction by a private mob. Thus in 1845 Mrs Morgan of Fishguard organized twenty supporters and physically removed Mr Nicholas, her tenant.[27] The Rebeccaites said, with a degree of truth, that they had received good tuition from their betters in direct action. For its part, too, the wider community had always been prepared to act together against wife-beaters, adulterers, slanderers, informers, and others who broke unwritten codes of behaviour. It 'set in motion a crude and primitive machinery for enforcing its own code of conduct,' said D. Parry-Jones, who knew the area better than almost anyone. 'It seemed at such times to sink down to subconscious regions where it made contact with an older civilization that knew only the tribal law.'[28] Recalcitrants were warned by threatening letters, and, if this failed, they faced the indignity of a beating, a mock trial, denunciation in verse and song, harassment, and, in the last resort, ostracism and removal. At times of serious economic and political tension, such community policing was adapted for other purposes, and it became, as we shall see, one strand of Rebeccaism.

Although it is now impossible to estimate the extent of illegal activities, we can form an impression of the kind of people who were usually associated with crime. About 80 per cent of people fined and imprisoned during the Rebecca years were male, and the average age of those men tried at the higher courts was some 27 years, with women a little older. However, these figures are misleading; the sexual division of all known offenders was slightly more even then this, and the most criminal age for such people was 19–23 years. Popular female crimes were stealing, assault, drunkenness, concealing the birth of, and deserting, children, and infanticide, the balance altering from urban to rural settings. In the countryside especially it seems that as many single as married women were taken into custody by parish constables. For this, and for other reasons, it was said that 'mother Rebecca' was a singularly inappropriate name for the leader of a great popular uprising.

The information on the class and occupations of known offenders is disappointing. A small number of them were clerks, respectable tradesmen, and large farmers, who had committed offences such as embezzlement, forgery, assault, and breaches of the game laws, but the overwhelming impression is that of working people passing through the courts. In the countryside two-thirds of the males arrested, and identified by their occupation at this time, were called 'labourers' and 'servants in husbandry', and the females were usually described as 'servants'. The rest of the village criminals were returned as 'farmers', 'blacksmiths', 'shoemakers', and the like.[29] There were several reasons for this occupational breakdown, not least, one suspects, the laziness of the court scribes. It reflected, also, the basic assumption that certain types of offences were the responsibility of particular groups of people. An example makes the point. When farmer Thomas Hutchings of Rudbaxton parish in Pembrokeshire lost a handkerchief and neckcloth in 1842, suspicion immediately fell on Mary Williams, who had been his servant. Like many of the other servants and vagrants who appeared in court for similar property crimes, Mary was found not guilty, but a word from the magistrates on the lack of integrity amongst servants reinforced the prejudice.

In the towns, of course, the pattern changed. 'Labourer' was still the dominant group of named offenders, but in places such as Haverfordwest and Carmarthen sailors, miners, colliers, weavers, fishermen, and others made up the remainder of the lists. The literary evidence suggests that children were important in the urban crime story; in the same two towns there were constant reports of them stealing, pulling up shrubs, throwing stones, breaking windows, and generally trespassing. In Haverfordwest some of the young lads were whipped around the town as a warning to others, while in Carmarthen special stocks were erected for them in 1827. When the evidence is given, which is not often, it appears that a third of the males and half of the females could not read or write, and two-thirds were without a previous criminal record. No one, not least in Wales, was sure of the value of this information, but it was in tune with judges' homilies on the need to give youngsters a better education, particularly of a religious kind.

One of the most popular claims throughout the nineteenth century was that crime in the Principality was committed by outsiders. It was widely assumed that the influx of 'foreigners', whether they were seamen, soldiers, sappers, or miners, was followed by rising statistics of disorder and robbery. When, as in 1843, 'strangers' suddenly appeared in the Llanarthne–Llandeilo area, the authorities were immediately on their guard.[30] Road builders and railway navvies were associated with drunkenness, fighting, and racial conflict; in the view of some Nonconformist ministers, they besmirched 'a Godly land'. But no one, perhaps, was more disliked than the Irish, English, and Scottish vagrants. 'By far the greatest number of those convicted of the highest class of crimes tried here, have not been natives of the county', said Henry Leach, chairman of the Pembrokeshire Quarter Sessions in April 1848, 'but persons who, in familiar language are called trampers; men who come to prey upon the fruits of your industry, and not to live upon their own—who preach about politics and not about labour, and who tell you that the idle, who toil not, have a right to a share, or to the whole of your property or your earnings, to supply the wants of idle and seditious politicians.'[31] For years Leach, and the middle class of Milford, had been concerned with the seasonal progress through their neighbourhood of the distressed Irish and all manner of other travellers. The perceived threat had made them converts to a measure of police reform, and grateful for the Royal Marines close by.

This attitude towards outsiders was, of course, commonplace whenever crime and protest was discussed in the nineteenth century, but it is difficult to evaluate. It seems probable that up to a third of those taken into custody for both serious and petty crimes in south-west Wales could be regarded as outsiders, though many of these were arrested in just a few towns, such as Haverfordwest and Pembroke, 'Trampers' were regarded by magistrates and police as a threat to property and public order, but the majority of their offences were breaches of the Vagrancy Act of 1824. People like Thomas Jones, a Monmouthshire man, and Thomas Williams and John Thomas, two Merthyr labourers, were convicted mainly for being 'rogues and vagabonds', for begging in the streets, and for deserting their children. Punishment consisted of a few weeks' hard labour, and a warning about future conduct, but the majority were acquitted of any charges and then moved on before they became a burden on the parish. When down-and-outs committed more serious crimes, they could expect little mercy. Lewis Henry, a local tramp, was given 10 years' transportation in 1843 for robbing a house in Abergwili parish. He was arrested because he had money in his possession, and found guilty because the size of his feet matched footprints near the scene of the crime.[32]

In the early 1830s, and again a decade later, when the numbers of people on the roads increased sharply, Carmarthen, Milford, Cardigan, Aberystwyth and other places ordered their constables to lead vagrants outside the town boundaries. The burden was thus passed to rural parishes, and in the summer months, when these vagrants were not being taken on as cheap harvest labour, they were regarded as a menace. At times of general depression the countryside was awash with destitute men and women, all with tales of dreadful accidents and personal tragedies. When they operated in small gangs they could frighten a little charity out of lonely farmers' wives. Their appeals to the Poor Law authorities for casual shelter and food were so demanding that a work test was prescribed for them in 1842, and immediately a further category of offences was added to the list. At Cardigan, Haverfordwest, and other workhouses angry travellers demanded relief, and then refused to break stones and stole and burnt Poor Law property. On occasions these outsiders were accused of more insidious behaviour, such as seeking to benefit from both the Swing and the Rebecca troubles. 'It is suggested that they [the large number of vagrants in Carmarthen gaol] have been attracted to Wales,' commented the *Welshman* in December 1843, 'in the hope of finding some booty during the disturbances.'[33]

In general, the more serious crimes in Assize, Quarter Session, and Petty Session calendars were committed by natives of the south-west. Thus of the felons registered in Carmarthenshire in 1844–5, 75 per cent were Welsh, 63 per cent were natives of the county, and 31 per cent had been born close to the place of the crime.[34] In more isolated country districts almost all the recorded cases were the responsibility of local people. Some of these, possibly the majority, were first-time offenders, but each village had its persistent poachers, petty thieves, and offenders against the Excise and Licencing Acts. Sarah Evans of Llandysul, who was committed to the Quarter Sessions in 1841, had, it was said, a forty-year history of pilfering. David Howells, a native of Meidrim, was one of the most infamous robbers of his day, and so was Mary Williams of Lampeter, who toured the region with her equally notorious husband telling fortunes and thieving. William Davies, alias Enoch Thomas, carpenter and sheep- and horse-stealer, William Jones, the tinker of Tregaron, and other well-known criminals from rural Dyfed spent time in Merthyr Tydfil before returning to practise their trade. For some, like Thomas Evans of Penboyr, a reappearance in their home parish was a sure sign that they were in trouble elsewhere. In the winter of 1841 Evans was captured by a Glamorgan constable and escorted back to Dowlais on a charge of stealing tools from Sir Josiah John Guest.[35] All in all, Rebecca's children were no strangers to crime.

The nature of the records, apart from any other consideration, renders it impossible to say whether areas within south-west Wales were more criminal than others. Nevertheless, in the mid-nineteenth century certain towns and parishes were associated with particular crimes. Llanelli had a reputation for the amount of stealing committed within its boundaries, Haverfordwest was to be avoided for its drunkenness and disorderly conduct, Cenarth was the place to be assaulted, and Solva had grown weary of vagrancy crimes. Despite the opinion to the contrary, there was no apparent connection between the extent of illegal activities in such communities and the presence or otherwise of the gentry, clergymen, and Dissenting ministers. Nor was the known incidence of crime closely related to that of Rebeccaism. Parishes like Llanarthne and Llangendeirne figured prominently in both fields of activity, but this was probably only a coincidence. Some

Rebeccaites were criminals, but most appear not to have had a criminal past.[36]

What kind of offences were committed in Rebecca country? On rare occasions the press gave revealing lists of unsolved thefts, killings, and other crimes, and hinted at half-hidden but popular forms of delinquent behaviour. It was always presumed that the pattern of criminal activity was somewhat different from that indicated by court cases. However, the legal evidence is the best that we have, and can be used with care. At the local courts one can distinguish between those crimes which were punished with fines, and those which usually resulted in imprisonment. Of the former, about a third were highway and turnpike offences, notably furious riding, riding on a cart, evading tolls, leaving a cart unattended, and having no name on it. Every week, for example, Abernant farmers and their servants raced their horses into Carmarthen, bypassing gates and terrifying pedestrians. Prosecutions against them were brought by the appropriate turnpike trust, and they received part of the fines of 5s. and 10s. Another third of the cases which resulted in fines were assault and drunkenness, the former being more common and resulting in heavier penalties of £1 and £2. The remaining fines were handed out for a variety of crimes, such as breaches of the Game, Fishing, and Excise Laws, evading market tolls, using fraudulent weights, and trespass on canals and railways.

Those offences which resulted in imprisonment can also be grouped into categories: stealing, vagrancy crimes, assault and disorderly conduct, which accounted for two-thirds of reported cases, and breaches of the Poor and Excise Laws, deserting one's service, damage to property, and the like, which formed the remainder. Gaol sentences for the lesser of these offences at Petty Sessions were commonly from seven days to one or two months in duration. At the Assize and Quarter Sessions the composition of calendars and the punishment were very different. At least two-thirds of cases involved stealing and related offences against property, and those found guilty expected a minumum of three months' imprisonment. In addition, there were at these higher courts civil cases, about which the Welsh were said to be particularly keen. Each Assize had its *nisi prius* proceedings, over libel, slander, breach of promise, land disputes, non-payment of debts, and other actions. These took up much of the time of the court, and were chronicled in great detail by the local newspapers. In their way, they tell us as much about society and the background to Rebecca as the more spectacular breaches of the criminal law.

It was no accident that offences against property formed a large share of recorded crimes. The defence of property had been enshrined in statute after statute in the previous century, and was now supported by a wide range of people and classes. Although the great landowners and industrialists initiated many prosecutions, so did shopkeepers and other ratepayers.[37] When she came 'to put the world right' Rebecca remembered this. The most hidden form of theft was forgery, fraud, and embezzlement; there were a considerable number of cases in the mid-century of forging notes, wills and signatures, of obtaining money under false pretences, and of providing worthless documents as security for loans, but there was no indication of the real extent of such crimes. Other forms of property crime were more visible. From the available evidence it seems that the property most commonly stolen was clothing and footwear: shirts, dresses, shawls, stockings, shoes, handkerchiefs, ribbons, and flannel disappeared from hedges, houses, shops, and workhouses. The clothes were often taken and then worn by the offenders, although a few items were passed to receivers and pawnbrokers like Harriet Thomas, lodging-house keeper of Carmarthen. Thieving from one's master and mistress was to be avoided. Richard Matthews, a Llanboidy farm servant, was different, not least because in court in 1844 he asked for seven years' transportation for his crime of stealing clothes. He received two months' hard labour.[38] Had he been a known recidivist like Evan Thomas, who took two blankets from the Carmarthen workhouse about the same time, the judge would have granted his wish.

Food and drink were also prime targets for thieves, as were, especially in the towns, money, watches, jewellery, silverware, and toys. Much of this theft was of a petty kind, its value being no more than a pint of milk, a loaf of bread, and a pound of vegetables. Elizabeth Davies and Mary Rees, spinsters who received a month's imprisonment in 1844 for stealing potatoes from William Evans at Llanelli, were representative of the many women and young people who engaged in this type of crime.[39] The motivation was frequently a mixture of greed and opportunity, exemplified by the case of a gang of six Cardigan boys, who made a small trade of stealing ropes and selling them to the marine stores in 1845. Nor should necessity be forgotten; 60 per cent of recorded food thefts occurred during the winter season. Penniless Lewis Henry and John Williams at Llanelli warned that they would steal so long as work and food were denied them. Williams, a man of Kent, put his fist through a window in 1843, and waited with a watch in his hand for his next prison meal.[40]

Less common but still important in the region were the thefts of rails, plates, shovels, copper, lead, iron, and brass. William Chambers, Richard J. Nevill, Alexander Druce, John Rees, James Child, and others prosecuted hundreds of people for these crimes, though perhaps the main complaint of these employers was the disappearance of great quantities of coal and culm. This was taken at the pithead, along the railway tracks, and in the docks, and, with few exceptions, women and children were the guilty parties. Their main excuse was that it was just 'lying about'. On the Pembrokeshire coalfield, in the hinterland of Saundersfoot and Tenby, where large amounts of coal were stockpiled in 1843 because of the lack of demand, the employers set watchmen to protect their property and demanded a minimum gaol sentence of two months for the thieves.

In a similar way it was found necessary in the early nineteenth century to give the new coastguards the task of preventing smuggling and wrecking. For generations villagers on the coast had claimed rights to cargoes washed up on the shore. They looted the contents of ships battered by high winds in Cardigan Bay and of those lying helpless on Cefn Sidan sands. On 19 December 1833, when *The Brothers* was lost on the sands, its cargo of hides and cotton had totally disappeared before customs officers arrived, and, in a huge storm ten years later, 'unfeeling brutes' ran from ship to ship, and ignored the pleas of customs officers, constables, and the gentry.[41] Smuggling was a related crime, and throughout the period wine, spirits, tobacco, tea, soap, candles, and malt were illegally acquired and sold. Along the coast officers carried out innumerable raids on lonely cottages, but when, for instance, they travelled inland to a Llanboidy farmhouse early in 1844, looking for smuggled malt, Phoebe David, Sarah David, and seven other women turned on them.[42]

Excise officers, who had the difficult task of discovering the unlicensed trade in food and beer, had a reputation for toughness. The restrictions on home brewing in a land of cheap barley were greatly resented, and illegal stills were to be found everywhere. Officer Rowlands, at Narberth, boasted that he had initiated forty prosecutions in 1844–5, but perhaps the most numerous and notorious cases came from the Teifi valley.[43] In one of them, at Cardigan, when excise officers took everything from the house of an old man because he was unable to pay their £25 fine, there was anger and violence. Such prosecutions, and fines of £5, £10, £50, and £100, brought ruin to unlicensed male and female traders, and must have been a severe warning to others. Perhaps contemporaries were right in the mid-century when they declared that illegal brewing, smuggling, and wrecking had been forced into a steady decline.

In the case of pickpocketing, highway robbery, and burglary, the reverse was possibly true. From time to time the local newspapers warned their readers of the dangers from 'swell mobs' who were roaming the countryside looking for gullible victims. In 1834 and 1835 there was a wave of hysteria over the arrival of a group of pickpockets, apparently under the control of 'Cockney Bill', a man who stabbed a constable and was eventually transported. His mob, which consisted of small gangs, was said to have numbered over a hundred young criminals from Glamorgan and Monmouthshire. Like others based in Bristol, Gloucester, and London, it operated mainly at fairs and markets, where it was comparatively easy to practise the arts of thimble rigging and stealing from the person. In 1835, John Devonald, James Richard, Thomas Lewis, 'Jack Straw'(berry), and others of Cockney Bill's mob were convicted of robbery and given the same punishment as their leader. Five years later, in the autumn, another gang collected £200 from farmers at the Newport, Capel Cynon, and Newcastle fairs. At Mathry the crown turned on them, and two of the 'light-fingered gentry' were taken by the constables. Sometimes these gangs were joined in the illegal work by Carmarthen prostitutes or Haverfordwest coiners. All of them feared the harsh sentences of magistrates, who became determined to curtail their activities. The men taken at Mathry, and John Brown, convicted two years later for the same offence of pickpocketing at Rhos fair in Cardiganshire, were each given 10 years' transportation. The coming of professional policemen, especially from the county forces, added to their difficulties, and there were protests in 1843 when superintendent Henry Westlake took five men into custody to prevent them attending two Carmarthen fairs.[44]

The crimes of robbery and burglary were always taken seriously, and they were committed by both residents and outsiders. There were almost sixty cases of highway robbery in the years 1830–48, many carried out close to Llangeler and Pencader, and on Nantycaws hill. The most likely victims were farmers travelling homewards in a drunken state, traders and wagoners on their own in the late evening, and anyone known to be carrying money or valuables. Sixty-year-old Margaret Williams, who travelled weekly from Llandeilo to Swansea market to sell farm produce, must have expected her night attack in the spring of 1843. So must colliery owners and others with rates, wages, and rents in their pockets. In November 1838 an old servant of Lord Cawdor, who had been collecting rents, was pulled from his horse two miles outside Carmarthen and robbed by a man with a white nightcap over his head. All manner of disguises were adopted, with smock frocks and blackened faces being a favourite. Robbery was frequently a violent business and David Jonathan, assaulted near Llandeilo in 1834, was one of those who lost his life. So did David Lewis, a farmer from the Lampeter area, who carried butter to Merthyr. On his journey he was set upon by Thomas Thomas, a 24-year-old native of Llanfynydd, and had his throat cut and pockets rifled. Thomas was executed at Brecon.[45]

Burglary, and breaking and entering, could also end in bloodshed. In an effort to shame both the public and the constabulary, it was repeatedly suggested that burglary had never been a popular Welsh crime, but the records do not bear this out. Before the days of regular policing, professional watchers, and good lighting, buildings were easily entered by the skilled burglar. By no means all homes were locked, and thatch could be cut open. The time of entry was carefully chosen, typical hours being in the early morning, during divine service, and when families were out in the fields. Sometimes, however, the business was conducted in a more open and, for contemporaries, outrageous fashion. Groups of robbers walked into Cardiganshire churches and removed the silver, or called, with hatchets in their hands, on widows in isolated farms. John King, an old labourer with white gauze over his face, threatened to knife Mary Sutherland of Laugharne unless she handed over cash.[46] Gangs of men and women, disguised in a variety of masks and clothes, battered their way into vicarages and small mansions and forced inmates to reveal their hidden treasures. Magistrate David Harries James of Llwyndwfr, Pembrokeshire, and the Reverends Lloyd of Cellan, Cardiganshire, and Picton of Iscoed, Carmarthenshire, were three of the principal sufferers. The takings were valued in hundreds of pounds, and large rewards were offered for their recovery. When such incidents reached unacceptable levels, as in the spring of 1828 about Carmarthen, and even more in the neighbourhoods of Pembroke, Haverfordwest, and Carmarthen in the years 1833–6, the men of property at last got together to demand better protection.[47] But resistance to intruders did have its hazards, as we shall see in the case of Shadrach Lewis of Clydau, Pembrokeshire.

In the countryside property crime had its own distinct and familiar pattern. Gleaning was one element in the story, if only because the taking of the last traces of corn, wool, and fallen wood had been, with the farmer's permission, part of the peasant economy. There was, everyone admitted, a vast amount of both legal and illegal gleaning in the mid-century, mostly by women. Turf, rushes, bark, sticks, and branches were taken for fuel; in this way hundreds of mothers and children kept their homes warm through the winter months. The collecting of underwood was the most common, and least discovered, crime in the Welsh countryside. If found guilty the offenders expected a fine or one-month gaol sentence, but in one case, during the winter of 1852–3, two old ladies got off with only two-weeks' imprisonment as Lord Milford wanted leniency for their first offence.[48] George Bowen Jordan of Pigeonsford, James M. Child of Begelly, Owen Owen of

Crime and Deviance

Cwmgloyne, John W. Lloyd of Danyrallt, John Jones MP, and the earl of Cawdor were less tolerant, especially when new plantations were decimated. Richard Gardnor, agent of John Jones, said that the people had to be deterred from such habits. The same point was made about the gathering of furze, leaves, and nuts, and the taking of sand from the Tywi and other tidal rivers. The *Carmarthen Journal* hoped that the arrest of the culprits would convince the 'lower orders' that they had no 'rights' in such matters.[49]

Other goods which vanished in the countryside were corn, hay, turnips, and cabbages, as well as eggs and milk. A large number of poor women were caught in the act of milking cows in the open, and, as they tended to do this more than once, they were frequently given between six weeks' and six months' imprisonment. During times of hardship this particular crime, and the taking of green vegetables, reached epidemic proportions. In 1847, when charity was being organized for the hungry, Richard Woodfall, forester of Abel L. Gower of Castle Malgwyn, came upon twenty people digging up turnips in one of his master's fields, and about the same time hundreds of cabbages disappeared from an estate near Tavernspite in Carmarthenshire. In one case, at the end of the Rebecca troubles, the magistrates took pity on Anne Davies, a single and destitute woman of Llansawel, who was accused of lifting potatoes valued at 2s. The Bench raised a subscription to supply her with adequate clothing.[50] One suspects that there were many like her; as always in the story of rural crime, the number of these court cases bore no relationship to the actual level of depredations.

Besides vegetables and fuel, other typical thefts in the countryside were of forks, knives, and shears, and of hens, ducks, and geese. One or two losses of this nature were probably ignored, but when small farmer David Evans, who lived near Carmarthen, had six hens stolen within a month in 1844 he accused neighbour Mary Davies of the crime and then, as often happened, failed to prove it in court. According to observers, crimes of this nature, like the pilfering of small items of clothing and household goods, took place when the families were out at work or at chapel. Garden robberies were common across west Wales, and poor and rich alike lost fruit and vegetables. Taking produce from the orchards and walled gardens of the upper classes was a tricky business; it needed a couple of people to deal with the dogs and trip wires. Eventually, people like Captain David Edwardes of Rhydygors found it worth while to employ armed guards.[51]

More serious, and more annoying for the farmer, was the disappearance of pigs, cows, oxen, horses, and especially sheep. Until the early nineteenth century the stealing of the larger farm animals was a capital offence, though very few of the criminals were actually hanged. It is hard to establish the impact of the change in legislation, for court records, which give low figures for cases in the mid-century, and contemporary comment about the crime were at odds. Cattle were driven off the moors by unscrupulous dealers, like John Thomas at Begelly parish and John Lewis of Llanegwad, and walked with others 'to the hills' of Monmouthshire and beyond. The success of this crime depended greatly on the alertness of the victims, as these two cases show. A mare was stolen late one night in 1842 outside the home of Morris Richards, who lived near Aberystwyth. His son traced the horse all the way to Shrewsbury, and helped to secure the arrest of the two young thieves. David Reynolds, a Carmarthen butcher, who stole an animal from farmer David Davies about the same time, only got as far as Neath; on his return in police custody, crowds surrounded the Swansea mail coach and he had to be rushed into the borough gaol. William Williams was another unpopular character; he made a precarious living in north Carmarthenshire out of borrowing horses and then selling them as his own property, telling their angry owners that for the moment he needed the money more than they did. From the legal evidence it seems that dealers, drovers, and butchers were more involved in this crime than labourers, despite the prospect of transportation for life.[52]

Sheep-stealing was regarded as one of the most common and least detected property crimes in the south-west. Thousands of sheep grazed on open hills and moorland, with only the company of shepherds and the Crown and manorial escheators. Each year hundreds of animals disappeared, some straying for miles, some killed by dogs, and some deliberately taken. There were many examples of farmers welcoming strangers into their flocks. Not all the sheep were marked, nor were registers of marks kept, so it was fairly easy to sell the stolen animals a few miles away. For the hungry, mutton was a pleasant change to their sparse diet, though they needed to ensure that they disposed of all the carcass. When Ann and Elizabeth Evans were arrested for stealing a sheep in the St Davids area of north Pembrokeshire in March 1839 the former claimed that she was hungry and that the animal had been found on the high road. The meal cost her two years' hard labour, though her case was not helped by a second charge at the same court. This time the mutton was eaten by Martha Jenkins, but Ann Evans had supplied the meat, and for this she was given another year's hard labour.[53]

On the mountains of Cardiganshire and Carmarthenshire sheep-stealing became something of a cottage industry. Charlotte Williams, who had a house and premises against the Black Mountain, asked her three young sons to bring down flocks of sheep. Theirs were separated from the rest, and the latter were shorn and killed. Thirty-seven carcases were found buried at the small farm, and, when the case was brought to the crowded Assize in 1831, all four defendants were sentenced to transportation. This was the common punishment for those with a previous conviction, or for those who, like farm servant Samuel Thomas of Llanarthne, were actually employed to look after the sheep. A close inspection, especially of the many cases in Cardiganshire, reveals that sheep-stealing sometimes formed part of wider conflicts over family honour, sheep walks, and grazing rights. Not all the animals were taken for food; some were killed and left in a conspicuous place. 'Ill will' was one explanation for twenty-seven sheep drowned in the river Usk not far from Charlotte Williams's home. The sheep, which belonged to John Joseph of Penyrallt in the parish of Llanddeusant, had their horns knocked off and their eyes eaten out, and two young graziers, and their dogs, were blamed.[54]

Certain breaches of the Game and Fishery Laws also fell into that hazy divide between crime and protest. The early nineteenth century witnessed the planting of woods and covers for raising birds, the importation of new types of game and shooting methods, as well as the licensing of fishing on many of the rivers of the south-west. Some of the legislation which both established and protected these new practices was resented and ignored by a broad spectrum of people. The most recent changes were introduced by the Game Act of 1831; this restricted the killing and selling of game to people who had access to land and a game certificate. Over the next twenty years, as the shoot and fishing grew in popularity as gentlemen's pastimes, the *Welshman* and other reforming papers complained of the class nature of this legislation, which turned ordinary people into poachers and poachers into criminals. Some of the places which experienced the worst depredations during the 1830s and 1840s were the Reverend Benjamin Lewes's preserves at Dyffryn, Charles Arthur Pritchard's land at Tyllwyd, both in the vicinity of Newcastle Emlyn, the Adams and de Rutzen's holdings at Middleton Hall and Slebech, and the bigger estates of Golden Grove and Picton Castle. Amongst those who appeared in court were gentlemen's sons who sported on farms where they had no property in game, tradesmen who had no game licence, and tenant farmers who resented the losses caused by birds and vermin. John Rees of Llanarth, and other farmers' spokesmen, thought that occupiers of land should have the right to shoot game living on it, and Sir James Williams, the Liberal landowner of Rhydodyn (Edwinsford) agreed to this request in 1844.[55]

Most of those arrested for poaching were small farmers, innkeepers, craftsmen, fishermen, servants, and labourers, though it was again rare for employees to be apprehended on their masters' land. A few of these men were widely regarded as exceptional criminals. In court they refused to concede that rabbits, hares, and fowls of the air were anything but God's creatures, a notion which was to be taken up by Nonconformist Liberals in their political battles a generation later. Two rebels of sorts were poachers George Morgan and T. Carew Phillips, the former being arrested in 1838 on a charge of setting fire to a valuable plantation belonging to Thomas Lloyd of Millbank Cottage. Phillips, a 30-year-old cabinet maker, whose house at St Dogmaels was full of guns, nets, and other useful tools, was noted for his intelligence, strength, and agility. This accomplished poacher, sheep-stealer, and burglar was, according to one newspaper, a sort of 'Rob Roy' figure, and his enforced

Crime and Deviance

departure for Australia in 1845 was long remembered in the area.[56]

The conies, fish, and winged game taken by these people were probably eaten by their families, but there was organized commercial activity by poaching gangs and a regular market in out-of-season fish and game. Gangs typically had about six or a dozen members, and when one of them at Cilrhedyn was trapped in a hay loft, they paid a collective trespass fine of £31 without delay.[57] For people on the banks of the Dovey, Aeron, Teifi, Tywi, Gwili, Gewndraeth, and the other main rivers, fishing had been a way of life for centuries, and they refused to give it up at the whim of landowners in Parliament and the Quarter Sessions. Attempts to introduce closed times, to outlaw certain nets, and to impose licences for line fishing were bitterly resented, and so were the fish traps that belonged to the up-river gentry families. In spite of everything salmon was still taken and sold out of season.

As we shall see in the last chapter, the arrival of the new Rural Police and small armies of river bailiffs in the mid-century allowed the conservators to press ahead with their plans. For their part the coracle men of the lower Teifi and the tenant farmers who fished along the upper Tywi protested peacefully and then vigorously; evenings were spent in noisy, and sometimes violent, poaching expeditions. This was the sign of things to come along the Wye and other Welsh rivers later in the century.[58] There was also conflict when the owners of Crosswood, Nanteos, and Llwyngwair estates experimented with rabbit farming, and when those with extensive manorial and common rights sought to control the taking of conies and birds on open moors, marshes, and waste lands. Some of the 'old offenders', men like Thomas Llewellyn of Slebech and John Williams of Narberth, whose cough was a precaution against the treadmill, must have felt that it was too late to change their ways, especially in places where watchers were thin on the ground.

Once in court, however, their chances of acquittal were slim, for the magistrates were landowners and the keeper or constable's word was 'law'. The punishment of poachers had become less vindictive by the Rebecca period, but it was still 'sharp and vengeful'. The crime was usually dealt with at Petty

Sessions, where hunting magistrates were naturally more hostile to offenders than their colleagues. Trespass in pursuit of game resulted in a fine of £1 or £2, or a month's gaol, and a similar fishing offence was even more expensive. Those actually caught in possession of game on their own land were penalized by a £5 fine, while those doing the same on someone else's estate were often given three months' imprisonment, and had to provide sureties at the end of that, or face another six months. In most of the poaching cases a moiety of the fine was handed over to the informer. Some informers made a good living, although there were dangers. When an employee of John E. Saunders of Glanrhydw wrongly accused two men of illegal fishing in the Gwendraeth, he was driven from the Carmarthen shire hall by a mob throwing rotten eggs.[59] Night poaching, which became the subject of a special Act in 1844, was a particularly serious affair, bringing longer gaol sentences and even transportation, and this is one reason why violence accompanied the crime.

The largest estates appointed as many as a dozen keepers and watchmen, often with English and Scottish surnames, and their conflicts with poachers split communities down the middle. A number of poaching farmers and labourers were killed in the 1830s and 1840s, and even in the case of J. Mabe, the butcher who drowned at Canaston Bridge in 1843 while escaping arrest, the local people were adamant that he had been pushed over the edge. William Jones, huntsman to Major Herbert Evans of Highmead, who killed poacher John Evans, a farmer's son of Brechfa, did appear in court and received a sentence of six months' hard labour. Some of the worst exchanges took place at Picton Castle, the home of Sir Richard B. Philipps, MP. One Saturday night in early February 1842 six or seven men disabled a keeper in front of the Castle, and two years later, in an even bigger clash, five keepers were badly wounded, two seriously, and another crawled away with a fractured skull.[60]

The extent of violence in rural counties was a topic of much interest to contemporaries. Defenders of Welsh peasant society spoke of peaceful communities, where religion and education had made a positive impact. The standard of private and public behaviour was said to have improved markedly in the second quarter of the nineteenth century. 'How fortunate the Welsh that they know little of the knife and the gun', exclaimed one admirer, who shared a common belief that brawling, cutting, and wounding were things that belonged to foreign seamen and drunken vagrants. Yet every now and then, admitted the *Carmarthen Journal*, faith in the peaceful nature of the region was shaken. On one such day, in 1829, a stream ran blood red for two miles down Pencader mountain, and at its source was the headless body of a woman.[61] A different perspective on the incident was offered by critics of rural society, and apologists of urban civilization. The Royal Commission on the Constabulary Force in 1839 suggested that much deviant behaviour, notably of a violent nature, was tolerated in backward societies, and, it was claimed, in Wales even the crime of murder had been ignored by, or hidden from, the inefficient forces of law and order.[62] The blood in the stream was a sign of the dark forces in the peasant psyche. In such a demoralized society, so the argument continued, the Rebecca riots came as no surprise.

No one who has studied the court and newspaper evidence can doubt that there was much disorderly behaviour in the three counties, both by natives and by newcomers.[63] Sailors, vagrants, navvies, soldiers, policemen, and other outsiders were always likely to inflame feelings amongst the local community; that is, when they were not fighting one another. Soldiers looking for a late drink clashed with publicans, and perhaps the bloodiest riots of the period were between the military and police officers. Few people were sad to see them leave, except, perhaps, said the newspapers with monotonous regularity, the young ladies. The arrival of English and Irish workers had a similar impact, and conflicts between them were a feature of the lead- and coal-mining villages. The building of the railways brought additional trouble, and in 1851 the Irish families about Ferryside were driven from their homes, much as their countrymen were in the mining communities of south-east Wales.[64]

Private, social, and political life in west Wales had always been impregnated with violence. As we saw earlier, members of the upper classes still settled disputes with the gun and sword. Perhaps the greatest irony was that the very magistrates who urged rational behaviour were themselves inclined to lose their temper. At one Pembrokeshire Assize two of them, William C. A. Philipps and James M. Child, had to pay fines and damages for striking an auctioneer and a farmer, Lewis Evans, David Davies, and Rice P. Beynon were bound over for threatening behaviour, and on occasions justices of the peace fought each other in open court. The Reverend Ebenezer Morris, who was fond of lecturing his Llanelli congregation on the need for improvement, once began kicking William Chambers in an effort to change his mind over a road-widening scheme by the churchyard. Other gentlemen were charged with attacking their servants and officials, and no doubt there were other cases settled out of court. Indeed, the judges at Assizes encouraged this trend, requesting that such business should be dealt with privately or at the lower courts, and setting damages deliberately low to keep people away. Thus only 20s. was awarded to Thomas Thomas, a 'very respectable' young farmer's son, when he was badly beaten and dumped in a quarry by the Griffiths family on their way home from Drwslwyn fair in 1843. Henry Griffiths the elder had a long-standing quarrel with Thomas's father.[65]

The fact that so many assaults were dealt with at Petty Sessions, and that offenders there faced a maximum fine of £5, indicates the seriousness with which society judged personal violence. Each week the courts at Carmarthen and Llanelli were full of men and women charged with drunken and disorderly behaviour, common assault, and attacks on, and rescues from, peace officers. Amongst them were prostitutes Elizabeth Burnett ('Betsy Cow'), Sarah Banner ('the Great Western'), and Mary Ann Sutherland. They caused disturbances by their drunkenness, their soliciting, and their ham-fisted attempts to rob clients. Their relationship, especially that of Sutherland, with the police was uncomfortably close. Other regular visitors to the police courts were the heavy drinkers, men like Carmarthen's John Evans ('John Wilkes'), who died of the disease, Margaret Evans, alias Mary Williams ('Peggy Clarach'), at Aberystwyth, and a few psychopaths. James Evans of Newcastle Emlyn had this unenviable gaol record in November 1842: one rape, four assaults, and two vicious attacks on constables. A year later he almost killed his wife with a knife, and was sent down for twelve months' hard labour.⁶⁶

Although such behaviour was exceptional, contemporaries were aware that violence was a part of urban and rural life. People might have walked in fear of James Evans at Newcastle Emlyn, but there were others in the town who filled their evenings with obscene abuse, primitive forms of wrestling, spitting and stone-throwing, and the smashing of shop signs and windows. One form of entertainment was the pitched battles between the youth of Llechryd and Cardigan, the newspapers reporting each defeat and victory.⁶⁷ Not far away, at St Dogmaels, life had a similar excitement, and no one there seemed shocked by the number of women fighting one another in the street, or setting about one of their number for some anonymous moral crime. Fishermen of this place, like their colleagues in Carmarthen and Ferryside, had a reputation for defending their interests with their fists. So had young farm servants and labourers, whether they were courting or enjoying the licence of harvest celebrations.

Market and fair days were always busy times for parish constables and the new professional policemen. In 1844 half the Carmarthenshire force attended one fair. As a result the drunken assaults by farmers, butchers, and drovers were committed on their way home. Most of the brawls were probably no more than exuberant rutting, but for a few, like butcher Lewis Lewis of Llanfynydd, the night of the Drwslwyn fair in 1843 was the time when his face changed shape. The *Welshman*, a little late, warned the lower orders against the common practice of head-butting.⁶⁸ Much to its regret, violence formed a part of traditional recreational activities and even some new ones. As we have seen, kicking a football and other forms of community mayhem were still practised and, in the aftermath of Guy Fawkes day, Shrove Tuesday, and similar festivals, tired bodies waited in line outside the police courts.

In general such disorder was unwelcome but contained. Yet, every now and then, matters got out of hand, and murder and manslaughter were entered on the Assize calendars. The killing of adults was, so far as we know, comparatively rare in this region during the second quarter of the nineteenth century. The execution of Thomas Price and John Evans in front of Cardigan gaol in 1822 for fatally beating and hamstringing Thomas Evans in 'an old quarrel' was one of the last public executions in the region.[69] Most deaths were the result of affairs of the heart, the enticement of sweethearts from boyfriends, husband-and-wife conflicts, pub fights, attempts at abortion and suicide, and the accidental outcome of armed robberies and poaching incidents. Manslaughter was more common, and more likely to have been the consequence of virility contests between young and drunken males. Amongst those who died in our period were William Grey, a carpenter, on his way home from the Rose and Crown, Llangadog. The trouble began when his party tossed to see who would pay the bill for sixteen quarts of ale. The local police stopped the subsequent fight in the public house, but missed the one that broke out on the road. Grey's lip was torn badly in the contest, and he died of blood poisoning. John James was similarly unlucky; in January 1840 he and four others were driving horses with loads of coal. When they reached a bridge at Llanfihangel-ar-arth a quarrel broke out about sharing the money which they had swindled from the collector at the Glangwili tollgate. A penknife pierced James's heart.[70]

Other murders, and attempts at murder, were more difficult to explain. In the 1830s there were several incidents when people fired guns into houses, either as a warning or as a genuine attempt to kill. Councillor Thomas Tardrew, the Carmarthen chemist, and a weather-glass manufacturer of Haverfordwest, whose bedroom was raked with shots, believed their lives had been in danger. The same feeling was generated when individuals were attacked by gangs or mobs. Eight ruffians set about stamping the life out of John Lazenby, the head constable of Carmarthen, in 1834, but he was saved in the nick of time.[71] Others, as we have seen, were killed by groups of armed robbers and poachers. More unusual was the experience of Shadrach Lewis, a retired soldier, widower, and woodward, who had become unpopular in the parish of Clydau, Pembrokeshire. On the night of 24 February 1840 he was sleeping with his three children, when sounds were heard in the thatch. On entering the loft, Lewis found several men, one or two of whom were dressed in women's clothes. They were apparently busy pulling down his cottage, with the support, in person or otherwise, of a large number of people. Before the pensioner could find help he was struck in the head with a hatchet. The savagery of the fatal attack, and even more the silence that came over the community, became legendary. With the assistance of police inspector Pugh of Carmarthen, and an informer, two men were charged with the crime, and both Benjamin Griffiths and Thomas Thomas were found guilty. The jury recommended mercy, petitions were drafted in their support, and the sentence of death was later commuted to transportation for life. The punishment, and indeed the whole affair, had a profound impression on the region, and convinced shocked observers that Rebeccaism had been all but encouraged.[72]

In some ways perhaps the most intriguing murder was that of a young female servant of Taliaris, north of Llandeilo, whose throat was cut while she was bringing the cows in for milking.[73] At first the reason for her death was a mystery, but rumours quickly spread as to who wanted a pregnant girl killed. Her murderer was never discovered, but her death opened a window on to the bleak landscape of female vulnerability. Although the rural community encouraged some pregnant women to marry their lovers, there were hundreds of women who faced the prospect of motherhood alone. After the Poor Law Amendment Act of 1834 life was made harder for these people. The bastardy clauses, said Lewis Evans of Pantycendy in one of his strongest speeches, 'have caused more crime, infanticide and general demoralization than was ever known in Wales before'. Pregnant women, unable to prove paternity and obtain the required maintenance payments, were increasingly obliged to enter the workhouse. The effect of this change was much debated, but the number of new-born illegitimates who suffocated in workhouse beds, and the unannounced departures of young mothers, hardly suggests that it was a pleasant regime. From Builth came the news in 1844 of Londoner Mary Furley, who fled the workhouse with her child, only to be caught and imprisoned. She attempted suicide and, when that failed, she killed the bastard. Mary received a lecture in court and the death sentence, commuted to transportation.[74]

Contemporaries found it hard to admit that infanticide was

the most common type of murder in Rebecca's society. 'Happily the crime of child murder is rare in Wales,' commented one newspaper in 1843, 'and in this instance the perpetrators have for some time past borne very questionable characters.' The case under discussion was the killing of illegitimate twins by their grandmother and mother. Margaret Hughes of Llannon in Carmarthenshire had lost her husband when he was transported for Scotch Cattle crimes more than a decade before, and she was determined that her daughter should not face a similar battle of bringing children up alone. The father of the twins was apparently a married man, a respectable farmer of the district. Four or five days after they were born, the twins were dropped down a 53-foot pit, only to be discovered by a hedger. Of the guilt of the two women, no one seemed in any doubt, but they were acquitted at the Assize, a familiar occurrence.[75]

The extent of the child murder cannot be known, although it had become one of the most common of recorded female crimes by the mid-century. There were scores of reports during this period of babies found dead, having been apparently deprived of food, care, and air. In the spring of 1845 the Cardiganshire Quarter Sessions expressed concern at the large number of inquests on children, and the chairman stated their intention to suppress the evil of mothers 'overlaying' babies. It was easier said than done; despite the evidence of suffocation given by a midwife and a surgeon, Mary Davies of Rosemarket in the neighbouring county was able to evade the charge of 'wilful murder' by proving that she had expressed concern about the health of the child before its sudden and unexplained death.[76] Coroners' juries, faced with doubtful, difficult, and conflicting medical advice about collapsed lungs, and inflamed and empty stomachs, returned verdicts of death by 'cause unknown'. Neighbours might have preferred 'culpable neglect', but there was little that they could do.

There was also the unstated problem of new-born children whose deaths were never registered with the coroner. Burying without licence or certificate remained fairly common after 1837, as sextons and churchwardens in Carmarthen, Haverfordwest, and Llanboidy knew only too well. Often it took place at night, as on 15 March 1842, when a one-month child was interred at the Baptist chapel in Llanfynydd. On this occasion, and on others, the matter came to light only because neighbours could not hide a secret. There was a feeling in the village that the child, which was dreadfully thin, had been starved to death by a mother who was anxious to return to service. Her aunt, with whom she was staying, strongly denied the charge, and stated, for good measure, that her niece had to work as poor relief had been denied her. A verdict of natural death was subsequently recorded.[77]

Most of the women who appeared in court after 1834 were charged with concealing the birth of children rather than infanticide. Those who had been born dead, or who died soon after birth, were found everywhere, in trunks, sacks, and baskets, trees, pools, and shallow garden graves. Their mothers were often female servants, like Elizabeth and Phoebe Phillips of Tenby and Amroth. The former, who worked for Colonel Ferrier, was sentenced to four months' gaol for hiding the body, apparently still-born, in a toilet. Phoebe received one month less, because her employer, magistrate William B. Swann, gave her a good character and because of unspecified 'extenuating circumstances'. Like many of these women, Phoebe had tried to hide her condition, but a fellow servant discovered her secret in a trunk. Another offender whose treatment was sufficiently lenient for embarrassing questions to be asked was Ann Jones, a servant at the White Horse pub, Aberystwyth. Her bastard child was pulled out of a pool in the Llanbadarn area, but the murder charge was changed to a lesser one, and she never served out her full sentence of twelve months' hard labour.[78]

After the Poor Law Amendment Act of 1834 the desertion of children received greater publicity and was the subject of more court cases.[79] Although there were conflicting views over the impact of the Act on child deaths, no one disputed the increase in desertion, and the Pembroke union was only one of many to petition the Home Secretary on the matter. A number of cases occurred within hours of a mother failing to affiliate her child. Some women left their babies in places where they were likely to be spotted. In Carmarthen, for example, they were found by the bridge, and at the doors of shops, chapels, and the workhouse. In the villages they were discovered playing in barns, with small bundles of spare clothes near the door. Others had fewer chances, being abandoned on dung hills, in dark woods, and on hillsides. One initial reaction when these babies were discovered in rural parishes was to blame it all on vagrant and gypsy women, but the mother, when traced, was usually a local resident, and her plea: 'I had no choice'. These were the precise words of Elizabeth Vaughan, a single woman living near Newport, Pembrokeshire, whose baby was found in 1838 by a stray cow some miles away on Treffgarne mountain.[80] A number of the women brought to court for this crime were in extreme poverty, and society recognized a difference between them and the people who strangled and mutilated the very young. So far as one can tell, this kind of child murder was rare.

Apart from this and the other violence described above, there were three types of violent conduct that took up much of the business of the local courts. Firstly there was enough family conflict recorded to indicate that life within the home was by no means always idyllic. Quarrels in labourers' families were over many things, including the lack of interest which they showed in their homes, as well as interminable arguments about money and food. Eventually they turned on their partners and children; even some of the Rebeccaites, like farmer Thomas Phillips of Topsail, Pembrey, who were supposedly upholders of morality, beat their wives 'severely'.[81] At the Carmarthen Guildhall on 26 January 1844 George Rees, David Morgan, and Henry Evans were all bound over to keep the peace towards their wives, and there were times when desperate women pleaded with magistrates to keep their husbands in gaol. Families were given little support by the courts; when Benjamin Lawrence of Haverfordwest, a sadist in anyone's language, was acquitted of brutal assault even hardened correspondents were surprised. People must have become used to seeing women with bruises and wounds across the neck, but when a Cilrhedyn horse-dealer killed his wife on suspicion of seeing another man he had to be rescued from crowds outside the Assize court.[82] Women, for their part, probably made fewer attempts to murder their husbands though they were frequently accused of stirring up conflict. An unusual case was that of the wife of John Davies, a labourer in Nash, Pembrokeshire. She left him to become a housekeeper to

B. R. Robertson, and sometime later was driven in his carriage past the home of Davies. The labourer, not the coolest of men, snapped; he managed to turn the phaeton over and threatened the life of its owner. In court Davies was treated as a hero, and his terrified wife was fortunate to escape the crowd of women waiting to duck her.[83]

In addition there was considerable strife within the wider family unit, and evidence of the persistence of personal vendettas and hostility between in-laws. Anger, like silence, was a common currency in west Wales, and at its source were problems of sex, marriage, and property. David Evans assaulted Elizabeth James in the Cardigan area, for naming his brother as the father of her child, and George Rees of Priory Street, Carmarthen, received a tremendous beating from his wife and his parents-in-law.[84] His attitude towards his wife and his remarks on the honour of her family had started the trouble, and the magistrates confirmed this by making him promise to keep the peace for six months. He took his revenge later in a public house, and there can be little doubt that many of the drunken brawls were over family matters.

Quarrels between neighbours were perhaps less common, but when they did occur the point of conflict was almost invariably property rights. It is hard to convey the bickering and violence that accompanied the crime of trespass; a woman was killed over a plum tree, and people were bitten by dogs and throttled over the ownership of boundary lines. In the parish of St Issells, Pembrokeshire, a dispute over a hedge gradually engulfed the life of ex-friends Thomas Phillipps and Evan Jones, until one January day a pitchfork was planted in Jones's foot. In the same county, Martha Evans of Nevern was gaoled for assaulting a neighbour who took her farmhouse after her eviction, and in the parish of Wiston, Rachel Rees proved as stubborn as Evans in defence of a field. She was eventually knifed in the back by Benjamin Davies and John Howell. They insisted, wrongly as it turned out, that her cattle could not graze on the land, and for their display of brute force they were given not the customary fine for assault but two and three months' imprisonment respectively.[85]

All the attacks on females have to be examined carefully, for it was customary in the courts to reduce the charge of sexual assault to a common one. Amongst the men who breathed a sigh of relief as a result were well-to-do farmers, ministers of religion, and young haymakers. One of the worst examples was that of Thomas Williams, who was accused of raping Ann Skone in Llanddowror in 1838, and given three months in gaol for assault only. According to the *Carmarthen Journal*, not usually sympathetic in these matters, virtually everyone in the court felt Skone had been greatly wronged. More fortunate still was the man imprisoned for two months. The jury seem to have accepted the man's claim that the 11-year-old girl herding cattle in Cardiganshire encouraged his advances. Mary Price, one of the many women attacked on lonely mountains, managed to prove her rape, but the offender escaped with just three months. What David Saunders Davies, the chairman of the Cardiganshire Quarter Sessions, defined as a 'bad case' of proven rape was worth a year's hard labour, or, in very special circumstances, transportation. At a meeting in 1844 the MP insisted that such offences were rare in the legal records of that county, though he could have added that many more young lads and gentlemen were accused of sexual harassment than ever appeared in court.[86]

One other form of assault which took up much of the magistrates' time was the attacks on officials. Excise officers, for example, were regarded with much animosity, and dogs and small mobs set on them. Like so many of the unpopular representatives of authority their job could not be done without a degree of force. John Mosely, excise officer at Cardigan, became convinced in 1842 that Jane Griffiths, the wife of a shoemaker, carried small amounts of malt on her person. His groping search earned him a penalty of £2. 10s. Three years later he was back in court, though this time his colleague Lloyd was the more guilty party, having broken into a man's house and pushed people about. According to one report the people of Cardiganshire showed an 'excessive interest' in the trial of the two men.[87] Keepers of pounds and tollgates, as well as inspectors of weights and measures and market toll collectors, were other targets of hostility, and the lives of Poor Law officers were made a misery by irate paupers and vagrants. Stephen Rees, the master of Llanelli workhouse, went in perpetual fear of Martha Beynon, and Evan Jenkins, the assistant overlooker at Carmarthen, was given the thankless task of trying to make fishermen break stones for their poor-relief loaves.

The police suffered even more, whether they were the old constables or the new breed of professionals. Farmers and tradesmen were understandably reluctant to act as parish or special constables when there were elections in the year, and no one relished being asked to assist with distraints. When Carmarthenshire constables Lewis and Evans tried to remove four cows in lieu of unpaid poor rates in March 1832 they found the shed locked and a mastiff on the loose.[88] In the largest towns police chiefs anticipated violence every weekend, as gangs of youths roamed the streets looking for a fight and the excitement of an arrest and rescue. Attempts to enforce the local licensing and highway laws only added to the excitement. In 1834 and 1835 the Carmarthen authorities, concerned by the effects of the recent Beerhouse Act, authorized a clampdown on after-hours drinking, games of chance, and skittles. The police lost hair, ears, and eyes in the campaign, and were accused of depriving the people of their rights to walk the streets and drink when and where they pleased. It was difficult to separate this drunken bleating from the wider hostility towards the police which was said to have existed during and after the Rebecca riots. Nowhere was this more apparent than in Newcastle Emlyn where, one weekend in March 1845, crowds kept up a frenzied battle with superintendent J. T. Hughes and his men. 'The police in this affray were beaten, their weapons forcibly taken from them, and they were only saved probably from a violent death by the interposition of the better affected portion of the townspeople,' declared prosecutor Edward Lloyd Hall at the Carmarthenshire Assize. 'It was a struggle of life and death; and the police were, under such circumstances, obliged to use their cutlasses in their own defence.'[89]

Perhaps the most abused men were the sheriff's officers who came to distrain on goods or to evict tenants for not paying loans, mortgages, rates, rents, tithes, and taxes. These bailiffs were seen by some as class traitors, given to 'over-levying' in an aggressive manner at unsociable hours. Evan Morris and James Thomas, when executing a legal warrant at St Dogmaels, took a cow, which was worth more than the debt, and then sold it back to a member of the family at an inflated price. In a case of riot and rescue at Llandeilo, seven prisoners were fined only 1s. because the Bench felt that attorney Thomas Lewis and his three bailiffs had levied too much on their visits to debtor Richard Williams.[90]

The arrival of these officers was the occasion for angry words and private and public protest, and this was particularly true in periods such as the post-war depression, and the last years before the Rebecca riots. At Aberaeron in January 1839 a large mob with blackened faces drove two bailiffs from a stationer's home, and kept them in a pub throughout the night. Two of the crowd's spokesmen, a man and a woman, were given short gaol sentences and a warning to keep the peace for three years. Two years before there had been a series of riots in the Llansawel–Cynwyl Caio area. In one of them Richard Peacock and six men came to seize the property of David Davies only to find a large crowd waiting. Two months later twenty-five sheriff's officers returned and took his goods as far as Llansawel where another mob forced them to hand the goods back. David Davies, the debtor, Esther Jones, and others who swapped punches and fired guns at Llansawel, appeared before the Assize and to everyone's astonishment were acquitted. The judge expressed his relief that the jury and not he had been responsible for the verdict.[91]

This offence and many of the other criminal activities described above arose out of legal disputes. The business of all the courts was partly taken up with civil matters. At the Petty Sessions decisions were made on the repair of roads and bridges, the weight and quality of goods sold, the non-payment of rates and tithes, breaches of the Highways Acts, cases of bastardy and desertion of the family, and conflicts over jobs and wages. As new legislation was passed, and the county courts established, it became possible after the mid-1840s to deal more expeditiously with small debts and similar issues, but before that time people settled debts, and cases of trespass, replevin, broken promises, libel, and slander, at the religious and manorial courts, and at the Great Sessions and Assizes. Sadly, the records of these civil cases have not survived as well as those of the criminal, but the proceedings at the three county Assizes in March 1835 give some idea of the range of complaints heard by the judge and jury. They included recovering damages for the loss of farm animals killed by dogs, establishing whether an excessive distraint had been made for rent owed, battles over wills and mortgages, the sale of diseased cows, the ownership to a piece of grazing land, the rights to tolls at Narberth market, and a dispute about not keeping a farm in proper repair.

From the records it is clear that a few of the inhabitants of Wales deserved their reputation for settling even minor questions by legal judgement. Judges commented unfavourably on the reluctance to compromise, especially in property disputes, and told the parties that they were simply putting money into the pockets of the legal profession. Farmers replied by suing attorneys over excessive bills, or, as in one cause involving Edward Adams of Middleton Hall, by not paying at all. Cases of property, debt, or libel took up considerably more time than criminal proceedings, and judges registered their annoyance by awarding reduced damages to the successful side. In one dispute, between Walter Rees and the family of John Richards over a slanderous remark about a Rebeccaite burglary, the judge was astonished that the injured person refused to accept an out-of-court settlement of £5. After a long and tedious wrangle, Rees won the verdict, and was given 1s. in damages. On occasions people took their cases to the highest courts in the land, and a suit won in London was celebrated like a naval victory.[92]

In explanation of this, it has to be remembered that questions of ownership, rights, and honour were important for the people of south-west Wales, and the history of land-holding was long and complex. Wills, deeds, mortgages, legacies, and debts were the very stuff of economic and social life. People were born with legal burdens on their shoulders, only to be relieved by a good marriage or an early death. Both circumstances figured large in the civil courts. The breach of a marriage promise was taken seriously, and damages of £100 and more were common. Similarly, the death of the head of a farming or business family was an important economic event, and those who stood to gain or lose often turned to the courts. David Herbert appeared at the Cardiganshire Assize in 1845, on the information of his mother, for stealing and destroying his father's will, and there were long degrading disputes in the consistory courts over the sanity of dead fathers and the division of their estates. In these, and other cases, people had to watch their words, for a person's standing in the community was important, and charges were brought over careless accusations of criminal, drunken, and immoral behaviour. Recantations appeared in the local press.[93]

The courts were used, when other methods failed, to resolve disputes over land and conflicts between landlords and tenants. Some causes, like the action taken on behalf of James M. Child and others against the vicar of Begelly for building a mansion across a public footpath, attracted much attention, but most were very ordinary affairs. Typical was the victory won by Harris, the occupier of a farm near Fishguard, who claimed that he had a sixty-year-old right to depasture livestock on Goodwick common and who resented the recent impounding of his stock for trespass. Even more representative was the attempt by a gentleman to recover £30 rent from a widow in the parish of Freystrop, in her case because she disappeared from the farm after a quarrel with her neighbours.[94] The courts, in so many instances, were called upon to decide exactly how much was owed to the landlord, and this could be extremely difficult, partly because of the practice of subletting. No one who has looked at the complicated property cases of this period can have anything but sympathy for the judges, and their desire for comfortable courtrooms; the documents, and the queues of old witnesses, must have seemed endless. Nor was this the end of the matter, for court decisions were evaded and challenged. Far better, perhaps, for rich and poor alike, to take direct action, and flood the mine in dispute or tear down the cottage that should not have built in the first place.

Only the most determined of landlords used the courts regularly. The de Rutzens of the Slebech estate, and their long-serving agent William Currie of Rosehill, initiated proceedings against several farmers in breach of their agreements. In 1835 Currie declared that he refused to receive the rents of all farms held under lease, and notices were given that, if they wished to continue, such tenants had to carry out the repairs expected of them. Minwear House was in a very poor state, but its occupier, John Lewis, insisted that this was the fault of the previous tenant, who had left him the lease in his will. The de Rutzens tried unsuccessfully to obtain an ejectment order, but a few months later they had a measure of revenge. Lewis had already been convicted once for cutting trees and underwood on his land, contrary to the terms of his lease, and he was now taken to court a second time for a similar offence and imprisoned for two weeks.[95] The courts did sanction a considerable number of eviction orders, especially on behalf of smaller landlords, but these, too, were not easy to enforce. The tenants of the Reverend Morgan Jones at Llanedi, James Davies of Llanfyrnach, and other proprietors were prepared to go to gaol rather than leave their premises peacefully.[96]

Of the civil matters dealt with at the lower courts, some of these, including the non-payment of rates, taxes, and tithes, and the many highway offences, have been mentioned earlier. Others which were of special interest to contemporaries were those associated with work, the family, and welfare. Conflicts between masters and servants were usually resolved outside the legal system, but there were times when the issues of wages, conditions, and contracts were brought before magistrates. At the ports scores of sailors were charged with jumping ship, and sentenced to a fortnight or a month's gaol. Servants in husbandry were treated rather more harshly, especially if they deserted their masters at harvest time. As a deterrent to others, one labourer, who appeared before magistrates Cullen, Dunn, and Leach at Pembroke town hall in August 1841, was put on the treadmill for three months, and a dairymaid of William Currie of Rosehill received one month less. As the pull of the industrial east grew strong, so the temptation to run away increased, but a few failed to escape. William Davies, farm servant of Thomas Rees near Carmarthen, was on his way to the mines when caught and ordered to return, and had money deducted from his wages.[97]

Desertion was also one of the most common offences under the Poor Laws. In 1839 the Narberth Board of Guardians decided to shame the people concerned by publishing their names. Each year a large number of men were charged with deserting their wives, children, and parents. Richard Lewis, a Carmarthen carpenter, David Griffiths, a Talley labourer, and

George Rowlands, known as 'the King of the Gypsies', were three who received the usual one or two months' imprisonment for leaving their families on the parish. They protested that, until they obtained a job elsewhere, usually at Swansea or Merthyr, they were in no position to help their families. Their village wives became used to such stories, and when one husband was brought back from London, where he had been living with another lady, the women of Pencader tried to kill him.[98] Women were both sinned against and sinners. Mary James, who left her aged mother chargeable to the parish of Llangennech, was placed in the county gaol until such time as she found £7 to cover their costs. By comparison, John Evans of Llanelli was treated lightly. On his appearance at the Petty Sessions on 6 November 1841 he said that his wife 'had intercourse with other men, so he separated from her'. Seventeen months later she had a child, and Evans, 'thinking he was at liberty, took another more virtuous wife'. The Bench ordered him to pay his legal wife only 1s. maintenance a week, instead of the more usual 2s. 6d., as she was 'very able-bodied'.[99]

The labour test, which seems to have been applied to this lady, was something which the courts had to enforce. The Poor Law authorities needed both the law and the police to impose the strict workhouse regime. The range of offences committed by casual and permanent paupers was wide. One crime was running away from the workhouse in clothes which belonged to the Board. When 15-year-old David William left Cardigan in this manner, he, like most of the others, was sentenced to three months' imprisonment, despite an appeal for mercy from the Cilgerran Guardians. When in the workhouse, refractory paupers destroyed property and refused to work or to obey orders. From the 1840s until the 1860s there was much of this in all three counties. The resident paupers were said to have been encouraged in their anti-social attitudes by vagrants in the casual wards. Travellers were repeatedly brought before the town sessions for refusing to break stones. On 31 January 1845, after months of trouble, the *Welshman* stated that the 'germs of incipient rebellion and treason, on a small scale, would appear to lurk in the stones of the Workhouse, in the Carmarthen Union'.[100]

Some of the criminal activities in Rebecca's society were regarded as special, both by the ordinary people and, to a degree, by the respectable inhabitants. Such offences contained an element of protest in them, or were an acknowledged part of the popular culture of the time. The relationship between crime and protest is a difficult subject, but there were offences that were committed for more than personal gain. In a society where other forms of action were difficult, the behaviour described in the rest of this book was a way of expressing resistance, bargaining, and imposing solidarity. Undoubtedly a few of the poaching cases, and of the other property crimes and assaults, were conceived as protest on behalf of a section of the community. Perhaps this was one aspect of the apparent increase during the late eighteenth and early nineteenth centuries in the 'malicious destruction' of property. Such destruction, so common in the post-war depression, included the breaking of gates, fences, and walls, the burning of houses, barns, and ricks, the cutting of trees, bushes, and plants, and the maiming and poisoning of animals.

The main reason for these malicious acts was personal pique over disputed ownership, trespass, and eviction, but the stabbing of horses in the throat, the cutting of their hamstrings, the hanging of sheep, the painful removal of cows' teats, and the poisoning of pigs, dogs, and game were sometimes part of more general conflicts.[101] The perpetrators of these horrible offences were rarely brought to justice, and this exasperated the local newspapers as much as it does the historian. The *Carmarthen Journal* was always disturbed when people committed depredations 'not so much with the intent of stealing as of destroying property'. It could offer no explanation as to why the large gardens of Thomas Lloyd of Glanafon, near Haverfordwest, and Dr Bowen of Carmarthen were so often vandalized, unless their formidable reputations as magistrate and Poor Law Guardian had affected the minds of the culprits. Similarly no comment was forthcoming when whole plantations of firs were levelled to the ground, long lines of fencing smashed, mill-ponds ruined, and bellows cut, and even when crowds of angry people invaded the homes of people at Aberystwyth, Llanarthne, Kidwelly, and other places in the 1830s and smashed windows and furniture.[102]

The same anonymity surrounded arson, the most despised and dreaded method of destroying property. Incendiarism was less common in this area than, for instance, in East Anglia, though prominent landowners like Lloyd Williams of Gwernant, Lloyd of Dolhaidd, and Picton of Iscoed were threatened with the torch in the years between the end of the Napoleonic Wars and the Reform crisis. Thereafter, during the subsequent decade, there were a few court cases and more than twenty newspaper reports of stacks, trees, farm buildings, homes, and workhouses deliberately set on fire, and insurance companies did a good trade. The burning of parts of the Llandovery and Narberth workhouses was, as we shall see later, one aspect of the anti-Poor Law agitation, but most fires were, almost certainly, acts of private vengeance. Despite substantial rewards, few offenders were identified and fewer still put on trial. William Lewis of Tre-lech, who was, faced a charge of two arson attacks on the farmhouse of Anne Howell, tenant of the Reverend Augustus Brigstocke. Lewis was taken into custody because his girlfriend had left him for Anne Howell's son, but he walked from the court a free man. New tenants suffered as much as anyone at the hands of the arsonists, and this hostility might have reflected the feeling, not just of the outgoing tenant, but of the wider village community. As always with this crime, proof was hard to come by. Even the letters which sometimes accompanied incendiarism were anonymous. One sent to Major Evans of Highmead in the winter of 1835–6 threatened to burn his house down and kill him and magistrates John Lloyd Price of Glangwili and John Lloyd Davies of Alltyrodyn. Only the Rebecca signature was missing.[103]

In other respects, too, Rebeccaism was an extension of well-established forms of popular action. David Saunders Davies and fellow MP George Rice Trevor told a rather bemused government during 1843 that the people of the three counties had, as part of their popular culture, always handed out humiliation and punishment to wrongdoers. 'It is an oft-recurring incident in a society where ignorance is prevalent, where the vanity and turbulence of youth are not kept in check by a decent self-respect, and where men of mature years have not learned to venerate the law even in its unessentials,' commented the *Spectator*. 'A grievance is felt and avenged, half in joke and half in earnest . . .'[104] The timing, place, and character of this community justice had been laid down centuries before, and variations on the main theme can be found in other European countries about the same time. In England, said the *Times* correspondent, it took the shape of 'rough music' and 'riding the stang'.[105]

In Wales, where the law of the English kings had taken generations to supplant that of Hywel Dda in the people's affections, the notion of direct retribution remained popular. Many people, including the radical lawyer Hugh Williams of Kidwelly, still believed that the community, and especially the victim of physical or moral crimes, had the right, in certain circumstances, to take matters into their own hands. Even the gentry and judges at the Assizes were slow to condemn it. 'The object of this proceeding formerly was to check gross immorality and to visit against the offending party . . .', ran a petition from JPs of the Teifi valley in 1837, 'and while it was confined to those limits, and conducted without riot, the sense of the country connived at it . . .'[106]

There were several forms of community disapproval, two of which were closely related and identified as the 'root of Rebeccaism'. The first of these was the carrying of effigies and portraits, with placards and flags for explanation. Such behaviour was common at times of political conflict, and heralded by the appearance of abusive notices. In these the objects of the people's anger were frequently ridiculed by association with animals and birds. The first citizen of Carmarthen became a 'Simple Mare', and David Davies of Greenhall near the town was a 'T. Y. Black Bull', a connotation well deserved, or a 'Jim Crow'. The effigies of these people were carried through the streets at night, with a small band of music leading the crowd. On arrival at the home of their victims, ringleaders of the mob subjected the man or woman to a loud public trial, or sold the person in a kind of mock auction, before burning the effigy. David Davies of Greenhall, in letters to the magistrates of Carmarthen and the Home Office in 1837, complained of the embarrassment caused to his family by the obscene language and accusations of the crowds, but that, of course, was the very point of the exercise. A few miles to the south-west, at

Laugharne, the effigy of Elizabeth Gibbs was also tried and then burnt on a gallows, after the lady herself had been found not guilty in a poison case at the Assize. As Gibbs left for Swansea in the spring of 1851 she reflected, as Davies did, on the half-hearted forces of law and order.[107]

The same criticism was made of the response to the carrying of the *ceffyl pren* (wooden horse). It was both a warning and a punishment. In the first instance, a man was denounced in public for the old offence of adultery, and the parish crier gave notice of three visits to be paid to his home, usually on the same night in three successive weeks. The crowd assembled about 9 to 10 p.m. with blackened faces and other disguises, and with guns, horns, and long torches in their hands. Somehow leaders 'were appointed', and these put on the horse's head and other masks, as well as women's clothes and bonnets. The procession often travelled considerable distances. The wooden horse was carried by four or six men, and another—'the preacher'—sat astride the horse. His task was to deliver a warning and a sermon, 'full of obscene abuse', outside the home of the wrongdoer.[108]

It is impossible to know from official records how frequently the *ceffyl pren* was carried, for magistrates had for years ignored or supported the custom. Rice P. Beynon of St Clears likened the processions to those of the Oddfellows and Ivorites, and stated that he had no power to stop them. One gains the impression that landowners were happy to remain silent so long as the practice was confined to sexual misdemeanours and within the 'lower classes'. When a married man of Newchurch, north of Carmarthen, sought the comfort of another man's wife in 1835, one of the magistrates was actually asked to sanction the community action. A year later, when a *ceffyl pren* was carried to Cilrhedyn, one report suggested that Henry William Howell of Glaspant was secretly pleased by the hostility shown to a rival party.[109] The judge, on sentencing John Williams, carpenter, for his part in a Llechryd affair in 1837, said, at the Cardigan Assize, that the prisoner was the first to be convicted of the crime of carrying the horse.

In 1837 the carrying of the *ceffyl pren* was at last taken seriously by the local and national authorities.[110] The custom may have become rather more popular about this time, particularly in the parishes between Cardigan and Carmarthen, and between the latter town and Llandovery, and young farm servants and labourers were said to have been its keenest advocates. In the spring of 1837 the mayor of Cardigan sought help from soldiers stationed at Brecon. In four parishes close to the town the people were accused of using the *ceffyl pren* at the expense of the common law. Those at risk now were not only adulterers, but also wife-beaters, mothers of illegitimate children, informers, libellers, bailiffs, overseers, unpopular new tenants, and all persons who exploited the economic misfortunes of their neighbours. Sometimes, to the embarrassment of Cardigan's parish and special constables, the *ceffyl pren* was brought into the town, and people were shot and injured, but most of the proceedings were conducted without hindrance in places like Llechryd, Cilgerran, and Bridell.[111]

Over the next few years, the lower Teifi valley continued to be a favoured spot for the 'lords of misrule', but in 1838 and 1839 there were more reports of the *ceffyl pren* from the parishes north of Carmarthen and Llandeilo. The information, drawn mostly from court records, has to be used cautiously, but the following three examples illustrate the growing solidarity, resolve, and openness of the disaffected population. In the first case the target of the crowd's anger was Mr Gordon, a Scotsman who worked for Abel L. Gower at Llechryd when Thomas Hazelby was convicted of cutting wood on his plantation. The verdict was attributed to the information supplied by Gordon, and so for two or three nights in May 1837 John Williams, carpenter, David James, labourer, Thomas Hazelby, and about 200 others visited the man's home in Pembrokeshire with a wooden horse. Hazelby, with his coat turned, his face disguised, and neighing like a horse, was the star performer, and when he was captured by constables, his release was quickly negotiated. Williams, the only man who stood trial, was given the minimum possible gaol sentence, but the judge warned that on the next occasion the rioters would receive the maximum.[112]

Within a year nine people were brought before the Carmarthenshire Quarter Sessions for the same offence, and treated more harshly than Williams. In view of the physical punishment meted out to their victim, the difference was not too surprising. The defence lawyer claimed that the activities were nothing more than 'fun and frolic', and very common in the Gwynfe area east of Llandeilo. They were intended to make Jane Davies 'a virtuous woman'. Jane was on her way home from Llandeilo market on 20 October 1837 when she was caught, beaten, and put on a 'hurdle or ceffyl pren'. She was carried through Pontarllechau several times by a small mob firing guns, and was ducked in water at three different spots. She claimed that she nearly drowned in this water ordeal, but two people managed to rescue her and barricaded themselves in a mill house. Morgan Thomas, a Llangadog labourer like the other eight defendants, later warned Jane not to take him to court, but this single woman had courage as well as stamina.[113]

In the third incident the victims stated that they were even more confused than Jane Davies by their apparent unpopularity. Perhaps, as Colonel Love suggested, the *ceffyl pren* was now being used to settle quarrels, pay off old scores, and express opinions about the rights to, and occupation of, land. In the autumn of 1838 the carrying of the wooden horse was publicly cried in Tre-lech a'r Betws, and several visits were made to David Thomas Lewis and his wife Elizabeth, of Maesycrugiau, and to neighbouring farmhouses. The mob was well supplied with guns, torches, and placards, and bonfires were built and lit in the fields of Lewis and his neighbour William Morgan. Whether, as alleged, the house of Maesycrugiau was also attacked, and Mrs Lewis and her sister beaten with torches, was not clear, even to the Assize judge, but a gun was shot down a chimney and a young lad injured. Some of the accused were set at liberty by the examining magistrate William Lewis of Clynfiew, and later by the judge at Assize, but William Davies and several other labourers of the area were sentenced at the Quarter Sessions to three and six months' imprisonment.[114] This was in April 1839, about the very moment when the practice of carrying the *ceffyl pren* was at its height, and only one month before the Rebecca riots began at Efailwen, ten miles to the west of Tre-lech. At the time no one seemed aware that a great popular movement was about to start.

5
The Rebecca Riots

THE very name of the Rebecca riots has become synonymous with the destruction of tollgates. The first recorded appearance of Rebecca was at the new Efailwen tollgate on the Pembrokeshire–Carmarthenshire border in the summer of 1839, a fact celebrated in verse, engravings, and stone memorial. For a few years she became 'the best-known lady' in Britain, and her name appeared prominently in newspapers and political discussions. In the cartoons of the time she was portrayed simply as the 'female destroyer of gates, chains and bars', and the waning of her support was associated with the falling incidence of such crimes and the reforming proposals of the turnpike commission in 1843–4.[1]

There was, it must be said, much popular dislike of road administration well before 1839. The reporter of the *Morning Herald* was wrong when he stated that magistrates had no warning that the tolls were unpopular. Evasion of tolls and other breaches of the turnpike Acts had always been common, and assaults on collectors and on their property were becoming a regular feature of urban and rural life. One of the reasons for these conflicts was the determination of the trusts and their officers to obtain a higher return at the gates. William Williams at the Water Street gate in Carmarthen was the leading exponent of this policy; in the early 1830s there were many complaints of overcharging against him. David Davies of Carmarthen and Trawsmawr, a mayor of the town, and John Evans, a Tre-lech carrier, successfully prosecuted Williams, but other angry victims were unable or unwilling to press charges. The same was true at the Royal Oak gate in the town, the Pensarn gate nearby, the Ffairfach gate south of Llandeilo, the Clarence Suspension Bridge gate at Llandovery, and several others. Not all the complaints were justified; James James of Cwmffrwd, Carmarthen, carrying a load of potatoes from one farm to another, was staggered to find in court that a toll could be demanded on a kettle carried on the return journey. There were other shocks, too, not least when magistrates who were themselves trustees revealed to litigants in the 1830s that the sharp increases in tolls had been legally, if quietly, sanctioned.[2]

Frustration led to disputes, and disputes turned into violence. Thomas Williams, son of the Carmarthen collector, and William Lewis at Llanelli were just two of the trust employees fined for assault. How much of this aggression sprang from self-defence one can only guess; some of their customers, like John Charles, the formidably tough Carmarthen carrier, David Davies of Blaenporth, Cardiganshire, and 'the Llanfynydd men', were renowned for their stubborn refusal to pay tolls. Sometimes they had reason to feel aggrieved, and received support for their direct action from the wider community. Since the late eighteenth century there had been sporadic attacks on tollgates. During the latter part of 1832 and in 1833 there were, for example, riots at St Clears and Carmarthen. On the first occasion a side-gate was destroyed, and in the other attorney Henry Lewis appeared before the Carmarthen Petty Sessions for inciting a mob to use their sticks and knives on all toll collectors.[3] These were the seeds of the Rebecca revolt; magistrates in the early 1830s admitted that some of the new side-gates and bars were of doubtful legality, and the local press grumbled that Carmarthen, amongst other towns, had become 'positively, beleaguered with Turnpike gates'.[4]

This popular opposition to aspects of road administration was therefore a permanent feature, which pre-dated and outlasted the Rebecca riots. In the years 1840–1, when Becca was supposedly quiet, there was much defrauding of trusts, forcible passing through gates, and violence against officials and their families. Such familiar illegality, which was common in many other parts of Britain, persuaded observers that the Rebecca movement had to be more than 'a turnpike affair' and that it drew inspiration from other sources. 'You know that I care nothing about the gates,' added Rebecca in a helpful footnote.[5] Thomas Bullin, the professional toll gatherer, and Archdeacon Venables, a Radnorshire magistrate, were agreed that there was nothing in the long history of the turnpike trusts to justify the frenzied attacks, especially against the old gates. In parts of south-west Wales there were repeated suggestions that the gates were no more than convenient targets for a disaffection that was deeper and more sinister than people realized.[6]

There was, we shall see in the next chapter, some truth in this view, but it would be wrong to ignore the animosity that was shown towards the turnpike trusts in the Rebecca years. No one will ever know the full extent of the Rebecca riots. One suspects that some incidents, which were Rebeccaite in all but name, were never reported, and even the evidence on known riots is sometimes vague and contradictory. Table 4 is a compilation of information supplied by Home Office papers, legal records, private and public reports, diaries, and newspapers. Only attacks by crowds on gates, tollhouses, bars, and chains which can be dated and placed precisely have been included in the table, and the accompanying map (Map 2). This comprises some three-quarters of all the reported incidents. It is the largest sample of its kind, and reveals some interesting chronological and geographical patterns.[7]

According to the table 293 attacks were launched between 1839 and 1844. This is considerably more than historians have generally indicated, and confirms the wilder contemporary estimates that several hundred gates, bars, chains, and tollhouses had been destroyed. The chronological sequence leaves

TABLE 4. *Attacks on tollgates, tollhouses, bars, and chains, 1839–1844*

Month	No. of Attacks			
	1839	1842	1843	1844
Jan.			7	1
Feb.			5	3
Mar.			12	1
Apr.			9	1
May	1	1	7	1
June	2		53	1
July	1		59	2
Aug.		1	52	
Sept.			35	2
Oct.			17	
Nov.		6	5	
Dec.		6	2	

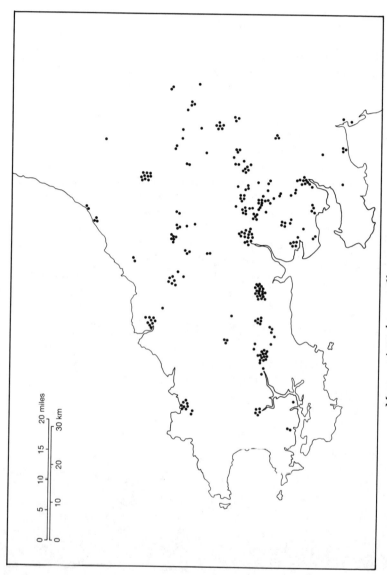

MAP 2. Attacks on tollgates, 1839–1844

Note: Thirty-one other attacks in Montgomeryshire, Radnorshire, and Breconshire.

no room for doubt; after a small outbreak in the spring and summer of 1839, there were few recorded Rebecca attacks until November of 1842. Thereafter there was a steady monthly incidence before the explosion in the summer of 1843 and the gradual decline in the autumn. From December 1843 onwards there were only one or two cases per month, and the movement was virtually over. Several people likened the riots to a meteor; for a time in 1843 there were on average two or three attacks every day, and this momentum continued until, in some trusts, no gate or bar was left standing. Others survived only by the permanent presence of detachments of soldiers and police. This was, to use a favourite phrase of the time, 'a rural war', if a short one. Like many such battles, the seasonality of the fighting was partly determined by the demands of farming, especially the harvesting, but this was less important than was sometimes suggested. The weekly pattern caused something of a surprise, for most attacks were carried out—in descending order—late on Friday, Monday, Wednesday, and Tuesday nights. Thursday nights were comparatively quiet and Saturdays and Sundays even more so. With fewer than a dozen exceptions Rebecca was true to her claim that she observed the Lord's Day.

The geography of the riots can be seen on the map. In general, Rebecca confined her activities to the three counties of the south-west, parts of Glamorgan, and the very heart of mid-Wales. Only a few Rebecca incidents were recorded outside that boundary. Indeed, people were intrigued by this, not least because there were gates in Breconshire, East Radnorshire, and Monmouthshire that were known to be unpopular and where tolls were high. The usual explanation was that the turnpike trusts in these counties granted concessions more easily than the 'stiff-necked' ones in Carmarthenshire, but the answer was obviously more complex than that. Even within the south-west there were regions, such as north Cardiganshire and areas of Pembrokeshire, where riots over tollgates were infrequent. Between Aberystwyth and New Quay only six attacks were recorded. The reasons for such disparities were many; in some places people were said to have been too prosperous or preoccupied with other issues to worry much about the gates. Across the hilly heartland of all three counties there were simply few gates, while certain trusts, like those of Aberystwyth, Tavernspite, and Milford, maintained that they escaped lightly because they responded quickly to the first signs of discontent.[8]

Other trusts were less fortunate: most of the gates and bars on the Cardigan Lower District, Carmarthen and Lampeter, Carmarthen and Newcastle, Llandeilo and Llangadog, Whitland, Main, Three Commotts, Kidwelly, and Swansea trusts were pulled down. The map indicates that there were five main lines of trouble: between Haverfordwest and St Clears, Cardigan and Lampeter, Carmarthen and Llandovery, Carmarthen and Pembrey, and Carmarthen and Pontarddulais. The map confirms the widely held belief that the riots were very much a Carmarthenshire phenomenon, for two out of three incidents happened in that county. Even so, over a short space of time the geography of the movement changed markedly; in the first period, up to the end of May 1843, the reports of trouble came from the western half of Rebecca's country, while during the next two months most of the attacks were located along the Teifi valley and especially the south-eastern quarter of Carmarthenshire. In the autumn the riots were more evenly scattered across the region, but over the following twelve months an increasing proportion of the outrages occurred in Radnorshire and its borders. When they at last came to an end, it was estimated that more than £10,000 worth of damage had been done to trust property.[9]

By tradition Rebecca's first appearance was at Efailwen on the Pembrokeshire–Carmarthenshire border in the summer of 1839. At its January meeting the Whitland trust decided, by a very small majority, to erect several new gates, which would add to the income of Thomas Bullin, the new toll contractor, and catch farmers carrying lime from the kilns about Ludchurch. The gate at Efailwen was set on a parish road, which the trust had belatedly adopted. Almost immediately the gate was destroyed and the tollhouse set on fire (13 May). Over the next few weeks the gate was re-erected twice, and each time the crowd returned in the daylight hours. Despite the swearing in of special constables, and the arrival in the area of the Castlemartin Yeomanry and twenty-five soldiers from Brecon, plans for the destruction of the Efailwen gate and that at

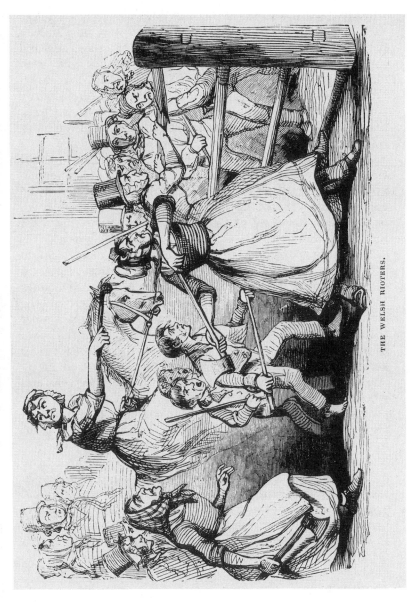

3. *The Welsh Rioters* (*Illustrated London News*, 1843)

nearby Maesgwynne were openly made and notices displayed. Already there was much of the ritual that was to characterize the movement. On the second visit to Efailwen, on 6 June, a crowd of 300–400, demanding the abolition of tolls on coal and lime, was led by people in women's clothes and with blackened faces. They slowly and deliberately dismantled property with large sledgehammers. Nine days later a slightly smaller crowd did the same at Maesgwynne, with all guns blazing, and, in its account of the next affair, at Efailwen on 17 July, the *Carmarthen Journal* recorded the presence of 'a distinguished' lady, known as 'Becca'.[10] According to local tradition 'she' was Thomas Rees of Carnabwth, a labourer and pugilist who lived on the edge of the Preseli hills. Quite why he chose the name 'Becca' has never been satisfactorily established; perhaps the 'Rebekah' of Genesis 24: 60, remains the most likely source.

This outbreak of rioting brought magistrates racing back to the district, and involved them in a rather fruitless search for ulterior motives and conspirators. William B. Swann and James M. Child, as well as an informed Dissenting preacher, told the Home Office that there was much feeling in the area against the new Poor Law and the county rates, and that two Birmingham Chartists had been seen in the vicinity of the gates.[11] No one knew what to make of this information, for no one seemed to know at this stage who the agitators were, and whether the people in the processions were willing or unwilling protesters. From the beginning of the movement intimidation played a powerful part; threatening letters were sent to magistrates, and notices were posted up denouncing the two 'sassenachs', Thomas Bullin and his brother Benjamin ('a little bull'), who kept the Efailwen gate.[12] Constables who helped the magistrates with their enquiries were attacked, and everyone warned of the dangers of talking too much. The only person committed to the Assizes, the blacksmith Morris David, never faced a trial, and farmers William Philip and Daniel Luke, taken into custody for non-payment at the Efailwen gate, later won damages for trespass and false imprisonment.

In the end the division of respectable opinion over the wisdom of the new gates proved decisive. A packed meeting of trustees, gentlemen, and magistrates at St Clears on 23 July advised the trust to revoke the order for the new gates. This meeting was the outcome of much behind-the-scenes activity by John Jones, the county MP, Walter R. H. Powell of Maesgwynne, Rice P. Beynon of St Clears, and the local Anglican clergymen, and its resolutions reflected local concern over the multiplication of gates and bars, and the use of dangerous chains on side-roads. Such an assembly might well have encouraged the opinion that some 'interested persons of a superior class' were behind the outrages, a recurring notion in rural protest movements.[13] In the view of Timothy Powell of Penycoed, George Rice Trevor, and his father Lord Dynevor, the 'extremely injudicious' concessions made after the meeting only invited further trouble, but for others the opposite lesson applied.[14] In their opinion the death of John Jones in 1842, and the stubborn determination of Bullin and his supporters on the Whitland and Main trusts, ensured that 'the first Rebecca war' would not be the last.[15]

Although the name of 'Rebecca' dropped out of the public headlines for two years, it is difficult to decide whether she actually ceased her operations during this time. There were violent clashes with toll collectors at a number of gates in the vicinity of Carmarthen and Llandovery, a chain across a road at Fforest, above Pontarddulais, was broken, and the building of the new tollbars at Penblewin and the Commercial Inn, to the east of Narberth, was sabotaged. Then late in 1842, no doubt helped by the sharp fall in agricultural prices, there was evidence of growing tension amongst the farming population and renewed resistance to any additional burdens. In a typical act of individual protest, at Aberystwyth on 19 and 21 December 1842, John Williams the carrier objected to paying 'return toll', crashed his horse and cart through the North gate, and rammed his fingers down the throat of Llewellyn Thomas, its keeper.[16]

Once again the spark that ignited widespread rioting was the decision in October to erect a new gate, the Mermaid, on the east side of St Clears. This attempt by the Main trust, under pressure from Thomas Bullin, to increase income and prevent evasion of tolls on the northern road between Narberth and Carmarthen set in motion a campaign of destruction, intimidation, and terror. Threatening letters and notices were issued, private property attacked, constables and soldiers abused and ridiculed, and possible informers again put in fear of their lives. If nothing else, this early period of the riots convinced George Rice Trevor and others that the movement was fully supported by the country people, and only a permanent police force could bring it to a satisfactory end.

Between mid-November 1842, when the Mermaid was first demolished, and mid-February 1843, some twenty attacks were made on gates and tollhouses within a short distance of St Clears. After a while the proceedings grew farcical; each time, for instance, the Mermaid and Pwlltrap gates on either side of the town were re-erected, mobs returned within hours to undo the work. The account of the destruction of the restored Mermaid and Taf Bridge gates about midnight on 12 December serves as a case-study. The attack was carried out by 100 persons who blocked roads and took contributions from people staggering home from Narberth fair. Directions were given by Rebecca and other leaders who had painted faces, horsehair beards, and women's clothes. They warned the population of St Clears to stay indoors.[17] Much to the astonishment of the authorities, the mob on this and other occasions behaved in a military manner, forming up behind mounted leaders and marching in ranks with firearms and hay forks over their shoulders. Pickets were placed to keep watch, and guns were fired at noisy sightseers. After the affair of 12 December an urgent request was sent to London for police aid, and George Martin and his two Metropolitan officers, arriving eight days later, were amazed by the scenes of devastation and the impossibility of obtaining information.

The magistrates, especially Timothy Powell and Rice P. Beynon who lived in the district, continued to have a difficult time, and were criticized on all sides, not least by the lessees of the gates. But farmers proved to be reluctant specials, and three London policemen and a few unstable pensioners could not stop the Rebeccaites. The authorities bowed to the inevitable, and on 6 January 1843, at Carmarthen, drew up a request for fifty regular soldiers from Brecon. At two further meetings, this time at St Clears, the Reverend John Evans of Nantyreglwys, and fifty other gentlemen condemned the cowardice of farmers and their servants.[18] John Lloyd Davies of Alltyrodyn, perhaps the most outspoken Tory present, described to the Rebeccaites the delights of transportation, and said he would invoke the County Police Act if the situation did not improve.[19]

Sir James Graham, the Home Secretary, responded to appeals for help by reminding their authors of the value of the Castlemartin Yeomanry, but when the justices at last got advance warning of an attack on 16 January in the Whitland district they called instead for the assistance of the Royal Marines at Pembroke Dock. These, as well as the London policemen and magistrate Beynon, arrived too late, and they vented their frustration by early morning house-to-house searches in nearby Cwmfelin Boeth, a hamlet 'notorious for lawlessness'.[20] The authorities were convinced that most of these 'daughters of Rebecca' came not from St Clears and Whitland but from parishes a few miles to the north, where the first Rebecca war had been fought. No one had any doubts that, by occupation, the culprits were small farmers—'the instigators', millers, farm servants, labourers, and other village craftsmen. Despite substantial rewards, very few of them were arrested, and only two, farmer Thomas Howells and miller David Howells, appeared at the Assizes. 'I shall show no mercy to anyone', was Rebecca's message to those seeking their conviction.[21] Their acquittal, and the crowd's denunciation of the chief witness, a drunken pig-dealer, increased the feeling 'that the people now reign'.

During this winter of 1842–3 the words and actions of 'Turnpike Rebecca' indicate that she had two main grievances: first, the accumulation of tolls on the boundary of two trusts, always a source of trouble, as they did not clear each other, and second, the appearance—at the behest of the 'sons of Hengist'—of side-gates and bars on roads maintained by the parishes through which they passed. Yet there was, it was feared, a deeper discontent, for on the night of 13 February when a troop of yeomanry was winding its disgruntled way homewards from an exhausting stay at St Clears, a body of 15–20 men, speaking mainly in English, removed the roof and bar of an old tollhouse at Trevaughan, in the neighbourhood of Whitland.[22] Like the Maesoland, Blue Boar, and a number of others destroyed about the same time, and the Penygarn, Pontarllechau, and

Pontarddulais which were attacked some months later, the Trevaughan gate had been a feature of the landscape for years, and its demise was a sign that no trust property was safe. The rumour that meetings of Rebeccaites had decided to remove all turnpike gates, and impose 'free tolls' and 'free laws', now gained credence. The Whitland and Main trusts, conceding a little ground, published a version of their accounts and chose not to restore a couple of the most unpopular gates.

In the spring of 1843 the riots began at last to sweep outwards from the close confines of Whitland and St Clears. Although any attempt to re-erect gates in that district still caused trouble, the letters being carried to London increasingly reflected the anxiety of magistrates who were facing Rebecca's vengeance for the first time.[23] Violence extended westwards, via the Narberth, Robeston Wathen, and Canaston Bridge gates to Haverfordwest, beyond Carmarthen to Kidwelly, and northwards in the direction of Newcastle Emlyn and Llandysul in the Teifi valley. Some of the hostility at this time was of a private nature, such as the annoyance of a wedding party over the tolls at the Llanddarog bar, the irritation of drovers passing through half-destroyed gates at Narberth and St Clears, and the mounting anger of regular users of the bridges over the Eastern Cleddau and the Tywi. As several gentlemen noted, with a touch of irony, success bred success and private initiative, and soon Rebecca was declaring her intention 'to set the world to right'.

In Pembrokeshire a number of the riots, including those in Haverfordwest and Fishguard, were in the nature of well-planned protests against the legality of gates within town boundaries and the non-renewal of a turnpike Act. Meetings were held to plan the attacks, and farmhouses visited for money and victuals to help the cause. Such was the organization and secrecy that toll keepers were quite unable to explain what had happened to the property in their care; at Prendergast, on the northern edge of Haverfordwest, the collector slept through one visit, while the old lady at Robeston Wathen gate talked vaguely about a gentleman caller and a splendid horse.[24] At Narberth, two miles away, the Rebeccaites made a springtime resolution to remove all three of its gates, and, as at St Clears, Rebecca processions through the town were led by men on white-robed horses and accompanied by the firing of guns. Plaindealings gate, to the north of the town, was a special target at this time, for it caught people coming for lime and culm from Llawhaden, Llandyssilio, and further north. On one famous occasion the wife of Colonel Colby of Ffynone got her servant to pull down the chain that had replaced the hated gate, and received enough support from the Grand jury to escape the legal consequences of her action.[25]

The activities of the Rebeccaites above Carmarthen and along the Teifi valley were marked by even greater intimidation and ritual. In the hilly landscape to the south of the valley, where John Lloyd Davies of Alltyrodyn kept a lonely vigil and where lawyer Edward Lloyd Hall was a constant sniper, the forces of law and order were notoriously weak. Their unenviable task was to protect those gates known to be under threat. These were the ones which the trusts had erected in the district to prevent people using parish roads when carrying goods or cutting fuel on the local mountain slopes. In the late spring and early summer of 1843, when more than a dozen attacks were made on these gates and tollhouses, parish constables were too terrified to act, and letters pleading for military and police assistance arrived on the desk of the poor Home Secretary.

From the angry tone of John Lloyd Davies and his brother magistrate John Lloyd Price of Glangwili, no one could doubt the extraordinary nature of the events. Public meetings of hundreds of people had been held, warning notices appeared everywhere, scores of unwilling farmers were brought from their homes at gunpoint, and rockets, horns, and gunfire echoed through the night. A few of the outrages seem to have been carried out by small gangs, as when, on 15 March, six or seven armed men forced a collector and his wife 'out in nudity into the road' before pulling down part of their house, but most visits were open demonstrations of 'Rebecca's awesome power'. At 11 p.m. on 18 April a large crowd wandered the hillsides south of Llangeler, led by a man with a woman's cap, blackened face, and loose striped shirt, and carrying a broad sword. At Bwlchyclawdd they forced the toll collector and his wife to leave the tollhouse before demolishing it and the adjoining gate. The man was encouraged to choose another profession. On the way home the Rebeccaites stopped at an inn to get ale, 'the leader of the party holding the sword across the door way to prevent the publican coming out'. On the following night the same people marched through Cynwyl Elfed and left a warning that if the magistrates at the Petty Sessions acted against 'our Lady' they would be besieged in the village and their property set on fire.[26]

The month of May was comparatively quiet, but June was one of the busiest, and during its first three weeks the Teifi valley was the focus of attention. On the evening of 19 June three gates were destroyed by the same crowd. It comprised 150 people, resplendent in elaborate masks, make-up, and women's clothes. At Llanfihangel-ar-arth the leading actors formed a circle with guns pointed outwards as a warning to intruders. One of the special constables, who had been appointed to defend the gates, was asked not only to strike the first blow but also to remove the stumps of the gate posts on the following day. His failure to carry out his promise brought a threatening letter and, eventually, an attack on his home. His experience was shared by the toll collector of Pont-tweli, Llandysul, who bravely appeared at a Petty Sessions to support charges against several people for refusing to pay tolls in the area. On the night of 12 June his house was set alight.[27]

The leader of the crowd on 19 June was said to have had soft hands, 'unused to work', but in reality it was probably a sixty-acre farmer, David Evans of Penlan, the tall middle-aged father of four children. He was later arrested for playing a prominent part in this riot, and in one on 10 July at the Gwarallt gate nearby. Several people claimed that he was their leader, the man who forced his tenant, employees, and others from their homes, gave directions about the destruction of the tollhouse in English, and fired his gun at will. He was taken into custody by police superintendent Hughes months afterwards, and charged with the two offences. He had no previous convictions. The Assize jury, to the obvious anger of the judge, acquitted him of the charge of destroying trust property while agreeing that he had been present at the scene of the crimes.[28]

The intention to destroy the gate and tollhouse at Llanfihangel-ar-arth had been proclaimed for days before about the neighbourhood, and indeed Rebecca carried out so many of her promises that by the third week of June almost all the gates on the Whitland, Fishguard, Lampeter, and Newcastle trusts had been demolished. Some trustees and magistrates no longer bothered to enforce the payment of tolls, but John Lloyd Davies took the fight to the enemy. He proceeded against the Hundred of Elfed to recover damages for the loss of trust property and continued to fine individuals who were either unwilling or afraid to pay the charges at the broken gates. He was convinced that, besides the immediate assistance of soldiers and Metropolitan policemen, the Hundred of Elfed, and the Hundred of Derllys, which contained St Clears, needed a permanent Rural Police. He was supported by MPs David Saunders Davies and George Rice Trevor, but when the matter of a county police was first discussed by magistrates in June 1843 they listened to the counter-petitions and postponed a decision.[29]

Despite this, a feeling persisted that Rebeccaism in early June was entering a new and more sinister phase. John Lloyd Davies declared that the destruction of gates 'is becoming a mere means of meeting to concert other schemes of a more dangerous character'.[30] What he meant by this was rather obscure, but he was repeating the words of his favourite local newspaper. The *Carmarthen Journal*, in its editorials of 9 and 16 June, warned that the government of the three counties was falling into Becca's hands, and that no person or property was safe from her children. Before the month was out, she had taken control of the Tywi valley between Carmarthen and Llandeilo, and could be seen nightly in the Llanegwad–Llanarthne area, where the roads crossed the valley bottom. The most vulnerable spot was the Llandeilo-yr-ynys gate attached to the new bridge, which was used by farmers travelling south for lime, coal, and slates. The gate and tollhouse, which had been destroyed on 10 May, was again visited on 13 June and 20 June. Not far away the Penygarn gate suffered the first of its five attacks on 26 June, and William Davies, a freehold farmer and Poor Law Guardian, with land worth several hundred pounds, later confessed to his part in the affair, and had to be rescued from his drunken slip by a Carmarthenshire jury.[31]

In the British context these events made remarkably little

impression on the people in power. Graham and his colleagues, who had Irish and Scottish affairs on their mind, appeared undisturbed by the rural war in the Principality, and national newspapers gave it little space. On reflection, it was only the threat to the largest town in the region that aroused the interest of the public, politicians, and the Queen. As we have seen the turnpikes in and around Carmarthen had been unpopular for years. Since 1837, when its tolls were raised at a poorly attended meeting of trustees, the Water Street gate on the old Newcastle road, close to the town centre, had been the main target of complaints, certainly from people journeying to the market from Newchurch, Cynwyl Elfed, and beyond. Threats, verbal and written, were made against Henry Thomas, the gatekeeper, and warnings were issued to those foolish enough to pay the toll.

Matters came to a head in the early hours of 27 May when two men in women's clothes, called 'Rebecca' and 'Charlotte', smashed their way into Thomas's home, and imprisoned him and his family while the roof, porch, and windows were removed. Thirty people, carrying tools borrowed from a smithy, were responsible for destroying this tollhouse and gate, while several hundred others, by their continuous gunfire, kept the townsfolk and constables at a distance. The rioters were apparently farmers and labourers from villages to the north-west of Carmarthen. According to Henry Thomas, the leader was a tall man with a local accent, and he was identified by one local historian as Michael Bowen, a young farmer of Tre-lech. Besides grumbling about the high level of tolls Rebecca expressed her opposition to informer David Joshua, keeper of Glangwili gate, and her hatred of the local workhouse. For the moment, however, the leader was content with the night's work, and her followers enjoyed their journey home, grunting like pigs outside Fountain Hall and firing a dozen shots near Greenhall, the urban residence of David Davies.[32]

Shamed by the whole business, and furious at the cowardice of their policemen, the Carmarthen authorities decided on firm action, but in their keenness they ignited a community rebellion. In the next few days heavy fines were imposed on John Harries of Talog Mill and two of his neighbours, Thomas

Thomas the shopkeeper and farm servant Samuel Bowen, for not paying tolls at the Water Street barrier. The men were only obeying Rebecca's latest order to ignore all such charges. Thomas, whose wife was about to bear him a child, eventually paid the fine, but not the other two, and distress warrants were made out by Edmund H. Stacey, the mayor, and William Morris, and endorsed by David Davies.

On 9 June four Carmarthen constables were given the dubious honour of executing these warrants, only to find that Bowen, who lodged with his parents at Brynchwyth above Cynwyl, had no property of his own, while Harries's home was protected by a mob disguised in the usual way. The people had been brought together by a white-clothed lad, blowing a bugle, and they intercepted the constables at Blaenycoed. The constables took one look at the guns, scythes, and turf-cutting equipment and wisely returned to Carmarthen and an irate Bench of magistrates. No one was more annoyed than David Davies, for in the midst of these troubles a mob from the parish of Abernant set to work and destroyed half an acre of young trees and an ornamental wall at the entrance of his country home at Trawsmawr. His letter to the government, and his sacking of employees, convey his fury.[33]

After a short delay the mayor and his colleagues rounded up twelve constables and specials and twenty-eight pensioners living in the area. Led by the fiery road surveyor David Evans, who was deemed more reliable than the police chief, they set off in the early hours of Monday morning (12 June), and arriving at Talog Mill without any interruption, except for the running feet and gunfire of Rebecca scouts. The mill house was locked, but eventually one of Harries's daughters opened the door and the levy was made. Already, however, the boy with the horn and answering gunmen had alerted the neighbourhood, and before the party had retreated far towards Carmarthen large groups began to appear. Thomas Thomas, as conciliatory as ever, now approached and persuaded the constables to return the four boxes of goods, promising himself to pay the fine for Harries.

The crowd of people, which had now reached a total of perhaps 300, was disguised in a multi-coloured collection of red masks, horsehair pieces, women's caps, and the like, but it was the large number of new guns in their hands which attracted the constables' attention. Amongst those present in the mob were Jonathan Jones, a 21-year-old labourer and farmer's son, John Jones, Jonathan Lewis, Howell Lewis, and David Lewis, four labourers of about the same age, James James, a bugle-playing labourer, and an older man, David Davies, a 60-year-old tailor. On this occasion there were several leaders, and they all required the constables to give up their weapons.[34]

After a short and bruising battle, PC Nicholas Martin and his colleagues agreed to dispose of their ammunition and handed over their pistols. The crowd then accompanied them as far as Trawsmawr, where David Davies was busy refurbishing the mansion and its grounds. Having been assured that he had endorsed the distress warrant, which was ceremoniously ripped to pieces, the constables were forced to finish the work of a previous evening and remove part of the new wall. After this, the pensioners, who were unarmed and harmless, were sent on their way, but the hated constables were kept back a while and then released with shot flying about their heels. 'Law is at an end', cried one town official when he heard the news, and the reports of such incidents at last persuaded the Secretary of State to send the military.

When orders were given a few days later for the removal of a troop of the 4th Light Dragoons from Cardiff to south-west Wales, Sir James Graham knew that the Rebeccaites had threatened a mass invasion of Carmarthen. When Water Street gate was attacked at the end of May, the leaders of the mob had talked of visiting the town and workhouse sometime in August. The prosecution of the Talog men brought the matter forward, and by the end of the second week of June people had expressed their determination to recover the fines which Thomas Thomas had paid on behalf of himself and John Harries. The latter, who was the moving spirit behind so many of the troubles, called on David Davies on Saturday, 17 June, and warned him of the great Carmarthen demonstration planned for two days later.

For several days the villages were full of public meetings. The reports of the proceedings in Abernant and Cynwyl Elfed are unsatisfactory, but it seems that resolutions were passed demanding changes to the tolls and gates of the Newcastle trust, and the removal of the fines on the Talog men. Nor did matters end there, for most commentators were agreed that the issues discussed ranged well beyond the financial cost of new roads. According to the *Carmarthen Journal* and other hostile sources, tithes, rents, the Poor Law, and the behaviour of magistrates were near the top of the agenda, and 'Justice' was the slogan. It was agreed that the people's grievances should be taken to Carmarthen and the town paraded as a demonstration of their strength. A few spokesmen wanted to pull down the Carmarthen workhouse, but for the moment this feeling was held in check.[35]

The gist of these discussions was relayed to meetings of Rebeccaites on the north and western borders of the county. On the eve of the march on Carmarthen reports came from Newcastle Emlyn, Cardigan, and further afield that simultaneous attacks were planned for the beginning of the third week of June, with gates as the first target and workhouses as the second. The planning involved was considerable, and there was talk of money being levied and weapons bought or collected. Those organizing the mass demonstration to Carmarthen sent anonymous letters and visited farms. The following, placed by Becca under the front door of 'John Wood Esquire' of Cwm, Meidrim, on 17 June, fairly represents them all:[36]

Sir,
 You are hereby strictly requested to meet the Hamlet procession at Henfwlch on Monday the 19th Instant by 10 o'clock A M. Be you personally present on horseback and every male in your employment must appear. No excuse will be taken. Non compliance will bring vengeance on your head and most likely you will be launched into eternity without the least warning and you shall see whether your threats, your loaded guns & pistols, your cursing & swearing at the people's good cause will avail you something in that day. Recollect that the County is quiet [sic] tired of bearing the heavy burden of maintaining your fraternity the burden must be shaken off. If you will exert yourself in the people's cause all well & good, if not Monday will decide your fate.

Yours &
Becca

Like the Reverend John Jenkins of Meidrim, and several others, this small landowner was asked for money to finance the popular campaign and to pay for powder and shot.

Notices demanding the presence of all males over 16 years at the Plough and Harrow in the parish of Newchurch on Monday morning were pinned to the doors of the parish churches. On Sunday, after the church service at Abernant, James Evans, the parish clerk, was forced by John Harries to announce the demonstration at Carmarthen and the perils awaiting those who stayed at home. Lewis Evans of Pantycendy, who tried for days to avert the disaster, advised the church-goers to ignore the call, and asked John Harries and Thomas Thomas to meet him at the vicar's house. Meanwhile, similar announcements were being made at the local chapels, and placards posted on walls and gates. Anxious farmers met secretly and sought help from Edward Lloyd Hall and friendly magistrates, but there was no escape. 'You must attend or expect the torch' was the message that accompanied them to bed.

It seems, from subsequent investigations, that no riot was intended on 19 June, for the people were told to attend without disguise and to act in an orderly manner. When they arrived at the Plough and Harrow those who had brought guns were urged to leave them at the inn. At this point the growing crowd resembled a political assembly, not least because the leaders were still drawing up an ambitious list of reforms, which included reductions in tolls, equitable rents, and the abolition of, or cuts in, church rates and tithes. Lewis Evans, who had been in the area since the early hours, failed to get John Harries to stop the march, but when the landowner arrived at Carmarthen about 10 o'clock that morning, he brought with him the message that the crowd would listen to Thomas Webb, a prominent reforming councillor. Unable to procure his assistance, Evans returned to Newchurch with John Lloyd Davies, but the latter's speech to the mob was cut short by people demanding his blood. The assembled throng were prepared to guarantee the safety of the workhouse, but their demonstration was legal and their hopes were high.

As always in the Rebecca story, estimates of the size of the procession varied, but there were at least 300, mostly farmers, on horseback at the rear, and 1,500 labourers, servants, and others before them. Right at the front were a hired band of music, and scores of women and children, as well as a splendidly made-up Rebecca who has since been identified as Jonathan Jones of Abernant, Michael Bowen of Tre-lech, and Rees of Rhydymarchog! About 11 a.m. the crowd set off, and within an hour was seen near the outskirts of the town, their tall Welsh flags of 'Justice' and 'Freedom' waving slowly in the warm breeze. They marched past Water Street gate, but then turned right to meet a party which had come from the direction of St Clears. As they moved down Lammas Street they were cheered by hundreds of townspeople, and some of these leapt on the farmers' horses and followed the procession on a tour of the southern half of the town. On reaching St Peter's church, the band turned down the narrow King Street, and eventually came to a tumultuous halt in Guildhall Square.

The geography of the march indicates that the first intention of the country people was to place their grievances before the magistrates at the town hall, and retrieve the money which they had wrongly taken from the Talog heroes. The mayor and other town magistrates were waiting for them, but they had done rather more to defend themselves than the crowd realized. Shops and pubs were closed, and police and specials placed at the town hall, county gaol, and the workhouse. The Staff of the Royal Carmarthen Fusiliers, under Captain Bankes Davies of Myrtle Hill, had also been called out, and were strategically stationed near the workhouse. Finally, during the morning, Thomas Charles Morris and a few friends were spotted near the bridge below the Castle, waiting to meet the dragoons who were expected at any moment.

What happened next was much disputed, though not entirely unexpected. Those townsfolk who had attached themselves to the procession declared that Becca had come to save them from the workhouse, and there were people from Abernant and Newchurch who were happy to answer their prayers. At the Guildhall Square these people, along with Frances Evans, an ex-inmate, Isaac Charles, a young tailor of the town, David Williams the weaver, and John Lewis the fisherman led a section of the crowd northwards up Red Street to the workhouse, where angry spectators had been waiting since an early hour. In the mob, which was constantly growing in size, were John Harries the miller, and Job Evans and David Thomas, 28- and 49-year-old farmers, and these forced their way past the porter's lodge and into the workhouse yard. Councillor James Morse of the Stamp Office tried to reason with them from a window of the building, and swore that the soldiers were about to descend upon them, but the master and matron were obliged to give up their keys and the alarm bell began to sound. People poured into the dining hall and board room, and danced on the tables. Some of the inmates were released, and the mob said that they would provide places for the dozens of orphaned and illegitimate children.

In the confusion that followed, beds and bedding were thrown out of the upstairs windows, and magistrates and constables on the spot believed that a fire was about to be started. Others suggested that the more respectable elements in the crowd, influenced by Morse's words, were already leaving the scene. At that very moment, however, Major Parlby and twenty-nine men of the 4th Light Dragoons arrived on the scene, their horses panting and their swords flashing. Once in the yard, the gates were closed and an unequal battle ensued. The strongest members of the crowd tried to bring the horses down, and David Thomas fought county magistrate William G. Hughes to the ground, but most of the men and women fled into corners and over the 12-foot wall. They crashed into neighbouring gardens and stumbled down side-streets, and a few lucky ones found their horses. Even Rebecca, the only horseman in disguise on that day, lost some of her curls and stately bearing.

Almost a hundred were left behind in the workhouse yard, cut, broken and soon under arrest. After a quick examination, those identified as being also prominent in the Talog disturbances were taken to the county gaol; another small group were escorted to the borough prison and about sixty others, protesting that they had been compelled to join the mob, were bailed to appear at the Assize court or discharged. They included 'several respectable farmers' and some of the musicians who had been hired for the day. In the meantime, guards were set upon the major roads into the town, for the fleeing mob had promised to return with their guns. 'This is a day to remember', crowed George Rice Trevor, and the *Carmarthen*

Journal gave thanks that the banks and the mayor's house had been saved. Only the *Welshman* refused to join in the celebrations, muttering to itself about a manipulated crowd and military brutality. No one doubted, of course, that the events of the day were a significant turning-point in the story of Rebecca.[37]

The information that the daughters of Rebecca came within half an hour of demolishing a workhouse, and perhaps taking over Carmarthen, finally spurred ministers of the Crown into action, but it would be wrong to assume that this reduced the level of popular protest. On the contrary, as the table shows, June marked not the end but the beginning of Rebecca's busiest period, and the evidence for this was not just a matter of better reporting. On the very night of the demonstration at Carmarthen three gates were broken, and on the following day another seven gates and bars were removed on the line of roads between St Clears and Castell Rhingyll in the south-east of Carmarthenshire. At St Clears, as part of a long-running saga, thirty men from the Meidrim district threw gate posts into the river, and the next day they were retrieved and replaced, and then destroyed again. Timothy Powell of Penycoed, who was beaten when he tried to apprehend one of these rioters, stuck bravely to his duty. In the Teifi valley it was a similar story; for days rumour and anonymous letters had predicted 'an explosion' to compare with that at Carmarthen, and nocturnal meetings chose certain gates and workhouses as their next targets. Once again, the mayor of Cardigan, Lieutenant-Colonel Herbert Vaughan of Llangoedmore Place, and Edward Lloyd Hall of Newcastle Emlyn pleaded for military help, but when it came it was too late.[38]

The descriptions of the destruction of gates at Newcastle Emlyn and Cardigan on 21 and 23 June were almost too good for the government's peace of mind. Magistrates and landowners, left to their own devices, were obliged to contact Rebeccaite sympathizers and hold informal meetings with parish delegations, and because of this they were able to give the Home Secretary accurate predictions of when and where 'the Lady' would strike. Edward Lloyd Hall watched the proceedings at Newcastle Emlyn from his home. The number of participants was fewer than expected, though there was a handful of friends and 'strangers' who had travelled some distance to be there. Many of the 250 people who passed near his house were young, and they were divided into three groups; advance and rear guards, armed with guns, and the main body. As on a number of other occasions, instructions were given in English, and no names were used. They destroyed three gates in the area, and some of them probably joined the attack at Cardigan two nights later.[39]

The large crowd at Cardigan was a mixture of spectators who had come from neighbouring parishes and were congregated on the common, and a smaller group which arrived later and was led by a dozen horsemen. When they set off on their evening's work, the people were armed with guns, swords, scythes, pitchforks, hatchets, and saws. Their first call was the newly built Pensarne tollhouse and gate on the Aberystwyth road, and these they dismantled together with an attached wall which stretched for thirty yards. The task took longer than expected. 'Damn me, mammy, it's hard work, send more hands up here', shouted a roofer, but it was soon time to march through the town to Rhydyfuwch bar. According to one report, which was at variance with other evidence, Rebecca was not present during the destruction of the bar, but 'Nelly' was, and she assured the keeper of the Rhydyfuwch tollhouse that all her personal belongings and furniture would be safe. So was the adjoining gate, 'the idea of the Rabble being', wrote Herbert Vaughan, 'that those roads, & those only, where the Mail runs, shall retain their Toll gates'. Like Richard Jenkins, the mayor of the town, he was convinced that these attacks were part of a much wider disaffection, organized and concerted across the three counties, and 'emanating either from the Chartist or Anti-Corn Law League and accompanied with . . . strong revolutionary feeling'.[40]

Although there was, as we shall see in the next chapter, some truth in this picture of general unrest on the borders of Carmarthenshire and Cardiganshire, for the moment it was the threats on the gates and workhouses that were taken most seriously. On 20 June Edward Lloyd Hall issued his first address to the inhabitants of the district, calling on people to give up violence and concentrate on redressing one grievance at a time. Three days later he attended, along with parish delegates, a meeting of magistrates and the Newcastle trust at the Salutation Inn. It was an important assembly, at which John Lloyd Davies, Edward Lloyd Williams of Gwernant, and Thomas Lloyd of Bronwydd (in Welsh) expressed their incredulity that Welshmen could commit such offences, and George Rice Trevor, in a statement guaranteed to please, promised that during an emergency he would order troops to fire on them. In the mean time, it was suggested that a committee should examine the affairs of the trust. With considerable skill, and a fearful eye on the angry crowd outside, the conveners of the meeting nominated Rees Goring Thomas of Llysnewydd, Thomas Lloyd, Lewis Morris and John Davies of Carmarthen, the Reverend Benjamin Lewes of Dyffryn, Lewis Evans of Pantycendy, John Beynon, and Edward Lloyd Hall as members of this committee. It promised an early report and action.[41] Other trusts followed suit; before the summer was out there were clear indications that tolls would be reduced in the region, and gates moved or given up altogether.

This was complemented by other responses less welcome to the local population. Within a few days of the Carmarthen demonstration a troop of the 73rd Infantry Regiment arrived in the town to supplement the 4th Light Dragoons, and their colonel, James Frederick Love, was given the command of the south Wales military district. The colonel was an experienced campaigner, having been stationed at Merthyr after the riots of 1831 and at Bradford in 1842. One of his first tasks was to review and deploy the Castlemartin Yeomanry Cavalry and the Royal Marines at Pembroke Dock, as well as the regular troops now arriving at Carmarthen. Within a month he had over a thousand men under his command. He was helped in his work by George Rice Trevor, and, a little later, by staff officer Major Rochfort Scott. Their strategy was plain: to isolate the region from discontent elsewhere, and to place detachments of soldiers where the danger was greatest. By the end of June there were troops of yeomanry at Pembroke, St Clears, Narberth, and Lampeter, and marines at Cardigan and Newcastle Emlyn. On the nights of 26 and 28 June it was only the presence of soldiers which saved the workhouses at Newcastle Emlyn and Narberth from attacks by mobs assembled for that purpose.[42]

Towards the end of June the magistrates of the three counties finally responded to the criticisms of lethargy levelled at them by the Home Secretary and press. They held meetings in all three counties, and published addresses calling on the inhabitants to give up Rebecca and lay their grievances before their natural leaders. The Cardiganshire justices expressed the hope that landlords would insist that their tenants and dependants became special constables, and offered to match the substantial rewards that the government had just sanctioned for the conviction of rioters. They also asked Parliament to increase the punishment for the destruction of trust property. The Pembrokeshire magistrates, who were unable to enforce tolls at Narberth or execute warrants against offenders, demanded more military help at their meeting, while the Carmarthenshire authorities, under relentless pressure from George Rice Trevor, made the first moves to set up a county police force. Trevor, like Edward Lloyd Williams, the Reverend Augustus Brigstocke, and others, published his own warning address at this time, and other landowners gave notice that they would sack workmen and leave the country unless things improved.

Neither threats nor concessions seemed to work and, outside the larger towns, Rebecca carried on much as before. Colonel Love grew cold and his superiors impatient. Nothing seemed to work as well here as it did in other parts of Britain. The random night patrols of the cavalrymen were soon abandoned, for Rebecca's scouts and decoys were everywhere. During July re-erected gates on the edge of Newcastle Emlyn and Haverfordwest were taken down although soldiers were within shouting distance. Even more galling for the authorities was the time wasted on false alarms and deliberate distractions. On the night of 25 July, for example, the sound of a horn and shots on the road to St Clears roused soldiers in Carmarthen, but in their absence, the gate and part of the tollhouse at Croesllwyd to the south of the town were pulled down.[43] Even when rare information of a planned attack reached Colonel Love, the movements of the soldiers were carefully watched. On 21 July troops set out from Llandeilo and Carmarthen to a destination between Porthyrhyd and Pontyberem where Rebeccaites were expected. 'Shortly after we had passed a village or hamlet,' reported Love on the following morning, 'we heard shots fired in our rear to let people know the troops were out, & to which was added occasional fires on the hills . . .' 'It is difficult', added the *Times* reporter, 'to fight against a united people.'[44]

All this humiliation took place in the glare of national publicity, for one effect of the prolonged troubles, and especially the demonstration at Carmarthen on 19 June, was the arrival of newspaper correspondents from across the border. The *Times* man, Thomas Campbell Foster, got to Carmarthen on 22 June, and was immediately impressed by the fact that 'every person I spoke to sympathised with the rioters'.[45] His reporting was to infuriate the authorities, and soon colleagues from the *Morning Herald, Morning Advertiser,* and *Morning Chronicle* were following in his footsteps. Together with local journalists, notably the impressive correspondent of the *Swansea Journal*, and a host of private literary adventurers, these men ensured that the Rebecca rioters had probably the best media coverage in the history of early nineteenth-century popular movements.[46]

From the start Foster showed sympathy for the peasantry of the region and revelled in the embarrassment of the civil and military authorities. On 22 July he told the nation that 'not a single outrage has been stayed, nor a single Rebeccaite captured'. This was not strictly true, for a number of men about Carmarthen had been taken into custody, and eight people held for one night on suspicion of destroying a gate above Llandovery. The magistrates there were emboldened by the recent arrival of a party of 4th Dragoons from Brecon, but the soldiers were absent when the examinations of the prisoners took place, and so were unable to protect the constables from an angry mob. At Aberaeron a few weeks later guns were trained on the room where another suspect was being examined, and the magistrates wisely decided to imprison the informer.[47]

These experiences highlighted the difficulty in apprehending Rebeccaites. Despite fines for a refusal, very few people were prepared to act as special constables when required to do so, and although there were £100 rewards, hardly anyone at first would give evidence against the rioters. Nor did it end there, for informers were notoriously unreliable, and Welsh juries could not be trusted. Grand jurymen received threatening letters from Rebecca, and several Petty jurymen gave up the job in terror. 'If any informed,' boasted one arrested man, 'they would never be found guilty at the assizes, by a Jury of this country.' This was not quite as bad as Ireland, but it does explain why serious consideration was given to removing Rebeccaites for trial outside the three counties.[48]

More soldiers and heavy artillery arrived in the region during July but the riots were now at their greatest intensity. On the fourth, seventh, fourteenth, and twenty-first days of the month at least half a dozen gates and bars were destroyed on the same night, sometimes by one or two mobile crowds. Re-erected barriers at Cardigan, Newcastle Emlyn, Prendergast, Narberth, Kidwelly, and Carmarthen were an inviting challenge, despite the widespread use of iron posts and chains, while at the same time private gates and bridges were damaged and new targets found as far apart as Llanon in Cardiganshire, Bronfelin south of Builth, Devynnock in Breconshire, and Rhydypandy and Three Crosses in Glamorgan. At the last place a courageous and prominent figure, William Eaton, walked out like a western gunfighter to meet the crowd, only to be bombarded with pieces of gate and shot.[49]

Most of the attacks in July and August were carried out either near the towns of Lampeter and Carmarthen, or to the south of the river Tywi and against the property of the Kidwelly, Three Commotts, and Swansea turnpike trusts. In the vicinity of Abergwili, just to the east of Carmarthen, there were three barriers within half a mile of one another, and although the one on the main road was ignored, the other two were repeatedly broken. Above all, perhaps, it was communication centres like Llandeilo-yr-ynys, Porthyrhyd, Llannon, Meinciau and the road junctions on the western fringe of Pontarddulais which became forever associated with Rebecca in the minds of contemporaries. The ferocious battles at these places revealed that the trusts were as keen to defend their best sources of income as the depressed agricultural and industrial workers were to remove them. The *Times* correspondent, on 25 July, declared that the outrages had assumed 'a much more serious aspect than they have hitherto', because of their prevalence now 'in the midst of a thickly populated district', and because of the threatening letters being sent to landowners and industrialists. Some of the gates destroyed at this time up the Tywi valley were indeed close to Middleton Hall, Golden Grove, Dynevor, Glansefin, and other mansions, and there were fears that the rioters were about to pay these a visit.

Three of the most famous Rebecca incidents convey something of the growing apprehension and violence that now characterized the movement. The attack on Llandeilo-yr-ynys bridge gate on 7 July was anticipated because William Lewis, the lessee of the tolls, had summoned people for non-payment. When the crowd approached the gate they spotted a young gentleman farmer and two constables. The young man had volunteered to act as a special constable, and he was now beaten for his impudence. One of the constables failed to escape and suffered the same fate. Then Lewis was brought out of the tollhouse, brandishing a horsewhip, but after a few strokes from his own weapon he fell on his knees and promised three times to sever all connection with the turnpike trust. News of the attack reached Carmarthen at tea-time, but when the dragoons arrived at the spot the crowd had vanished.[50]

The second incident took place at Pontyberem, in the mining district of southern Carmarthenshire, where colliers and quarrymen lived side by side with the farming population. The former had their own special grievances and held large evening meetings to discuss them, but they also contributed to the Rebecca movement. About a quarter of the attacks on trust property during July took place in the mining villages, and some observers noted comparisons between the outbreaks there and those of the Scotch Cattle on the coalfield in Monmouthshire. One group of 'collier Rebeccaites' operated in the neighbourhood of Cilcarw, the home of Stephen Evans, who was one of the spokesmen of the movement. After destroying the tollhouse at Meinciau on 14 July they turned their attention to that at Pontyberem on the following night. About fifty men marched from the direction of the lime kilns, walking two abreast with three gunmen to the front and rear. Besides the usual equipment they carried large sledges and bars, for the pillars of the gate had been manufactured at the Gwendraeth foundry, and even the colliers found them difficult to shift. The work took an hour, and, as the gate was situated in the centre of the village, virtually all the inhabitants saw the Rebeccaites.[51]

Such audacity presumed total loyalty, but in the third example of Rebecca outrages a witness proved very willing to talk. His name was John Jones, and he was present at the attack on the Bolgoed and Rhydypandy gates, both in the county of Glamorgan. Bolgoed was one of a number of gates about Pontarddulais which became of interest to Rebecca during the summer. According to Jones, a crowd collected on the lower Goppa just before midnight on 6 July, and put on all manner of disguises. The leader or 'mam' for the evening was Daniel Lewis, who was dressed in cap, bonnet, and shirt, and rode a white horse. Under his directions and those of Griffith Vaughan of the Red Lion Inn, the gate was smashed and the tollhouse undermined. The crowd numbered several hundred people, about half of whom had guns, and they fired these off as they returned to their place of rendezvous and their home villages west of Pontarddulais. The pattern was repeated five miles to the east at Rhydypandy gate on 20 July, though there were fewer participants and a different Becca, who addressed her followers by female names. Once again Jones was in the crowd, but this time he went with Matthew and Henry Morgan of Cwm Cile, walking uneasily in their bedgowns.[52]

The reasons for Jones's betrayal were fairly obvious; he was a loner, with a criminal record, who was disliked by his wife and friends and who had come into conflict with the farming family of Cwm Cile. He was looking for a home and an income, and the offer of a large reward came at the right time. On Saturday 22 July, he walked to Swansea and implicated, William Morgan, two other members of farming families. The arrests were far from easy; Captain Napier of the Glamorgan police, inspector Rees, and two other constables infuriated the community by arriving with their warrants first thing on Sunday morning. Napier was set upon by the Morgans, but his athleticism saved him from a reaping hook, an iron bar, a fish spear, and boiling water. As John Morgan struggled for the captain's pistol, the police chief directed a shot into his stomach, and earned the gratitude of the Prime Minister and Her Majesty. By Monday night the whole family was in custody, along with several others, including Daniel Lewis and Griffith Vaughan, and a box of guns and ammunition bound for Pontarddulais was seized. The manner of the arrests, and the secrecy and delays of the examinations at Swansea, produced a secondary protest movement, and 400 soldiers were held in readiness to deal with the crowds shouting 'Rebecca for ever'. Most of those apprehended, and later tried at a special commission in Cardiff, escaped a gaol sentence, but the episode was a timely reminder to Rebecca that not everybody joined her movement out of conviction.[53]

The period of the late summer and early autumn proved to be an interesting one in the Rebecca story. Contrary to most accounts, it is evident from Table 4 that there were almost as many successful attacks on trust property in August as in June, while even in September they numbered on average one a day. Yet the geography of the outrages was changing; in August they were still widespread across the three counties, but in September Rebecca's attention switched to the periphery of the region. Most of the destruction now took place in the neighbourhood of Llandeilo and Llandovery, within Radnorshire, on the coalfield between Kidwelly and Pontarddulais, and at Fishguard. Such a change was not entirely unexpected, for many of the gates in the old areas of rioting were down, and meetings were being held to consider leaving them in that state. In August and early September half a dozen trusts promised to reduce the interest on tallies, the rate of tolls, and the number of tollhouses. The Kidwelly trust, for example, wanted to abolish thirteen side-gates and bars, and the Three Commotts, under Edward Adams of Middleton Hall's fiery chairmanship, ten, but they found it difficult to agree compensation with William Lewis, the lessee of the tolls. Edward Lloyd Williams of Gwernant, chairman of a meeting of the Lower Cardigan trust, which decided on many concessions, told a huge audience on 3 October that the turnpike grievance was no longer an issue.[54]

When trusts discounted Rebecca's warnings and put chains where gates had been, trouble was always likely. Toll collectors at these places became extremely unpopular, especially if they made notes of those who refused payment. In this way riots broke out at or near Fishguard, Haverfordwest, Narberth,

Whitland, Llandysul, and Carmarthen. At Bwlchyclawdd, in mid-August, two hundred Rebeccaites in straw hats, long white gowns, and sword-belts amused locals by indulging in a mock fight once the chain had been broken. The collector, anticipating their advance, had fled. At Fishguard, three weeks later, when two gates were smashed, the keepers promised Rebecca that they would take no tolls in future. When they went back on their word, they were advised to put their furniture out on the streets, and their houses were pulled down. On the same night the mob also destroyed the wall of the road surveyor, and fired guns at the homes of two other unpopular residents. One of these Fishguard Rebeccaites later turned informer and, in one of the most spectacular raids of the period, more than thirty people were arrested. The man, Thomas Williams, and his wife had to be lodged in Haverfordwest gaol, for their own safety, and when the prosecution eventually withdrew the charges, the couple had already left the area.[55] The Home Secretary found it too much to bear, and warned magistrates of the 'displeasure of Her Majesty' if they failed to protect public-minded citizens.

Perhaps no one suffered from the collapse of law and order more than David Joshua, the bookbinder and keeper at Glangwili gate, who had been regarded as an informer since Chartism was at its height in Carmarthen. Unlike some collectors, Joshua was prepared to name people who refused payment and even guess at Rebecca's identity. When about 200 people arrived at Glangwili in the early hours of Friday 25 August, they threatened to tar and feather Joshua, and they pushed the tollhouse in on his beloved furniture, tools, and books. He was hit several times with a hatchet, and obliged to start the work of destruction himself. The constables, who were there to assist him, fled at the sight of the mob and one took the news of the attack to Carmarthen. The dragoons arrived too late to catch anyone, but an angry Joshua gave the names of four Rebeccaites. After their arrest and examination at Carmarthen on Friday and Saturday, a mob attacked the informer in the streets of the town, and on the following day set fire to a temporary shed that had been erected for him at Glangwili. 'Behold!' began the handbill which he printed at the time, 'disastrous days have come . . .'[56]

While these old battles were being played out in the heart of Rebecca country, the Lady was finding new targets to the north and south. At the beginning of August all five gates in the neighbourhood of Lampeter were destroyed, to the sound of very loud music and synchronized gunshot, and in the next weeks there were similar riots at Aberaeron, New Inn, and Llanon in Cardiganshire. Colonel Love thought it necessary to station troops in this region, but not everyone appreciated their unruly behaviour and the costs of their accommodation. Rebecca, meanwhile, did her best to ignore their presence, and continued to hold meetings on the hills and intimidate those who enforced the tolls. At New Inn the wife of a collector was temporarily blinded by gunpowder, and a repentant Rebecca ordered some communion wine to help her recover from the shock, while a few miles down the Cardiganshire coast a more serious shotgun attack was made on the house of a Tory magistrate who disgreed with the concessions being offered to her followers.

Even the Whig newspapers, like the *Morning Chronicle* and the *Swansea Journal*, were inclined to agree that Rebecca had grown strong on other people's weakness. Accounts of her activities in the north-east of Carmarthenshire during August and September contained words like 'flaunting', 'daring', and 'arrogant'. 'By far the most daring act [to date]', was how one correspondent described the removal on 7 August of the Llandeilo Walk gate, near one of the entrances to Dynevor Park and under the very noses of the sleeping soldiers. The troops at Llandeilo and Llandovery seemed quite unable to cope with 'the elusive Lady'; the pen portraits of the period show them leaning exhausted outside their lodgings, try to ignore the Welsh taunts of passers-by and dreading the next piece of false information.[57]

Griffith Jones and his wife, the collector at Pontarllechau tollhouse, near Llangadog, saw them regularly on their patrols, but never when it mattered. Their house was located on the road towards the lime kilns and the Black Mountain, and they had lived there in comparative peace for eight years. Then, on the night of 1 August, a gang of Rebeccaites dragged Jones and his wife out of their home, and made them watch their furniture burn. This spiteful act, which was unusual on a first visit, was Rebecca's reward for Jones's public statement that he would inform if attacked. True to his word, he named Thomas Hughes, a sawyer, Benjamin Jones, a farmer's son, and 'Captain' John Jones, a farmer who was very active in the mob. All three were given sentences of one year's imprisonment, and the last, outside the magistrate's room, warned the toll collector to leave the country. On 17 August and 13 September his gate was again attacked. The last incident was part of a small 'war of destruction', for besides the Pontarllechau, the Waunystradfeiris and two other gates and tollhouses in the neighbourhood of Llangadog fell to the Rebeccaites that night.[58]

Joseph Downes began his 'Tour of the Disturbed Districts in Wales' for *Blackwood's Magazine* on 9 September, and he chose as his starting-point Llangadog. In this lovely 'village-town' he found it almost impossible to imagine the reality of 'Rebecca terrorism', and so, like many others, he moved quickly south to a blacker country, to a land of 'vindictive rebellion, arson, and spiritual oratory'. In this semi-industrial landscape there were broken gates, shattered windows, and half-dismembered houses everywhere.[59] Porthyrhyd, a village at the crossroads, was like a place under siege, for here in the late summer and autumn of 1843 Rebecca fought her longest campaign. There were at least seven nightly attacks on the village, and Richard Williams (Dick Morgannwg) and his friend Evan Thomas ('the Lion of Porthyrhyd') bore the brunt of most of them. The former was a gatekeeper, and the latter was a 56-year-old parish constable who had sworn to defend Williams and bring the Rebeccaites to justice. At midnight on 4 August five parties arrived at the tollhouse, and, having destroyed it and a gate, set off in a vain search for 'the Lion', who was hiding in a cowhouse. At Thomas's home they smashed the door, windows, crockery, a clock, and even trees in the garden. On their journey they also took powder and shot from Davies the shopkeeper, broke into a smithy, and damaged two public houses. The breaking of the windows of one of them was attributed to boisterous young recruits and Rebecca sent money to cover the cost of their replacement.[60]

In one of those fruitful historical accidents Thomas Cooke, the agent of Edward Adams, came upon the above parties as he travelled home via Porthyrhyd late on that Friday night. In a dark narrow lane he met forty men, dressed in white, with veils over their faces and with poles and guns in their hands. They stood as quiet and motionless as statues, and Cooke was allowed to weave his shying horse through the crowd. 'Several of them were so near to me,' he informed his anxious mother, 'that I could have thrashed them with my stick, but I thought it more prudent to thrash my horse instead of them, and was glad to get off so cheaply.'[61] So was Evan Thomas, the owner of the smithy, and several other special constables who gave up their staves of office a few days later.

Five further 'midnight visits' to Porthyrhyd occurred before the month was out. On 7 August, for example, the half-repaired tollhouse was pulled down, and Dick, its keeper, was abused. If his evidence can be believed, the rioters were this time without disguise, and the most prominent of them were young stonemasons and servants of the neighbourhood. Eleven days later, following a decision to re-erect the tollhouse and a gate, a crowd arrived at the parish constable's home, and forced Evan to march before them in his night-shirt to the gate. There he was ordered to begin the work of destruction, and afterwards sank on his knees and swore never again to meddle with Rebecca and her daughters.[62] Then, firing guns and cheering loudly, the party moved on. Amongst them were John Jones (Shoni Sgubor Fawr), who, like his compatriot David Davies (Dai Cantwr), had recently travelled from the mining district of south-east Wales. Jones was a prize fighter, who had served in the army and given the authorities information on the Scotch Cattle a decade before. Together, as we shall see in the next chapter, he and Davies raised the level of personal violence in this part of Carmarthenshire.

The destruction, in August and September, of tollgates to the south of Porthyrhyd, around Llanelli and Swansea, was a new departure and a direct challenge to the town authorities and their professional police forces. At the beginning of August the Furnace Lodge and Sandy lime-kiln gates at Llanelli were pulled down and burnt, and one of the keepers received a beating and a shotgun blast in the face. The subsequent arrest of three people and their examination in the town hall were almost as traumatic as the riots themselves. Not for the first time, the informant, Jenkin Hugh of the Sandy gate, appeared a little drunk when telling his story and was quickly disowned by his friends. One man, George Phillips, was discharged immediately, and the others allowed out on bail. Francis McKiernin, innkeeper and coach proprietor of the town, was carried home in triumph, and he and George Laing, another publican, were ultimately released without prosecution.[63]

During the late summer it was the turn of Swansea to experience a small epidemic of attacks, and in one of these an old female collector was badly beaten by a collier. The lady ran the Tycoch tollbar, on the Morriston side of the Tawe river, and the mob which came to destroy it consisted of one horseman and thirty 'working men' who came from the Llansamlet direction. The collector was woken from her bed early on that August day by the sound of an iron bar on her door, and her screams as David Jones broke her arm could be heard a good distance. She was well treated by the magistrates and the Swansea infirmary, and did her best to ignore the small crowd which later tried to punish her for informing on her attacker.[64]

Sarah Williams of the Hendy tollhouse was less fortunate. In a way she was the perfect martyr, a 75-year-old woman who spent the last years of her life making a few pence at one tollhouse after another. Whether she agreed to increase tolls at Hendy or spoke too freely about the Rebeccaites is not clear, but she had already received warnings about some aspects of her work. On 9 September she had been in the village no more than a week, when her home was fired by a mob. Sarah rushed to neighbours for help, but they were too afraid to give assistance. When she returned to pull some of her furniture from the burning tollhouse, someone shot her in the chest. Joseph Downes, passing through the village, found the body 'preserved just as it fell, for the inspection of the jury ... All was naked, ugly horror. An old rug just veiled the corpse, which, being turned down, revealed the orifice, just by the nipple of a shot or slug wound ... Another wound on the temple had caused a torrent of blood, which remained glued over the whole cheek. The retracted lips of this poor suffering creature, gave a dreadful grin to the aged countenance ...'[65] At the coroner's inquest two surgeons certified that shots had lodged in her lungs, but the jury decided that the precise cause of her death was unknown. This astounded the higher authorities, who offered a £500 reward for her killer, and convinced the wider world that Welsh community feeling and deceit knew no bounds.

It was obvious by early September that the use of firearms, by both sides in the conflict, had reached a dangerous level. Only a few days before Sarah's death, a full-scale battle took place in the neighbourhood of Pontarddulais. On 6 September news reached William Chambers at Llanelli and Captain Napier at Swansea that Rebeccaites were about to destroy gates at Hendy and Pontarddulais. Chambers took with him a small party of soldiers, and a reputation for fairness that was to come under the closest scrutiny in a few hours' time. Napier collected two county magistrates, Messrs Llewellyn and Dillwyn, as well as Matthew Moggridge, a Monmouthshire colleague, and about eight other constables, all armed to the teeth. In a field close to Pontarddulais bridge and the gate, they lay in ambush. Noises of horns and firing filled the air, but when the sound of a gate breaking and glass smashing reached them at about 12.50 a.m. it was time to move. The tollhouse was half-down, and a crowd of 150 people were being supervised by three mounted men, who now turned and tried to ride down Napier's party. In the mêlée three young men, David Jones, John Hugh and the leader, 24-year-old John Hughes, a farmer's son of Llannon parish, were seriously wounded. The rest hitched up their skirts, mounted their horses and rode back into the west, taking with them, it was said, several wounded friends. Some of them walked into the hands of Chambers and the soldiers who had journeyed beyond Hendy gate, and there was a little brutality and more arrests.[66] The *Carmarthen Journal* was in ecstasy; here at last was blood and the certain prospect of transportation for those captured. When, three weeks later, Shoni Sgubor Fawr and Dai Cantwr were taken as well, the victory seemed complete.

By late autumn the gaols were rapidly filling with those awaiting trial for the destruction of gates and related offences, and there were more people out on bail. The magistrates, stung by the constant carping from Westminster, had at last sprung to life, and were receiving more help from the middle and lower ranks of propertied society. These were worried about the scope of Rebecca's activities, and hopeful, perhaps, of the results of the government enquiries now in progress. Partly because of the investigations of visiting commissioners, open daylight meetings became more common, and received greater publicity. Resolutions were passed promising an end to nightly outrages, and gradually the call for special constables was answered. In October 'the most respectable parties' were summoned to be specials and, as November gave way to December, there were even volunteers for constabulary duty from residents in the Teifi valley and the Llannon and Llanelli district. This startling development was a tribute, said the *Times* reporter, to the rethinking that followed the violence at Hendy and Pontarddulais, but it was also a sensible accompaniment to their petitions on behalf of prisoner John Hughes and to their loud protests against the county police.[67]

In September the number of attacks fell by almost a half, and in October and November the rate of decline was even faster. The government never fully appreciated this, which probably accounts for their growing obsession with the apparent ineffectiveness of the military and the cautious approach of Colonel Love. 'The continuation of the disturbances in South Wales and the failure of the efforts to check them and to detect them,' Peel told Graham on 19 September, 'baffles all speculation.'[68] In a studied insult to their commander in south Wales, the Home Secretary sent Major-General George Brown to the region early in October, and he and local magistrates were given the aid of some 150 Metropolitan policemen. As Map 3 shows, Brown kept large detachments of troops in the major towns, but, unlike Love, he was willing to risk small units of soldiers and policemen in some thirty villages. Thus 21 men were put at Pontyberem, and were thanked for bringing peace to the neighbourhood, while 6 were placed at Llanboidy, where they were held to be unnecessary and arrogant. There were problems over their reception, accommodation, costs, and desertions, but Brown was convinced of the value of small detachments. Very soon, however, questions were raised about the policy behind the strategy; was west Wales to become another Ireland, 'its villages [to] be converted into barracks, and its cities into garrisons?'[69]

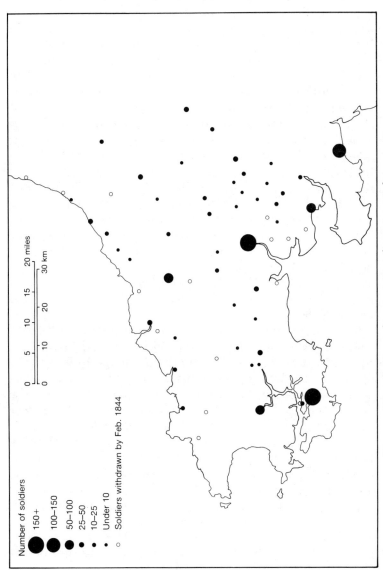

MAP 3. Distribution of troops, October 1843–February 1844

The presence of these units, and the nightly patrols, helped to check the operations of the Rebeccaites, but most of the successful detection work was the result of secret information and financial bribes. At the beginning of October a Royal Proclamation was issued which offered £500 rewards for information leading to the conviction of those who fired property or took part in a murderous assault, and much less money for those who trapped the destroyers of turnpike gates. George Rice Trevor, before he left for Christmas in Bedfordshire, offered money to just about anyone who would talk about the troubles, and in his county at least £1,500 was handed out in rewards by the Quarter Sessions. Amongst the lucky recipients were Richard Williams of Porthyrhyd, who received £100, and Thomas Phillips of Topsail, who informed on John (Shoni Sgubor Fawr) Jones and David (Dai Cantwr) Davies. Others were awarded money for unspecified assistance to the authorities, and Captain Napier's services earned a grant of £500. In a period of depression, these were good payments, but, as John Jones of Lletty, William Zacharius of Llandeilo, and many others discovered, informers had to endure long periods in custody for their own safety, and often a journey to a new land.[70]

Such were the problems with witnesses, and Welsh juries, that the government decided to appoint a special commission fifty miles away at Cardiff to try the Glamorgan prisoners.[71] It opened on 26 October, appropriately one day after the enquiry commission began its work in Carmarthen. The proceedings were conducted before Baron Gurney and Sir Cresswell Cresswell, and the guilt of the chief offenders was never in doubt. One sentence of twenty years' and two of seven years' transportation were handed out to John Hughes, David Jones, and John Hugh, an exercise in selective severity that was applauded by the government, and repeated at the Carmarthenshire winter Assize two months later. There John Jones was given transportation for life and David Davies for twenty years, and their contrite confession was immediately rushed into print. Another eight men subsequently joined them on the convict voyage around the world, and the heavily signed petitions on their behalf were left to gather dust. Altogether well over a hundred suspected Rebeccaites appeared for turnpike offences at these and other courts in 1843–6, and two-thirds of them were discharged or let off on promises of good conduct. The evidence against many of the lesser characters in the drama was weak, but there was also a feeling that the law had already been sufficiently vindicated. This was a measured response, said an admiring Irish newspaper, and indicative of the careful way in which the British state reduced tension on the mainland.[72]

'Rebecca is finished' was the message of the late autumn and winter of 1843. During October there was on average one attack on a tollgate every two days, but thereafter the rate fell to one a week and less. There were now few attacks in the heartland of the region, where many gates and bars were still down. The incident at Llannon on 15 September, when 300 'strangers' demolished a tollhouse, was one of the last of its kind in the district, while the same could be said of the removal of Tirfran gate near Llanelli at the end of the month. The fact that a group of local farmers rescued the Llanelli gate from a coalpit and replaced it was applauded as a sign of Rebecca's growing impotence. Colonel Love, as ever, refused to be swayed by such good news. Even six months later, when the economy had improved and gates were safer, he felt that 'the feelings of the people are not as yet much altered for the better, nor is the disposition to mischief much abated'.[73]

If trustees and toll collectors slept more peacefully during the winter of 1843, life was not easy. Hundreds of travellers ignored requests for payment on the road, magistrates sometimes refused to punish offenders or defend certain gates of doubtful legality, and any attempt to re-erect unpopular side-gates, bars, and chains precipitated a Rebecca visit. The inspector of police at Llangadog, speaking for many officers, said that he could still hear horns sounding through the night and was convinced that if he were to depart bad times would return.[74] At Fishguard, Narberth, St Clears, on the outskirts of Carmarthen, and at Penygarn, Porthyrhyd, and Pontarddulais there was a legacy of tension which never disappeared, and there were attacks months later on comparatively new targets like Brynlloi gate, Cwmamman, and the Steynton gate near Milford. At Robeston Wathen the Reverend James William James, one of the staunchest defenders of the gates and perhaps

Rebecca's keenest pursuer in Pembrokeshire, was shot at his home just before Christmas 1843.[75]

Even established gates, re-erected during the following year, were not safe. When Plasbach bar, in an isolated spot on the Lampeter–Llandeilo road, was set up again in February 1844 with a new lessee, Rebecca appeared within twenty-four hours, and a similar experience befell the female collector at Pontyates. A month later it was Cardigan's turn, when the long-standing grievances of the people of Verwig and Mount over a gate on the Aberaeron road finally exploded into violence. The gate was an old one, and extremely unpopular with everybody in the town, and it had only just been rebuilt. An angry government conducted a rather pointless investigation into the inactivity of the Cardigan authorities and the forty marines under their control, for the incident occurred at 4 a.m., when only ghosts and a few old ladies were awake.[76]

Those searching for Rebecca after the corn harvest of 1843 might have been forgiven for thinking that she had migrated north of Llangadog. The tollhouse and gate at Dolauhirion, about 1½ miles from Llandovery, was visited several times in the late autumn, people coming from the direction of Cilycwm and leaving notices warning Benjamin Evans to give up taking tolls. On 2 October, again at 4 a.m., about a hundred people were caught in the act of destruction by George Jones, a well-armed and much-hated Llandovery policeman, and he arrested two farm servants hiding in a cowhouse. A few days before this, Charlotte, Rebecca's daughter, had taken 250 followers to Cwmdwr, near Llanwrda, burnt the toll collector's furniture, and fired a volley at a clergyman's house.[77] During the next week there were more attacks east of Llandovery, and another arrest. Afterwards there was a lull in activity, although isolated incidents were reported in the district through the winter months. On the night of 9 December, for example, Waunystradfeiris gate near Llangadog was pulled down for the fourth time, despite the guard of soldiers close by, but on this occasion it appeared that Rebecca was not present. As the press lost interest in the movement, so it became more difficult to establish exactly what was happening, and to distinguish between public and private acts of vengeance.

In the last period of rioting, Rebecca seems to have 'moved

her headquarters' to mid-Wales, especially along the main roads between Llandovery, Builth, and Llanidloes. Discontent had been simmering in Radnorshire for some time, not least because the Welsh farmers resented the number of gates recently erected on parish roads. There were several attacks on tollgates in the southern half of the county in September and October. One was destroyed near Builth, two near Presteigne, where a large collector squeezed out of sight under his own bed, two at Cefn Llanddewi in Breconshire, and one at Newbridge on Wye. Each attack brought meetings of magistrates and calls for more police and special constables, and, in an enterprising development at Cefn Llanddewi, between Llandovery and Builth, residents decided to purchase the right of free passage for themselves from the contractor of the gate.

Sir John Walsh of Knighton, the lord-lieutenant of Radnorshire, prayed that these isolated incidents did not constitute a real extension of Rebecca's influence, but higher up the Wye valley, close to Llangurig and Rhayader, large numbers of people stopped paying tolls, threatening letters were received by magistrates, and gates were attacked on at least sixteen occasions. Perhaps the most frightening episode was at Botalog, an isolated mountainous spot to the south-west of Llangurig, where, on 9 October 1843, an old female collector was blasted in the face for trying to discover the identity of her attackers. She knew some of them, but despite threats and rewards never disclosed their names. The same anonymity was achieved at Rhayader, 'a wild, and miserable town'. It was blessed with six gates and a bar, and it came as no surprise when, late in September, Rebecca appeared there in full costume, with her supporters, many on horseback, blowing horns and firing guns. Two or three gates were pulled down.

The affair which caused most alarm and which changed the minds of gentlemen about the wisdom of military help and a permanent police force, was the entry into Rhayader of three companies of Rebeccaites in the early hours of 3 November. In a trailer for the astonishing poaching riots of a later generation, ranks of sword- and pikemen marched slowly in full regalia. Armed guards to the front and rear kept police and on-lookers away, and in two hours four of the most obnoxious gates were down. The noise of firing was so deafening that inhabitants gave up trying to sleep, and promised their families that they would flee the town in the morning. The arrival of dragoons and policemen changed all that, and the town settled down to a kind of peace. 'If we make a stand here,' wrote a chastened lord-lieutenant, 'we may keep it [Rebeccaism] out of the rest of the county.'[78] Through the winter, and into the spring of 1844, other gates along the northern and southern boundaries of the county were broken down, but the summer was warm and quiet. Then, in the autumn, the arrival of threatening letters heralded further attacks on gates at Builth and Rhayader, although this time they were on private roads. 'Is this the last outbreak?', asked one observer. The answer was 'not quite', but the movement against the turnpikes was virtually over.[79]

Even before this moment contemporaries were busy wrestling with some of the more difficult questions posed by the riots. In particular, everyone from Queen Victoria downwards wanted to know the identity of the rioters and their leader. Charlotte Guest, wife of the ironmaster, had a meeting with John Hughes at Cardiff and shed a tear for the convicted men. When they arrived in London on their way to Australia crowds turned out to see 'what such creatures as Rebeccas were'. The legal and newspaper evidence offers some clues. Those punished for refusing to pay tolls and other offences against the turnpike Acts were often carriers, drovers, butchers, and farmers, whereas the known and convicted Rebeccaites tended to be farmers, farm servants, and labourers. When occupations were given, in 86 cases, two-thirds of the arrested Rebeccaites fell into these three main categories. A quarter of the sample were farmers, and there were three times as many labourers as servants. In the village communities millers, blacksmiths, carpenters, and publicans were added to the list; in towns such as Carmarthen and Haverfordwest there was also support for Rebecca from shoemakers, fishermen, tailors, and weavers; and in places like Pontyberem and Llannon colliers, masons, and ordinary labourers played a part. All this modifies David Williams's statement that 'the riots were entirely an affair of small farmers,' though his main point remains valid.[80]

Of the instigators of the riots there were few doubts; in August 1839 John Allen of Cresselly said they were the occupiers of smallholdings, and William Day, assistant Poor

Law Commissioner, writing four years later, concurred: 'the leading actors . . . are the farmers and their labourers'. The owners of Alltyrodyn, Glangwili, and other great houses insisted that they knew which of their tenant farmers participated in the troubles, but Thomas Cooke, Edward Adams's agent, was a little surprised when one or two of Middleton Hall's tenants were examined. Sometimes, it was claimed, these farmers and their employees worked together at night, as they did during the day, and for this reason the authorities hardly expected to obtain inside information on rural Rebeccaites until the hiring season arrived. This was not, everyone seemed agreed, a rising of the most ignorant and lawless members of society.[81] So far as one can tell, comparatively few of those convicted of breaking tollgates had previous criminal records; not for the first time in popular movements, informers like John Jones and William Zacharius were the people most likely to have been in trouble before.

The ages of the suspected and convicted Rebeccaites were not always given, but the evidence indicates that, although some of them were quite old, the typical offender was a farmer's son or labourer in his early to mid-20s. In a few instances, as when six persons faced trial for the demolition of Dolauhirion tollhouse on 7 August 1843, no man was more than 21 years of age. Similarly most of those arrested for an attack at Porthyrhyd in August were in their early teens, while incidents at St Clears, Newcastle Emlyn, and Fishguard were also blamed on idle boys and youths. Jane Walters of Glanmedeni, whom we shall meet in the next chapter, regarded the Rebecca movement as a form of juvenile rebellion, but with a few exceptions the named leaders were from a slightly older age group and the description 'of all ages' was applied to the larger processions of gate-breakers.[82] Whole families were present at the great demonstrations at Carmarthen and Fishguard. Women were, it seems, rarely involved in the smashing of the gates, but they were often enthusiastic spectators and took part in other forms of protest. As we shall see in the next chapter, Becca was the channel for the moral force of females as well as for the physical force of males.

The court evidence indicates that some of the Rebeccaites lived fairly close to the scene of the crime. The people arrested for the destruction of the Fishguard gates in the autumn of 1843 were inhabitants of the town, and the landlord of the Blue Boar, St Clears, and Eleanor Jones of Botalog admitted that they recognized neighbours amongst the crowds. But the usual pattern, seen in the smashing of gates at Llanon, Cardigan, Whitland, Carmarthen, Pontarddulais, Dolauhirion, and Rhayader, was for the gangs to be composed of country people who had travelled a few miles to be there.[83] They were guided to the precise spot by residents of the town or village, although it was common for those actually destroying the gates to refuse all help. The advantages of such an operation were obvious; people in the vicinity of tollgates could deny all knowledge of the affair.

It was customary in the country riots to place responsibility for the troubles on 'strangers', and this was one brief, and comforting, response to the Rebecca riots. Edward Lloyd Williams, George Rice Trevor, and other landowners and clergymen suggested that the riots were not the work of Welshmen, but were stimulated and organized by 'busy meddlers' who were afraid to show their faces. The *Times* reporter was informed that 'an Englishman' had led the attacks on the Lampeter gates, and 'a stranger to the neighbourhood' had held a surgeon of the town prisoner during the proceedings. In Cardigan there was information that one or possibly three agitators had stirred up the 'quiet and inoffensive' people. Elsewhere in the same county, and in Carmarthenshire, men with seditious ideas had apparently travelled from 'a long distance' to plan the attacks on the gates, a notion encouraged in the summer of 1843 by the arrival of hundreds of unemployed miners from south Wales. Colonel Love believed that these migrants were 'active in persuading the people to mischief', but the evidence hardly supports the claim that they were 'at the bottom of the riots'. Edward Lloyd Hall, who was besieged by rumours at Newcastle Emlyn, agreed that a few English agitators were probably in the region, but only to learn the secrets of a successful movement.[84] In fact, the anxiety about 'outsiders' in these years was part of a wider concern over the presence of hundreds of vagrants and other newcomers. As we shall see in the next chapter, there were tramps and robbers across the region who said that they were working on behalf of Rebecca.

The leadership of the movement was a difficult matter. One of the theories, common to all such movements, was that 'intriguing men of some knowledge and consequence, have instigated or organised these deluded persons'. According to Captain Pell of the Royal Marines in Pembroke Dock, at the very least 'resident clergymen . . . and respectable farmers of property' had connived, by their inaction, at the proceedings, and there was a persistent rumour that a gentleman, or even a magistrate, gave the Rebeccaites his full support. The identity of this person was never revealed, though George Davies Griffiths of Berllan, a magistrate, was apparently sympathetic and related to a disaffected blacksmith. Captain Napier suspected for a time that the 'gentleman Rebeccaite' was Lewis Evans of Pantycendy, and Love, too, accused him of seeking popularity and tampering with the evidence of witnesses. William Edwardes of Sealyham found it necessary to deny that another suspect, his son Owen Tucker, was involved with the rioters. Like John Jones MP, Colonel Colby of Ffynone, David Rees, the minister of Llanelli, and other outspoken critics of turnpike charges, Edwardes believed that some of the people's grievances were justified. Such sympathizers were immediately accused of 'fomenting' trouble by Alban L. Gwynne of Mynachty, amongst others, but there is no evidence of gentlemen being seen or arrested during the Rebecca attacks. Of course, there was the usual talk of a refined leader with soft hands, superior bearing, and Latin scholarship, but no one took that too seriously.[85]

The Rebecca mob had one or more leaders who were, on a few occasions, the only people on horseback and in disguise. As in other movements of a similar character, it was suggested that these people were 'of the better sort', 'in good circumstances', of some property and standing. The *Morning Chronicle* reporter maintained that these people had the authority, influence, and contacts to command and organize. He was thinking of Rebeccaites like the 34-year-old Thomas Thomas, the farming son of a prominent freeholder near Porthyrhyd. It was to his home that the wife of Evan Thomas ('the Lion') went, seeking immunity from further Rebecca visits. David Evans, William Davies, and Thomas Howells we have already mentioned, and there was Lewis Davies of Llannon, a tenant farmer worth £35 a year, and a Penboyr tenant of Lord Cawdor who was called to account by agent R. B. Williams for threatening John Lloyd Davies at Carmarthen.[86] Their precise relationship with the remaining Rebeccaites is interesting; a formal division was made between farmers, on horseback, and their employees in the march to Carmarthen on 19 June 1843, and gatekeepers occasionally identified a small and separate group of 'respectable people' amongst the crowd, but in general informers did not make such distinctions.[87] Rioters were bound together by shared grievances and beliefs, and also by ties of work, kin, and community.

When taken, those deemed to be of reputable families pleaded that they had been forced by threats to join the gate-breakers, and one writer has said that Benjamin Evans, a solicitor of Newcastle Emlyn, was only taken along by the Rebeccaites as an insurance in case of trouble.[88] Such interpretations became more common as the months and years passed; as with so many popular movements, people of status later found it embarrassing to admit publicly that they had been active participants. Yet there were 'respectable Rebeccaites', and they were often organizers and advisers, the people who chose and hired the Beccas and composed the anonymous letters. Men such as William Davies of Pantyfen, Thomas Thomas of Porthyrhyd, Stephen Evans of Cilcarw, and John Lewis of Llangadog usually remained in the background, but they were none the less important for that. Magistrates sensed that these people were carrying over their community leadership from one side of the law to the other. If this were true, it must have caused a strain, for although they had led local protests before the actions had been peaceful. Colonel Love and Edward Lloyd Hall both felt that every effort had to be made to detach such people from the movement, and were delighted when in the autumn of 1843 this seemed to be happening.

The search for the mother figure of the movement occupied little time. Edward Lloyd Hall, when addressing a meeting of farmers on 23 June 1843, was informed that 'the real "Rebecca"', a stranger, was in the audience, but he remained sceptical. 'It is however now evident', he told the Home Secretary, 'that there are various different "Rebeccas" in different parts of the country . . .' Ironically, the Newcastle Emlyn attorney was himself suspected of being 'the elegant Lady', mainly because he acted as a spokesman for the aggrieved and had an exceptional knowledge of their activities. The same could be said of Hugh Williams, the Kidwelly Chartist and lawyer who, like Edward Lloyd Hall, gained popularity as a courtroom defender of the Rebeccaites, and was closely watched by the police. According to his friend W. J. Linton, Williams was the 'undiscovered leader of the "Rebecca Movement" ', but there is no evidence that either the Kidwelly man or Edward Lloyd Hall joined the night-riders. Williams had been frightened by the violence of the Chartists in 1839. 'I am ready to stand in the cause of liberty at any time or any where,' he declared in November 1843, 'but I do not pretend to be a leader only a follower. I will not advance before the crowd.'[89]

In the beginning, in the district around Whitland and St Clears, people had information that a very tall and strong man, perhaps a blacksmith, played the part of Rebecca on several occasions, but it soon became apparent that the impersonators varied from place to place and even from night to night. Thomas Rees, the pugilist of Carnabwth, has been long recognized as the first Becca at Efailwen, and, although it has been said that he played no further part in the movement, one of the later Rebecca letters was addressed from his home parish. The Rebecca at Carmarthen's Water Street gate on the night of 26 May 1843 was a tall man in worn clothes who also claimed that he was from Rees's part of the world, but it was soon another, short and thickset, Rebecca from Talog who was stealing the headlines. One of her letters was found some miles away on the house of a coalowner in Cwmmawr.[90]

It is astonishing how many people later claimed to have been the Rebecca on the great march into Carmarthen on 19 June 1843, and it is a reminder of the pitfalls of all evidence, literary and oral. With this in mind the following is a list of some of those arrested, and accused of being the female leader: David Evans of Penlan, William Owen of Fishguard, Thomas, a gamekeeper living near Haverfordwest, William Hopkins, the shoemaker of Dyffryn Ceidrych, John Hughes of Llannon, and Daniel Lewis, the weaver and word-spinner of Pontarddulais. The last issued his directions in Welsh, but more commonly

Becca addressed her children in 'good English'.⁹¹ This was a sensible precaution, and a faithful reproduction of the language of the gentlefolk. The disguise emphasized the detachment; almost without exception, the Rebecca figure was dressed in a white shift, a bonnet and false beard, and carried both a sword and a gun. Like the Holy One in Revelation, or a latter-day Captain Swing, she rode a splendid white horse.

There were times, especially in the final weeks of the riots, when Rebecca was absent, or when, according to the press, she sent a deputy, like 'Charlotte', or 'Nelly', 'her eldest daughter'.⁹² Why there should be such a distinction is unclear, but the use of female names was widely adopted in the planning and execution of the raids. David Evans of Penlan and the Becca at Rhydypandy on 20 July 1843 did it to avoid disclosing the identity of their comrades, and so Nelly, Phoebe, Susan, Lucy, and Mary became part of the mythology of the riots. In letter writing Rebecca was sometimes helped by Miss Brown, Dorothy, Lydia, Eliza, and Miss Cromwell. The last was Rebecca's most favoured lieutenant, designated as her successor, and honoured at meetings and on one of the rocks near Llangadog. 'Miss Cromwell' wrote almost as many letters as her more famous sister, and in them she and the 'Youngest Rebecca' referred favourably to the exploits of the Puritans during the Civil War, and their use of the gun and the torch.⁹³

Perhaps the most intriguing questions concern the contacts between these leaders, and the organization of the movement generally. Within a short time, it had been established to the satisfaction of the government that there was no central committee for the whole of south-west Wales, but there were references to 'a Privy Council' and 'a Rebeccaite association', with federations and lists of members. One informer insisted that he had attended a district meeting of Rebecca delegates six miles from Llandovery. Most places had small cells, or 'committees', which met in barns, public houses, and coal mines, and planned actions, administered oaths of secrecy, and kept in touch with neighbouring groups. Precisely how communication was maintained remains a mystery, but there was constant talk of 'strangers' arriving at farmhouses and pubs with disguises, guns, and messages.⁹⁴ Some of their information was about the time of an attack, for on occasions, as at Prendergast and Porthyrhyd in August 1843, three, four, or even five groups from different villages came upon the target simultaneously. According to John Jones (Sgubor Fawr), on the visit to Porthyrhyd Stephen Evans of Cilcarw brought one of the main groups, and John Philip of the Mill brought another from Pontyates.[95] On other nights people were told to cross the border from Carmarthenshire into Cardiganshire, and from Radnorshire into Montgomeryshire, to assist the local residents, though not all of them appeared. Contacts were difficult to sustain, and in Fishguard, and possibly other small towns as well, inhabitants preferred to rely on their own 'little Rebecca'.[96]

The work of the Beccas and their 'deputies' began nights before any attack. Spies and scouts fed them information on the movement of magistrates, police, and soldiers, and on the gates and their keepers. When the local committee chose a target, they sometimes called a meeting late at night to inform the 'children', to swear them to secrecy, and to make preliminary preparations. At one such assembly, near Pontyberem, attended by 200–300 people, the disguised leaders talked of the next attack, extracted a promise from an innkeeper not to serve William Lewis, the toll collector, and warned special constables to give up their staves immediately.[97] Not all these meetings were secret, for those on the hills were accompanied, like those of the Scotch Cattle in Monmouthshire, by the sound of bugles and gunfire. Nor was the decision of the 'committee' always kept to known Rebeccaites, for notices of their intentions were posted on bridges, public houses, and chapel doors. Another favourite method of publicity was the proclamation, with official and unofficial criers touring the area at Becca's behest. This method saved more than one tollgate keeper who had the good sense to depart on hearing the message.

The threatening letter also carried news of imminent attacks. Mayors, magistrates, specials, and even soldiers and jurymen were teased by anonymous notes, warning them to expect a visit. Notices were left outside the homes of landlords and farmers, requiring their support or that of their servants on a particular day. Thus Rebecca, in a note postmarked 21 June 1843, called on Edward Lloyd Hall, as someone 'on my side', to send his men to meet 'my daughter Charlotte' on the following

Wednesday, or face 'some unpleasantness'.⁹⁸ Sometimes the letters asked for donations to finance Rebecca's expeditions, and farmers, vicars, and even one or two magistrates handed over subscriptions to armed men. It was policy to send threatening letters to contractors and collectors of the tolls, and to those who foolishly paid them. Keepers like Jenkin Hugh of Llanelli admitted that attacks had followed upon the receipt of written warnings from Becca or one of her daughters. When the Cenarth Bridge gate was destroyed on the night of 21 July 1843, without prior notice, it was duly reported as being an exceptional event.⁹⁹

The preparations for the 'midnight visit' were a matter of great interest and speculation. Herbert Vaughan of Llangoedmore, John Lloyd Davies of Alltyrodyn, and others argued that Rebecca had ample resources, including sums of money and guns dispatched from sympathizers elsewhere in Britain.¹⁰⁰ In fact, the money was obtained locally by door-to-door collections, regular payments from certain farmers, and armed robbery from men of property. Landowners were accused of sending money on demand, aware, no doubt, that the property of their friends had been set on fire for less. The cash was used to pay those labourers and miners who worked almost full-time for Rebecca, and to buy the drink consumed during the busy nights.

Most of the money went on guns and ammunition, because several hundred shots were fired in a raid. New guns were, it was said, purchased at Talog, Llanelli, and Pontarddulais, but, as the *Times* correspondent discovered when he attended secret meetings, most farmers already possessed such weapons. Although it is hard to believe, there were reports of people on the hills between Newcastle Emlyn and Llandovery being given military training by 'strangers' and ex-soldiers.¹⁰¹ George Rice Trevor kept a suspicious eye on the army pensioners in the three counties, but failed to discover anything about this or the trade in arms.¹⁰² Firearms and powder were taken on the nights of an attack from farms, smithies, and shops, as were whittles, sledges, and saws. The weapons were purchased, or borrowed for the night, but at Tangiers, above Haverfordwest, in August 1843 the tools of a smith known to be hostile to 'the Lady' were simply stolen. Those who were without guns, pistols, and swords obeyed Rebecca's instructions and brought pitchforks, sickles, hay-knives on poles, and the like.

Colonel Love, who had much experience of these matters, was enormously impressed by the planning and timing of the outrages.[103] Scouts and spies were out everywhere, watching the movements of magistrates, checking the rebuilding of tollhouses, noting the patrols of soldiers, and generally feeding information to the committees. Edward Lloyd Hall reckoned that it was impossible to surprise the Rebeccaites, and he was proved almost exactly right. When the authorities' defences were weak, the visit of a large number of Rebeccaites took on the character of a military invasion or a respectable procession from village to village, with drums and music, but when soldiers and police were guarding the gates, the task required greater secrecy and cunning. As the degree of difficulty increased, so the timing of the visit moved from the day or evening to the early morning.

When the appropriate hour arrived, guns, rockets, fires, and, most commonly, horns drew people from a wide neighbourhood. They frequently met at a crossroads or on mountain slopes. Some of the crowd arrived already disguised in bedgowns and smock frocks, and with a mixed cocktail of fruit juice, flour, soot, cork, coal dust, and chalk on their faces, but others simply turned their coats or put on the clothes brought in a bundle from the womenfolk. Not all the mobs bothered to hide their identity in this way, but their leaders did. The nature of the disguises certainly intrigued contemporaries, but it was not that different from those traditionally worn by country revellers, those perambulating the boundaries, the Mari Lwyd collectors, Charter-Day celebrants, *ceffyl pren* participants, and even some robbers. There were other comparisons, too, with Irish Whiteboys and, more pointedly, with Bristol gate-breakers of the early eighteenth century who went about their business in women's clothes and tall hats.

Inevitably it was the impersonations of those in the front of the Rebecca crowd which caught the eye, with their straw hats, turbans, caps, and bonnets, decorations of twigs, leaves, and feathers, wigs, beards, red masks, parasols, coloured faces, bardic gowns and dresses, sashes, belts, and the white sheets draped over the horses.[104] A couple of these leaders used the intoning voice heard in chapels and eisteddfodau, but most of the theatrical performers mimicked the female voice, and the rolling gate of the Welsh 'mam', and there was, when circumstances permitted, a *ceffyl pren* type banter with those on the fringes of the crowd and with the tollgate keepers. In all of this there was a delight in the deference temporarily shown to a female aristocratic pretender, and a recognition by the latter that female disguise gave him a security, authority, and licence which he did not possess in normal life. The association of sexual inversion and rebellion already had a long history.[105]

Typically, the Becca of the night said few words. Instructions were monosyllabic and delivered in military style. The number of people who waited for her commands varied considerably, though the average size of a crowd was 50–100. About a fifth of the Rebeccaites had horses, and the rest marched in twos or threes, with scouts and gunmen to the front. On arrival, sentinels were placed on guard, while a small band set about the gate first, and then the tollhouse. The stumps of the former were usually left standing, as were the bottom two feet of the house. On the first visit the belongings of the keeper were ignored, although his or her books and tollboard might be destroyed, but if Rebecca were obliged to come a second time all the property was set ablaze. Then, as part of the drama, the keeper was encouraged to bend the knee, kiss the Bible, and swear to give up the trade. If, like David Joshua, the man were a stubborn trust employee, then he was also required to strike the first blow for Rebecca. When the visit was over, three cheers might be given for the Lady, and, to the sound of horns and gunfire, the 'damnable crew' disappeared into the night.

The degree of intimidation and violence in the Rebecca riots was a matter of much debate. The *Morning Chronicle* reporter was struck by the ferocity of the attacks on property and persons, and it was this as much as anything which, so we are told, persuaded the farmers to support the movement.[106] Rebecca took a special delight in forcing the servants of William Lewis of Clynfiew and other gentry families to join her in nightly adventures. On the evening of the attack on a gate at Llanfihangel-ar-arth (19 June 1843) tenant farmer Evan Evans, David Jones, and others were taken from their homes at gunpoint. A farmer of Cwmcoy in Cardiganshire told one

The Rebecca Riots

meeting that he had been attacked in his bed and his wife threatened because he would not go out, and a servant years later recalled how his master had trembled like a leaf when he learnt that Rebecca required his presence. On the last adventure the Lady promised her children that they would be shot if they turned back, and warned them not to try to recognize her or one another. 'It is enough,' she said, 'that you know yourself.'

Of the work of the recruiting parties we know little, but they seem to have acted with the minimum of violence. In fact, most people were impressed with the restraint shown by the Rebeccaites, so long as collectors and keepers themselves behaved 'modestly'. When tollhouses were raked with gunfire, humans were not the direct targets, nor were all the weapons loaded with shot. The keepers temporarily 'blinded' by firing of blanks were those who, having been told to stay indoors, peeped through the windows in an effort to identify their attackers. The one fatality that occurred during the destruction of gates has never been satisfactorily explained; the oral tradition was that she ignored repeated warnings, and there was little sympathy for her in the Hendy. Where injury, to persons and property, was inflicted by mistake or by indisciplined supporters, Rebecca offered apology and some reparation.

Those subjected to the worst intimidation and violence were 'my enemies and oppressors': the parish and special constables, who defended the gates and helped with arrests, the braggarts and informers, and the people described in the next chapter. The night after Griffith Jones gave evidence against the people who attacked Pontarllechau gate, his smithy was destroyed, and he and his wife received over £200 from the government to compensate them for the loss of their livelihood. Some of the beatings handed out by the Rebeccaites, with horsewhips and sticks, left such an indelible impression that the recipients immediately gave up their badges of office. Evan Thomas of Porthyrhyd, the most dogged of Becca's opponents, was driven out of two homes, and was still on the run in 1845. David Joshua was not safe on the streets, and policemen in St Clears, Newcastle Emlyn, and Llandovery were set upon when their guards were down. Others who suffered were John Evans, the Gower blacksmith who repaired the Three Crosses gate, and carpenters elsewhere who did likewise. According to some accounts the objects of Rebecca's vengeance were saved from imminent death only by the hand of fate, though, on their recovery, they remembered that it was the mother figure which stopped the baying pack. For their part, informers and witnesses were agreed that without police and military protection they would have been savaged by crowds of angry women. Even so, this was not Ireland, and the government appreciated the difference.[107]

Despite the talk of community fear, the support for Rebecca was exceptionally strong. 'They were all Rebeccas', one arrested publican was reported to have said, and English journalists were astonished by the solidarity shown even after the defeats at Carmarthen and Pontarddulais and the Hendy murder. In some of the attacks on town gates, people had to be stopped from joining the work in progress. Even in the last days the stories of an isolated Rebecca, acting secretly and with little support, have to be treated warily. An investigation of the destruction of a Cardigan gate in March 1844 revealed that a good number heard the sound of wood breaking, but preferred not to tell the soldiers. When one correspondent journeyed to Penygarn after the tollhouse had been burnt to the ground he found sixty spectators 'who seemed to be delighted with the disgraceful events'. Without willing co-operation the movement could not have achieved its high rate of success. Of the thousands who participated in or witnessed the riots, only a handful came forward to testify against the men on horseback. Metropolitan policemen and soldiers who had served in various parts of riotous Britain could scarcely believe the silence and the lack of assistance which they encountered, even from respectable people. Such was the stalemate in the early autumn of 1843 that there was talk of imposing military law in the region.[108]

Yet, as Table 4 shows, the ending of the riots came quite suddenly, and the concern over the character of the movement quickly changed into an enquiry into its disappearance. The *Carmarthen Journal* and other Tory papers offered the simplest answer; the people had at last been intimidated by a magistracy, judiciary, and government which had shown hitherto a lack of courage and direction. Soldiers, policemen, and specials were everywhere, and the prevention and detection rates climbed upwards. This new determination had, it was felt, opened the divisions within the popular movement, and encouraged farmers to change sides. Of course, the improvement in the economy made this a little easier; the rally in stock prices had 'drawn the attention of the agriculturists from Beccaism'.[109]

For Whig reformers peace on the roads owed more to the removal of gates, and the enquiries and concessions won from trusts and government. The *Welshman*, the self-appointed leader of the Liberal cause, indulged in a little self-praise. From the beginning of the struggle it had urged the Rebeccaites to hold peaceful protest meetings and to widen the scope of the movement.[110] During the last months of 1843 this had happened and, so it was claimed, turned a growing number of people away from the destruction of gates. It was a cheering prospect, at one with the half-restored myth of Welsh passivity, but it ignored other forms of illegal action which had always been a part of Rebeccaism and which were used to justify the extensive police and military presence in the winter and spring of 1844.

6
Rebecca the Redresser

In the autumn of 1843 the use of the torch and the destruction of farm produce and private property seemed to be as popular as the removal of tollgates. Rebecca had, it was reported, become more ambitious, holding open as well as secret meetings, and seeking to be 'the government' of the three counties. 'Ill-disposed and designing men', namely politicians, outsiders, and criminals, had entered the movement, and were driving out the respectable farmers who for so long had given it direction. Rebecca the 'farmers' friend' was becoming 'Rebecca the Redresser'. 'There are many things required to be altered in the country,' the Lady is supposed to have said to a gatekeeper in May 1843, 'and as the gentlemen and those in power refused to do it, they would do it themselves.'[1] In July the *Times* correspondent warned of 'the deep feeling of hate and vengeance' which he found amongst the people, and about the same time Rees Goring Thomas expressed his conviction that the disturbances had 'a deeper root' than mere dissatisfaction over gates.[2] Within a matter of weeks, Walter R. H. Powell of Maesgwynne and other landowners concurred with this view, privately acknowledging 'the lamentable and appalling change in the character of our countrymen', who were now displaying class antagonism and individual malice over other grievances.[3]

The contemporary view that, in its later stages, the people's cause was 'perverted' for other purposes has become part of the official history of the Rebecca riots.[4] Writers have used terms such as 'deviation' and 'disintegration', when describing 'the more unpleasant' aspects of the once 'pure' movement. Henry Tobit Evans, who was trying to give credibility to the troubles and to set them in the context of a certain view of Welsh history, took this line naturally, and Pat Molloy also writes of a departure 'from the true spirit of her movement . . . [as it was] perverted in the pursuit of self-interest and private revenge'. Even David Williams was saddened by 'the sordid events' and the change 'for the worse', as 'the lunatic fringe' came on to the stage.[5] Rebeccaism, when set against the tollgate riots, was slightly disreputable.

This approach confuses as much as it illuminates. The illegal forms of direct action, so condemned at this time, were after all not new, but rather an integral part of a society where grievances were common and class solidarity difficult. Arson, threatening notices, and night marauding by property-destroying mobs had disturbed the region at various times in the late eighteenth and nineteenth centuries, as we saw in the earlier chapters; Rebecca the Redresser was a direct descendant of several community heroes and heroines. Even when the demolition of gates was making all the headlines, other protests were in progress, and some of these were sanctioned in pen and person by 'the elegant Lady'. Significantly, the best reminiscences of the riots, such as those in *Tarian y Gweithiwr*, gave prominence to Rebecca as the feared guardian of public morality and upholder of 'the people's law'. In a number of districts within south-west Wales such activities were always more important than the destruction of gates, and commissioners who visited these places reported that 'tolls are not much dwelt upon'. We shall be discussing the other grievances in the next chapter, but for the moment it is important to establish that Rebecca was often regarded as a spokeswoman for a general community revolt. 'I am averse to every tyranny and oppression', was perhaps her most characteristic remark.[6]

A recognition of this fact helps to explain the alarm shown by so many of those with economic power and social and political influence. The destruction of gates alone could not have accounted for the tone of paranoia in certain letters to the Home Secretary, for the requests for guns and hand grenades, for the sudden conversion to the value of professional policemen, and for the rush of landowners, clergymen, and even magistrates out of the three counties. The public letters of the departing leaders convey a sense of anguish and betrayal now that Rebecca was the 'governor' of the region. Of particular concern was the fact that she seemed determined to impose her own laws, which the peasantry, and even a few of a higher social status, were ready to obey. This was 'master turned servant', a world upside-down. Nothing, not even private property and life itself, was safe. 'Society will fall apart if this carries on,' ran one editorial on 1 September 1843. 'Every man with an atom of influence, no matter what his politics are, must repress and discountenance the agitation which is now sapping the very foundation of the social structure, and threatens to engulf all law, order and personal security.'[7]

In London the news of the non-tollgate activities had a profound effect. 'The infection from Ireland has spread to S. Wales in its worst form', declared Sir James Graham on 15 September 1843, as reports of a boycott of rents and poor rates reached him from the three counties. On the next day he admitted to Lord Stanley that 'the burning and the assassination are more formidable than their open riots'. The duke of Wellington and Queen Victoria were equally worried, not simply for south Wales but for the example which the rebellious Celts were setting to the rest of the empire.[8] This anxiety helps to explain the Royal Proclamation in October, promising very large rewards for information on arson and murderous assaults, and why Major-General George Brown was sent down when the tollgate riots were nearly over, to reorganize the defences and cover the region with detachments of soldiers and Metropolitan policemen.

It is impossible to give an accurate estimate of the actions, other than tollgate riots, done in the name of Rebecca. Few people had the courage of Abel L. Gower, who tore up a threatening letter at a public meeting. Edward Lloyd Hall at Newcastle Emlyn knew friends who were frightened to admit receipt of such warnings or even to report the destruction of their property. They had been told by Rebecca not to take the matter to the authorities. Those accused by her daughters of crimes such as adultery were naturally too embarrassed to talk. Rumours abounded, and Lewis Evans, amongst others, strongly denied that he had suffered at the Lady's hands. Our information on the more private and secretive activities of Rebeccaites frequently comes from informers, busybodies, and angry truthtellers. The result is patchy; we cannot be sure whether southern Cardiganshire, for example, was as susceptible to Rebeccaism as the map suggests, or whether we know more about it because of the frank exposures of Edward Lloyd Hall and the ravings of the Walters sisters of Glanmedeni and Dr

Walter Jones of Lancych. For other districts the historian is heavily reliant on the military correspondence, the breathless requests to magistrates for protection at certain mansions and vicarages, and the less frenetic applications for plain-clothes detectives and claims for fire insurance. Perhaps the most difficult problem is establishing the motive behind poorly reported offences which could have been simply random acts by opportunist amateurs or professional criminals. Half of these have been discounted, mainly for lack of information, but the rest seem to have had some connection with the Rebecca movement.[9]

This last group can be found in Table 5 and on Map 4. Of the 235 incidents, some 45 per cent were offences against the person and 55 per cent against property. Over a third of the former were attacks by gunmen and mobs on unpopular individuals, for reasons not always stated, while the remainder were identified as assaults on bailiffs and pound keepers, *ceffyl pren* type punishment of anti-social behaviour, riotous assembly against police officers, and similar outrages. Arson, with 52 cases, is the main class of offences against property, with other forms of damage to houses and outbuildings not far behind. The destruction of fences, walls, weirs, corn, trees, and equipment constitutes much of the rest.

The chronology of these activities is striking; after a thin sprinkling of cases in the period down to May 1843, there is an explosion during the summer and autumn which almost parallels the tollgate affair. Although the sharp rise in Rebeccaism came a little later, it hardly bears out the common contention that only when the gates were down did the people turn to other forms of action. Intimidation and attacks on people and property were already being reported from Cynwyl, Newcastle Emlyn, and Llandysul by the early summer, and in August and September 1843 the authorities were dealing with a many-faceted onslaught. From October onwards the situation changed; there was now a greater incidence of Rebeccaism than of tollgate crimes, especially in the heart of Rebecca's country. This difference remained through 1844, though in this later period the number of offences and the 'spirit of mischief' were never quite as bad as Colonel Love suggested. Altogether, however, it is possible to argue that Rebeccaism, in

TABLE 5. *Rebeccaism; attacks on persons and property, 1839–44*

Month	1839 Attacks on persons	1839 Attacks on property	1840 Attacks on persons	1840 Attacks on property	1841 Attacks on persons	1841 Attacks on property	1842 Attacks on persons	1842 Attacks on property	1843 Attacks on persons	1843 Attacks on property	1844 Attacks on persons	1844 Attacks on property
Jan.	1										2	3
Feb.			1	1	1	1		1		2	2	1
Mar.	2			1	1				1		1	1
Apr.										2		
May										2		
June									6	4		
July	2								5	11		1
Aug.					1				18	7		
Sept.	1				1	1			29	31	1	
Oct.				1		1		1	15	25	4	1
Nov.								1	3	16		
Dec.				1	1			3	6	4	1	

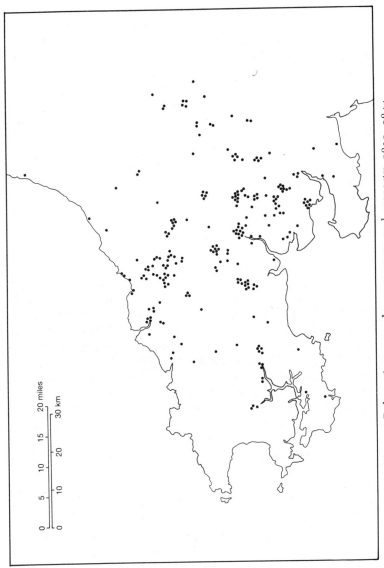

MAP 4. Rebeccaism: attacks on persons and property, 1839–1844
Note: Three other attacks in Radnorshire and Breconshire.

all its varied forms, almost matched the riots themselves in frequency and seriousness.

Map 4 bears a close resemblance to Map 2 and thus undermines the claims that Rebeccaism was 'most prevalent and formidable' in districts unaffected by road tolls. Most of the 235 offences were located in Carmarthenshire, though they were rather more scattered than the tollgate incidents. Pembrokeshire seems to have lived up to its reputation for being a fairly peaceful county, but it appears that north Pembrokeshire and Cardiganshire were more affected by these troubles than by those described in the last chapter. This point is strengthened by Map 3; soldiers and policemen were sited in the winter of 1843–4 at, for example, Eglwyswrw, Newport, and Mathry in Pembrokeshire, some distance away from the nearest turnpike gate.

Each category of recorded offences had its own particular geographical pattern. That of incendiarism is shaped like a big pear, with Llanarth at the top, Maenclochog and Gwynfe at its widest points, and Llanelli at the bottom. The most common form of arson was the burning of ricks, stacks, and outhouses, but there were, perhaps, proportionately more cases of inflagrations at dwelling-houses and workhouses than in the rural protest movements of south-east England. As some of the worst cases of violence against property and people took place within a few miles of Llanelli, it became known as the 'home of arson and murder', and was heavily garrisoned. Resistance to distraints and assaults on bailiffs and pound keepers were, by contrast, rather more common to the west of a line through Carmarthen, while the majority of known cases connected with sexual and family misconduct occurred within a six-mile radius of Newcastle Emlyn and Carmarthen. In summary, if one places Map 4 alongside Map 2, one can begin to understand why one contemporary asserted that the Queen's writ was being gradually replaced by the 'law of Rebecca'.

The dangers for people living in the lower Teifi valley, or in other heavily dotted parts of the region, must have been only too apparent, but Rebecca chose to spell them out in over 100 threatening letters and notices. A large proportion of the letters came to light in the vicinity of Cardigan, Newcastle Emlyn, Llandysul, Cynwyl Elfed, Abernant, Meidrim, St Clears,

Carmarthen, Llanddarog, Llandovery, and Llanelli. Written threats had been a part of the movement from the beginning in 1839, but reports of their use were especially numerous in the late spring and October of 1843. Most of them were sent to magistrates, their clerks and jurymen, Poor Law Guardians and workhouse officials, landowners, agents, farmers and tenants, clergymen and tithe impropriators, publicans, and of course turnpike trust employees. Over half the known Rebecca letters had nothing to do with tollgates. In addition, there were at least a dozen strange notices, usually posted outside Wales and sent to magistrates, mayors, workhouse masters, and toll collectors by other religious and secular redressers. These sometimes refer to Richard Brothers's egalitarian and millennial vision of rivers of blood, but sadly the evidence on this phenomenon is too thin to take the matter further.[10] Like all anonymous letters in the countryside, those of Rebecca had to be taken seriously; the threat they contained was frequently carried out. 'Becca', she boasted, 'always keeps her word.'

The content of the letters provides clues to the remarkable range of Rebecca's interests. She was always more than just 'a farmers' Rebecca', or an anti-landlord one. Those close to the protest movement were surprised by 'the multifarious subjects to the redress of which Rebeccaism is turning its attention'.[11] Her name was used to sanction a few individual disputes over pay and the truck system, and her followers drove 'outsiders' from the Rhandirmwyn lead mines and from collieries and ironworks. The anonymous letters accompanying these disputes contain echoes of both the Luddite and Scotch Cattle movements, and it is interesting that there were Rebeccaites in Swansea and the surrounding area who joined attacks on both rural and industrial targets. On one evening in September 1843 the same mob attacked the harbour-master of Llanelli and fired off their guns outside a copper works and tollhouse.[12]

Nor did it end there. Rebecca gangs were involved in moral, religious, and marriage conflicts, attacks on tax and excise officers, and battles over market tolls, the sale of alcohol, and many other matters. Two examples illustrate what was happening. Thomas Cooke, on 13 October 1843, described how fifteen Rebeccaites visited a female shopkeeper at Llanddarog who had a bad reputation for selling her groceries underweight, and Dr Charles Morgan, a Carmarthenshire gentleman, was given twenty-one days, on pain of losing his plantation, to return some of the excessive profits he had made on cutting timber for 'poor men'.[13] Even some fairly anonymous cases of burglary, animal maiming, cattle rustling, corn stripping, and the removal of private gates, fences, and houses have the 'Rebecca' hallmark on them. Action such as this was often sanctioned at the night meetings, and carried out by small parties in disguise. Of course, it was tempting for anybody with a grievance to use Rebecca's name, but it was not always a wise policy. 'My eyes are on you all', was the message for friend and foe alike.

Behind the growing concern about Rebeccaism was an old fear, that of obedience to an alternative code of laws and behaviour, known variously as 'lynch law', 'Welsh law', and 'Rebecca's law'. 'The public mind is perverted in its notions of justice,' declared Edward Lloyd Hall the lawyer, 'which in a political point of view is a thing much more difficult to be dealt with than a mere marauding banditti.' For Joseph Downes this clash of laws was 'the systematic waging of war against the very element that binds men in society—it is a casting off of civilization'.[14] 'I have got the world in my hand,' Rebecca told one auctioneer living near St Clears, 'to do justice, justice, justice.' 'I may quash a person that disobeys the laws of Rebecca,' she told another victim, 'because she wants very much to reform things that are out of the way.' When Margaret Thomas, the wife of Evan of Porthyrhyd, protested about the brutal vengeance meted out to her family, one of the mob on 5 September 1843 said that the 'law is now nearly all in the hands of Becca and her daughters'. 'Rebecca is more than ONE HUNDRED THOUSAND strong,' she boasted. 'If God spare the life of Rebecca she will work out their redemption ... We must be Free. I say it, I who command. I the Rebecca and Regenerator.' As late as December 1843, the message remained the same: 'before there can be any permanent peace for the Principality, Rebecca must her rights obtain.'[15]

Contrary to some reports, Rebecca did not completely supplant the Queen's law. People who committed theft and assault, for instance, were still taken before the Petty and Quarter Sessions. However, the Redresser did review cases which had been dealt with at the Crown courts, reversing some decisions and supporting others. The 'people's champion' attacked prosecutors and witnesses who had long been unpopular, especially landowners who had taken their servants to court. This was the reason why Captain Bankes Davies of Myrtle Hill, near Carmarthen, lost part of his plantation by fire in July 1843, and why a number of his friends suffered in the same way.[16] In southern Cardiganshire the Walters sisters of Glanmedeni were also brought to account. Their crimes were not paying a twenty-year-old wage debt due to an old female servant, and the assault case which they had initiated a few years previously against Thomas Lloyd, a young lad who had flirted with them in the presence of a small crowd. In anticipation of Rebecca's visit, poems were circulated in the area enumerating their wrongdoings, and mocking their physical attributes, their inflated ancestry, and their new family mansion. On 4 September 1843 a mob arrived at the big house, and fired into the building. The crowd were determined to regain the money for back wages and the fine imposed on the boy, and asked for just a little more for the cause. The mission proved unsuccessful, but on the following evening, under orders from Rebecca, young trees on the estate were cut down, and the ladies relented. They left immediately for a safer home in Newcastle Emlyn, and begged the government to build a barracks on their land.[17]

Lesser folk, too, were reminded of their errors; one old lady, after her furniture was smashed by a party of disguised men, promised, on her knees, to unswear her information against a friend for petty theft. At Caecrin mill, to the east of Llandovery, on the county border, a mob on its way to break a gate threatened a farmer with dire consequences unless he stopped legal proceedings against a late tenant.[18] There were many similar midnight visits along the Teifi valley, as mobs executed 'the people's justice' over debts, wills, and trespasses, often under the directions of a man in a long red coat. At Llanfihangel-ar-arth on the night of 11 August 1843, when disguised men with reaping hooks called on elderly Daniel Harris, it was worn by John Powell, though on this occasion David Thomas, a 37-year-old farmer and gate-breaker of Llanllawddog, was probably 'the prime mover'. With his 'smattering of law', he insisted that Harris sign a promissory note for £20 in connection with a legacy under a will. There were almost identical incidents elsewhere; in the parish of Llanarthne five men took a note for £14 from the treasurer of a friendly society, while at Tre-lech a larger mob opened the coffin of William Jones of Croesifan, in a search for the deeds of a farm.[19]

Slander had always been an offence which affected the village communities, and there were a number of people in these years who must have regretted their outspoken opinions. Of them all, Mr Jones of Esger Graig, who lived a few miles north of Newcastle Emlyn, underwent the most frightening experience. On one unforgettable evening at the end of August 1843 he went downstairs early in the morning and was blasted by a gun. In the dark he could see about forty people. The conversation went to and fro like a courtroom interrogation. He was accused of reducing the harvest allowances of his workpeople, of being mean to the poor, and of calling his neighbour an adulterer. He was given one week to make reparation to his neighbour, and the Rebeccaites then moved on to three more farms, firing as they went and conducting similar interviews.[20] This was very close to the mock trial which, as we have seen, was a prominent part of popular culture. Rebecca extended the practice; agents, tithe collectors, and Poor Law officers were tried in their absence at nightly meetings, or sold at mock auctions. If they were found to be justifiably unpopular these men were promised a beating or their property set on fire.

Although this behaviour was embarrassing to the authorities, their greatest frustration was over a more direct interference with the machinery of law and order. In parts of the region the Rebeccaites prevented the execution of writs and warrants. Early in 1843 a judge warned of the dangers of such a practice, but it grew so rapidly that, by the autumn, sheriff's officers and parish constables found their work almost impossible. 'Civil processes are at a standstill', said one oberver; it was an exaggeration, but no one doubted that it was very difficult to recover rent, rate, Poor Law, and other arrears. Proceedings against debtors became comparatively rare. 'There are no suitors,' said one of Rebecca's victims, 'no one dare complain.'

'No solicitor can now recover a debt for no sheriff's officer will perform his duty', grumbled another.[21]

When the central government demanded to know why charges had not been brought against both known Rebeccaites and ordinary criminals, the magistrates of Narberth, St Clears, Llandysul, and other embattled communities replied that no one could be found to execute the warrants.[22] The furious Home Secretary insisted that special constables should be used to oil the creaking legal machine, but Rebecca ensured that these were few and ineffective. Even magistrates were intimidated from doing too much; at crucial moments when, for example, Rebeccaites met noisily on the hills near Tre-lech and Cilycwm, and paraded Fishguard for night after night, justices were absent or pleaded illness and 'an aversion to night riding'. The newspapers, always looking for nice images, compared Sir James Graham to Canute and Rebecca to Boadicea and Alexander Selkirk.

As one would expect, Becca's attention focused on the preoccupations of peasant life: the family, poverty, debt, land, and customary rights. One of the best-remembered and most colourful forms of Rebeccaism was the punishment of those who breached the established rules of family life. As we have seen, the *ceffyl pren* had been used for this purpose in the past, and this continued during the early and latter part of the period 1839–45. An effigy of James Morse, the outspoken and amorous councillor who had tried to disperse the workhouse crowd in the town on 19 June 1843, was burnt at least three times in the autumn of 1844, and on other occasions the ritual humiliation of enemies was accompanied by the sound of horns and burning of hay ricks. Several of the most spectacular episodes took place in 1839, within a short radius of Cardigan. In the spring of 1839 a mob of about a hundred people, led by a Rebecca figure, carried a wooden horse to the home of a woman in that town, and a few months later, at Aberporth, they smashed their way into the cottage of a single woman, Sarah Davies. She was found hiding in a neighbour's house, dragged out by her feet, and beaten severely. Then, after a short journey to a nearby well, Sarah was obliged to swear three times that she had told a lie about the father of her baby.[23]

Sarah Davies claimed not to have seen the *ceffyl pren* during the incident, and it is possible that people no longer felt the need to legitimize their action in this way. '"Rebecca" has superseded and, as it were, stepped into the place of the old-fashioned Ceffil-prenn,' said a correspondent in May 1843, 'and is thus favourably looked upon by' the population. Sometimes, as at Caerae near St Dogmaels in December 1841, crowds of armed and disguised men just attacked houses with whatever came to hand. In this affair the people were chasing a 71-year-old man who had left his wife and family for a 27-year-old woman.[24] Unfortunately, it is not always possible to establish whether the Redresser gave her blessing to all these attacks, though Edward Lloyd Hall at least was convinced that they were planned at her nightly meetings. Certainly, she was renowned for being an advocate of family care, both for the old and the very young, and she reprimanded those who did not take their responsibilities seriously.

Rebecca had firm opinions about violence within the home. When husbands were infamous for their aggressive behaviour, or when it came into public view, it was time to hand out a warning. Those found guilty of beating their wives 'without cause' received retribution from 'a hard fist'. In September 1843, when Rebeccaism was at its height, Thomas Morris, shopkeeper, and Herbert Lewis, shoemaker, of Newcastle Emlyn were called together by a party of six of the Lady's daughters and told that if they beat their wives again they could expect 'the midnight visit'. An old man in nearby Llangeler, who went his own domineering way, was thrashed by a mob until his body was 'covered in blood'. What part the women played in this intimidation is kept from us, but it is known that late in this year about forty of them, led by a woman blowing a horn, gathered outside the home of a Methodist tailor near Pontarddulais and threatened to duck him in the river unless his domestic behaviour improved.[25]

Promiscuity, sexual deviance, and adultery were also frowned upon. Innkeepers were warned about the wisdom of allowing too much sexual licence on their premises, and a mother and daughter were hounded from two lodging-houses for 'keeping bad company'. Young girls who were openly indiscreet, and old men who should have known better, were other targets of

Rebecca's wrath. When gentlemen farmers and maids were discovered in compromising positions, the latter were ordered to leave their jobs, and the former were obliged to renew their marriage vows. Several men were punished for deserting their families for alternative female company, and David Jones of Troedyraur in Cardiganshire, who had committed no crime, was told to stay at home and curb his wife's venomous tongue.[26] Rebecca's interference even reached into the families of the clergy and the gentry; anonymous letters publicized their personal weaknesses, and on one famous occasion a warring couple were reunited at her behest. In October 1843 fifty people appeared outside the home of the sister of Mrs Walters, the estranged wife of the vicar of Bangor Teifi. Mrs Walters was taken from Llwyndafydd to the vicarage, where she and her husband had to swear to be loving and dutiful partners.[27]

The care of the abandoned mother and her illegitimate offspring was for a time a matter of special interest to Rebecca. These mothers were, as we have seen, placed under a disadvantage by the new Poor Law; it was more difficult and expensive after 1834 to establish paternity and secure maintenance payments, and more women and children entered the workhouses. As a first step Rebecca tried, by threatening letters and visits, to ensure that the male seducer married the pregnant girl. It was not an easy task. When a large crowd collected a woman and child, and took them to John Griffiths of the parish of Llangunllo, the farmer said that he would not give consent to his son marrying the young mother as he was under age. But the lad promised to wed the girl in due course and, when he and the mother had kissed the child as a sign of their affection, the gathering broke up. In another case, near Nantgaredig, a farmer who had been unfaithful to his betrothed was warned to carry out his promises to the girl.[28] One report describes how Rebecca took a man to a river bridge and read the burial service 'feelingly' over his bound body; on being given the choice of a watery grave or marriage to his pregnant Mari he wanted to rush her to the altar.

When marriage was impossible Rebecca took illegitimate children to their putative fathers. The commission of enquiry, when it called at Cardigan in November 1843, was told by the

Times correspondent that the practice was 'quite common', and Thomas Thomas, executed for murder two years later at Brecon, admitted this was the only Rebeccaite activity in which he participated. There can be no doubt that some of the parish authorities welcomed this aspect of Rebecca's work, and there was a feeling that it was in line with old Welsh law which gave equal status to a man's legitimate and illegitimate children. 'In this there is nothing morally wrong,' muttered a rather uncomfortable *Carmarthen Journal*, 'but it is a supersession of the ordinary law of the land and, as such, is mischievous and dangerous.'[29] Some of the dangers were apparent in the case of Jane Jones, who lived with her bastard child some three miles from the county town. One night in September 1843, after the dreaded knock on the door, she was requested by an 'elegantly attired' Rebecca to swear on a fine gun that she had told the truth about the identity of her lover. The man had so far contributed nothing in the way of maintenance and, when the mob arrived with his burden, he had the sense to be away from home. Not even Rebecca's threats could make him look after the child, and its distraught mother tried unsuccessfully to place it in the workhouse.[30]

Although it was reported that many bastards were returned by Rebecca to their fathers, detailed case-studies were understandably rare. From a dozen instances in the vicinity of Newcastle Emlyn and Carmarthen it appears that Rebecca was sympathetic to those poor female servants who had been seduced by respectable men, often comparatively well-to-do farmers like the Bowens of Wernmacwydd and their sons. In their cases the ritual was deliberately elaborate and imposing. One farmer of Abergwili, who had fathered three bastards, awoke one night to a white biblical vision, and early the next morning hurried to reclaim one of his neglected children, while a freeholder from a parish to the north, and another farmer from Abercothi, were given the special attention of 'the black footman'. The freeholder, who had seduced a female servant two years previously, was disturbed at twelve o'clock one August night by the noise of carriage wheels. He jumped out of bed, and on opening the door saw a black footman taking down the steps of a carriage. Out came Rebecca, 'very gaily attired with a child in her arms'. She entered the house, and introduced herself to the farmer. The child was his, and after he had fondled it, and promised to treat it 'as one of the family', the carriage departed. At Abercothi a crowd of two hundred people found their 'respectable farmer' hiding in a field, but eventually David Moses and his wife were persuaded to maintain the illegimate child which, until this moment, had been nursed on the parish.[31]

Not all the transfers of responsibility were as smooth as this. Some of the mothers were reluctant to reveal the names of the fathers, child-minders resented the loss of money and affection, and the bravest fathers ignored Rebecca's requests. When one farmer of the Narberth area received a threatening letter from Rebecca and Nelly telling him to maintain the child of an old female servant, and wishing 'to see you without Bread and Chees as Martha is at present', he immediately informed the authorities, who arrested the woman, the writer of the letter, and the man who had carried it. The writer, a teacher at the local chapel, declared that he had only been doing God's will.[32] At Nantgaredig another farmer, without the advantage of such a warning, tried to reason with Rebecca in person. His illegitimate boy, now 4 years old, had been brought up by a woman in Carmarthen, and when he and his real mother arrived at the farm, the father refused to kiss the son. After another request, and threat to his life, he changed his mind, and, with a sword hanging over his head, even agreed to caress the child. A suspicious Rebecca warned that if he abused the child, he could expect quick vengeance, and, to the sound of cheering, the mob moved off. Perhaps this farmer would have appreciated the choice offered to Jones of Rhydfach, Llangeler, owner of two farms and an inn. He was asked to take his bastard into his household or pay a £25 fine.[33]

This interest in the plight of the illegitimate child and its mother held a vicarious fascination for contemporaries, but it was only part of a wider concern for the poor. If Rebecca was the 'farmer's friend' she was also, to a degree, the representative of the village community. In the summer of 1843 gentlemen, farmers, and other employers were requested to be more charitable towards her 'oppressed children'. Letters and speeches reminded protesting farmers that their poorest brethren, and their labourers, were ill-fed and ill-clothed. Farmers were told to keep up their complement of servants, especially once their rents had been reduced, and threats were made against cutting back on hiring at harvest time. According to the Reverend Henry Lewis Davies, the curate of Troedyraur, the Rebeccaites tried to insist, by threatening visits, on the old customs of a supper and extra food for the families of harvesters. At a meeting held near Capel Iwan there was a call for an increase in wages for corn harvesters, as well as demands for more potato planting and free dung.[34]

One group from this meeting of cottagers and labourers actually visited a nearby farmer, and threatened him with dire consequences unless he threshed his corn and sold it at a fair price. He was not the only one to suffer in this way; in the autumn of 1843 two farmers from Llanegwad and Haverfordwest, who had large quantities of corn in their haggards, were visited by Rebecca and persuaded to send their produce to market. The second victim was a gentleman farmer, who was greatly impressed with the Lady as she stepped from her carriage in gown and long ear-rings. In view of the slump in the market this interest in the withholding of corn was a little surprising, but there was talk in the Preseli district, in Llanelli, and in other places of Rebecca issuing recommended price levels for farm produce, as well as expressing a general concern for the quality of food received by people at the farms and in the workhouses.[35] 'Freedom and better food' were the translated words of one of the placards carried into Carmarthen on 19 June 1843.

When the economic crisis was at its height more of Rebecca's children joined the unhappy band seeking poor relief. They immediately came within her 'caring arm', for it was her declared intention that no pauper would starve, either inside or outside the workhouse. Those Guardians, clerks of unions, and overseers who were identified with reductions of relief and work tests received threatening letters or, worse, assaults and armed attacks on their homes. One evening in the early spring of 1844 someone set fire to the uninhabited cottage and hay ricks of a relieving officer of the Narberth union. According to these victims, paupers and beggars simply called at their homes and, if their requests for help were turned down, promised to 'tell Becca'. Widow Bridget Williams of

Cenarth, steeped in poverty and ignorance, was convicted of composing a threatening letter to Benjamin Evans who had, amongst his other crimes, been responsible for stopping or reducing the poor relief of her and her neighbours.[36]

On one occasion Rebecca helped a pauper family and solved a tollgate problem. Late one night in October 1843 200 people called unexpectedly on one-armed James Thomas and his wife at St Clears. They packed their belongings and told the old man that it was time to leave his hovel. He was removed to Pwlltrap tollhouse, and promised a rent-free existence so long as he took no tolls. The crowd then visited the home of Rice P. Beynon, a local magistrate and an opponent of the Lady. Horns were sounded, and gunshot raked the bedroom of the absentee gentleman. Before departing they pinned a note on his door telling Beynon of the night's exploits. There was also a warning, subsequently relayed by the crier or clerk of a local chapel, that the person who removed the paupers would be dragged between horses.[37]

According to Colonel Love, in a confidential letter of 30 September 1843, 'the hostility of the rural population to the working of the Poor Laws has much increased', and he placed part of the blame for this on the *Times* correspondent.[38] In truth, as we have seen, there had been resentment against the building of workhouses from the beginning, and James Lewis Lloyd of Dolhaidd and other Guardians had been the targets of much criticism. In July 1838, just as it was about to be completed, Llandovery workhouse was deliberately set on fire, and the same misfortune befell that at Narberth. At the latter place a £300 reward was offered for information on the incendiaries, and constables guarded the rebuilding day and night. Although the Rebeccaites failed to destroy Carmarthen's hated workhouse in June 1843, popular feeling expressed itself seven months later when cheering spectators watched the lodge succumb to an accidental roof fire.[39]

Rebecca always maintained that poor people should not be confined in these places, and in February 1843 she warned the master of Narberth workhouse three times about the quality of food given to the inmates. Other threats followed and, by the summer months, the Rebeccaites had made plans to destroy this and other workhouses. One of the Lady's letters describes how, in her assault on the Cardigan bastille, she and 1,100 other 'liberals' would ignore the few soldiers and drunken constables. William Chambers, chairman of the Llanelli union, received a letter from the Lady asking him to remove the poor from their institution so that they would not be injured in the attack. In June and July, in response to desperate pleas from magistrates of all three counties, detachments were stationed either in the vicinity of, or more commonly within, the poorhouse walls. Chambers and George Rice Trevor were convinced that the continued presence of troops at these places represented the government's greatest triumph over Rebecca.[40]

To understand their point it is necessary to glance at the Home Office and Poor Law correspondence. There were many complaints from parish vestries and Guardians about the cost and damage of billeting soldiers in workhouses, and a concerted campaign was launched to remove them altogether during the autumn of 1843. In the Aberaeron, Cardigan, and Llanelli unions farming Guardians protested that their workhouses had never been in danger, and therefore the soldiers were unnecessary. How much of this was due to Rebeccaite intimidation cannot now be estimated, but about this time mobs visited the homes of scores of Guardians, encouraging them to vote against the misuse of Poor Law property.[41] The Poor Law Commissioners were cautious about the matter, but military leaders and the Home Office condemned the resistance of Guardians as a sign of cowardice and evidence that they preferred anarchy and the old system of poor relief. Of course, the number of paupers in the new workhouses was exceptionally small, but their behaviour when the soldiers were out chasing shadows left much to be desired. One old man in Newcastle Emlyn workhouse broke forty-two panes of glass and said that, had he the strength, he would pull the whole building down.[42]

The Guardians of the three counties were in an unenviable position. Changes in Poor Law policy were not popular, and the costs were a heavy burden at this particular moment. Alban L. Gwynne of Mynachty and C. J. Wrigley, the clerk of the Aberaeron union, received warning that unless the expensive new officers were quickly dismissed, the workhouse and their own homes would be destroyed. Wrigley thought of resigning himself, but changed his mind after Rebecca required him to do so. He did, however, like the clerks of the Cardigan and Newcastle unions, volunteer a 10 per cent cut in salary.[43] Other lesser officers, in all three counties, offered to hand back 20 per cent, which was the Lady's recommendation.

As the economic crisis deepened, so the resentment grew over paying the poor rate in cash, in full, and on time. Rebecca threatened those who collected the rate and those who paid it. In many parts of south-west Wales the arrears mounted, and neither overseers nor constables could make any impression on the problem. Wrigley received an anonymous letter saying that, besides his other failings, he was not a Welshman and would not be receiving the people's money. At Pembroke a mob knocked in the windows of an official, and the overseer of Penbryn was told that his crop would be burnt if he tried to enforce distraint orders. In July the Cardigan union considered taking contributions in kind, thus assisting both the farmers and the paupers.[44]

By early September this union alone was over £1,000 in debt, and there were similar difficulties in Aberaeron, Narberth, Newport, and other unions. A number of parishes tried to opt out of the system altogether, by managing their own paupers and finances, but brought the wrath of Edwin Chadwick and the Poor Law Commissioners down upon their heads. The Home Secretary, tired of the squabbling, and mindful of a recent boycott in southern Ireland, called upon the magistrates to act decisively, and to use the troops for this purpose. 'The Poor Rate', he declared on 15 September 1843, 'must be collected at the point of a bayonet if necessary.' In spite of mob intimidation, the debt was reduced and court cases and accounts of distraints for non-payment began to appear in the press, but it was a slow business, deserving of comparison with events in the county of Waterford.[45]

The poor rate was only one of the many burdens that weighed heavily on the rural population at this time. According to the *Carmarthen Journal* Rebecca was against the payment of all taxes and debts; it was an understandable mistake. At their nightly meetings resolutions were passed requiring resistance to rents and tithes which had not been reduced in line with prices, as well as to poor and police rates, and, as a necessary corollary, the Rebeccaites swore vengeance against anyone who took the land or goods of defaulters. The government, which had its hands full with the problems of Ireland, dreaded the prospect of a second country full of boycotts and bankruptcy, and the Queen reminded her Prime Minister that financial instability was the reason why empires fell.[46]

Resistance to the payment of rents naturally caused most concern to the landowning class, both inside and outside the government. Rentals were vital to the encumbered Welsh estates, and it was a matter of great regret to Sir James Graham that the gentry as a body could not be persuaded at such a time to adjust their demands. The *Times* correspondent, too, was adamant that the question of rent was at the heart of rural discontent, and some observers believed that only the slight improvement in prices at the time of the second delayed rent audit of 1843 saved the region from complete disaster. Certainly the peasantry of south-west Wales insisted that of all the 'hardships' they had to bear, none was greater than 'the enormous rents we have to pay . . . the oppression which actually reduces us to ruin'. Rebecca, in a long address to the parish of Penboyr, pointed out that lower rents were crucial, because they brought other benefits, including a greater willingness to take on labour.[47]

Although, as we have seen, the price of land was not exceptionally high, there was evidence that, in two counties at least, efforts were made by 'emissaries' of Rebecca to establish a fairer 'standard'. The notice at Penboyr asked parishes and tenants to meet with landlords and agree on arbiters. Whether they were actually appointed is unclear, though farmers were informed by letter and word of mouth that the good Lady had revalued their holding at a lower price. They were told to pay that amount on rent day, and ignore the agents' demands for more. A boycott was the only way to bring the great men to their senses. As for those who feared that they might lose their farms as a result of listening to Rebecca's advice, the Lady urged them to stick together. 'I will take your part, were we forced to burn the bodies of those that dare to take your land.'[48]

The standard set by the Rebeccaites appears to have been 20 or 30 per cent lower than the existing rents. Landlord and tenant received letters informing them of the fact, and warning them of the consequences of non-compliance. We shall be dealing with the pressure on tenants a little later in the chapter. Typical of the threats to landowners was this extract from a letter sent on 19 June 1843 to William Peel of Taliaris, sheriff of Carmarthenshire:

You know very well I dare say that every article the farmer has to sell is of a very low price in this county and you know too well that your rent is as high as ever ... you will do well to prepare a secure place for your soul we will do well with your flech we will give to the Glansevin hounds and your bones we will burn with those of Sir James Williams and Lewis Gilfach ... Down with the Rent all will be good.

Phillips of Aberglasney and the Lloyds of Danyrallt were also condemned in this letter, and a promise was made to 'visit Dynevor & Gelliaur palaces to see what they do there this bad time'. Peel's reaction to the note is not recorded, though a water-colour painting of him leading troops in search of Rebeccaites suggests that he was not a man easily intimidated.[49] On the Cwmgwili estate a similar letter was found, though this was signed by Charlotte and Lydia, on behalf of their 'respected mother'. It informed Grismond Philipps that he 'must lower his rents as other respected great men do, if not we shall see him according to our promise'.[50]

Landlords were thus faced with a delicate decision. They could concede the difference between the valuations, allow arrears to mount, and be singled out for praise at the nightly meetings, as David Pugh of Greenhill did, or they could face Rebecca's anger. When no rent payments at all were forthcoming the response was even more difficult. Several landowners admitted privately that their instinct was to take legal action to enforce rents, but they feared the threatening letters, and even more the incendiarism and visits from armed mobs.[51] Amongst those who refused to compromise, and suffered at Rebecca's hands, were the de Rutzens of Slebech, Edward Adams of Middleton Hall, and lesser characters who obtained reductions in rent themselves but were reluctant to pass on the benefits to their under-tenants. Only a few landowners, including Edward Lloyd Williams of Gwernant and William Chambers of Llanelli, came out openly against the insidious pressure on private transactions, but both diplomatically offered concessions 'without being asked' to their own tenants.[52]

The same difficulties were experienced over tithe payments. At the height of the Rebecca troubles, the process of commutation was delayed because of threats and violence; at Clydau people drove out the apportioner before he began his labours. Titheowners therefore speeded up the business by making conciliatory gestures. Where commutation had already taken place, Rebecca's directive was that a rent charge of 2s. in the pound on the rental was as much as anyone should pay. As a result of meetings in Penbryn, Abernant, and other parishes the titheowners did offer cuts of 10–20 per cent. Elsewhere farmers decided unanimously that unless they, too, had similar concessions they would refuse all payment at the September tithe pay-day. The reaction was mixed. The agent of the titheowners at Llandeilo gave the protesters an extra month to reconsider their demands, but other tithe proprietors responded more aggressively. The result was predictable; several Carmarthenshire vicars were horrified to find hay ricks on fire and gunshot marks on their doors.

The tension over tithes was at its height in southern Cardiganshire and northern Pembrokeshire. Incendiary letters were sent to clergymen and lay impropriators, their collectors mocked and attacked, and tithe-haggards threatened.[53] The Reverend John Hughes, the scholarly and 'inoffensive' incumbent of Penbryn, and his neighbour Eleazar Evans, vicar of Llangrannog, suffered as much as anyone. In the former parish Rebecca was annoyed because a Bible had been taken in a distraint for non-payment of tithe: 'I give you warning to return that poor man's Bible . . . I warn you to send back to them all that they have paid this year, more than they paid before . . . I will break two of your limbs one leg and an arm and I will put all your goods on fire . . . REBECCA.' Similar notices were sent to Evans, and both men required temporary military protection. 'I merely wish to state', Evans told the commission of enquiry, 'that I have been exceedingly annoyed and my life threatened in the parish, for a long time . . .' For six years tenants had paid the commuted tithes very reluctantly, and now they contributed nothing or only the 2s. in the pound agreed between them. The letters sent to this 'Minister of the

National Whore' were extremely unpleasant, but unlike his curate he escaped physical assault.[54]

Of all the midnight visits made to titheowners and their agents none was more spectacular than that of 22 August 1843 at Gelliwernen near Llannon. The major tithe impropriator of the parish was Rees Goring Thomas, who, despite disclaimers, had done rather well out of a local commutation Act some years before. During 1843 his agent, the arrogant John Edwards, received many complaints from angry farmers about the high level of payments and the actions against defaulters. On 21 August, presumably in an effort to defuse the situation, the curate of the parish chaired a public meeting at Llannon, but the representative of the tithe impropriator was quite unable to satisfy those present. Edwards, who was ill at the time, had unwisely refused military protection at Gelliwernen, for within twenty-four hours well over 300 heavily armed people surrounded the property.

They had travelled through the village, in full Rebecca regalia, behind their leader, riding a bay horse, but because of transport problems had left their splendid coach on Morlais bridge. At the house Mrs Edwards bravely attempted to speak to the mob from a bedroom window, but was driven backwards by rounds of gunfire. Five windows and a glass porch were shattered, before her daughter had the presence of mind to promise an early meeting between the sick man upstairs and their leaders. The crowd dispersed. Before doing so, they destroyed the large walled garden and much of the property in the gamekeeper's house. Edwards, all resolution gone, now talked of leaving the country, and the soldiers came. His master, at a meeting in Llanelli town hall a month later, announced his decision, made before Rebecca's attack, to give a reduction of 2s. in the pound. For Sir James Graham and Robert Peel, reading accounts of the affair in London, this was all depressingly familiar; Wales was becoming 'a second Ireland'.[55]

No one was too surprised when, in such conditions, arrears of rents, rates, and tithes rose sharply, and when the amount of indebtedness reached proportions not seen since the post-war slump. Magistrates admitted that it was extremely difficult to recover money owed to the State, the county, and private individuals. Landowners and their agents were warned not to contemplate action against debtors, and those involved in legal proceedings were paid a midnight visit or sent an anonymous letter. In the case of debts owed to her children, however, Rebecca took a different stance. She asked S. P. R. Wagner of Manordeifi, a deputy lieutenant of Cardiganshire, for the return of £41 which, she alleged, he had wrongly taken from ex-tenant John Thomas for lime and repairs to a barn. 'I don't like these dirty tricks I find so often in people that professes themselves gentlemen . . . I shall find more again of their savageness towards their tenants,' ran one of the letters Wagner received, and 'as you know my name this long time and also some of my works' it was sensible to close the account quickly. Other old and new debts were also settled on Rebecca's instructions; the wife of a small property owner of Llannon repaid several pounds loaned from a shoemaker 'many years' before.[56]

In the subsequent trials it became apparent that Rebecca had directed much of her hostility towards landowners, solicitors, auctioneers, and others who had, at a time of depression, foreclosed on the cash and stock loans which they had made to the peasantry. Mr Kynaston of Blaencorse, near St Clears, was a good example. During 1843 he leased some cows to a tenant who later fell behind with his rent. Kynaston took back the animals and, as a security for the outstanding payments, acquired the farmer's furniture. The tenant had neither the will nor wherewithall to continue. Rebecca wanted the goods returned to her child and declared, in a letter to Kynaston, that he 'should give them the cows this following year hoping the times will come better instead of driving the poor people out'. . Kynaston asked for, and got, a little police protection.[57]

Amongst other creditors harassed by the Lady were John Williams, a Carmarthen solicitor, several of his colleagues in Newcastle Emlyn and Llandeilo, attorney Henry Lloyd Harries, a member of a prominent Llandovery family, and auctioneers Thomas Williams senior of Llanwrda, Thomas Williams junior of Llandovery, and James Thomas of Llandysul. Williams senior loaned money to several farmers during 1843 at, if Rebecca can be trusted, the interest rate of 30 per cent, and, when the time came, was present at the sales of the bankrupt stock. In one case, that of John Jones of Danygarn, he was able to buy the whole farm itself cheaply. Jones protested about the auctioneer's behaviour, not least the fact that the proceedings were conducted by Williams's son. Rebecca demanded that Williams return the deeds to the farmer, in exchange for the paltry purchase price, as well as the goods bought on the day of the sale.[58]

Another farmer, David Jones, who lived close to Williams senior at Llanwrda, also found himself in debt during the summer to both this auctioneer and his landlord. At the sale of his effects, Jones became convinced that the Williams family were exploiting him. Corn and other goods were sold to friends of the auctioneer at well below the disastrous market prices. 'I hear you will not let my Children alone,' Rebecca informed Williams. 'She does not like to see them abused or injured.' The warnings of fire and death had some effect on the man, but later he tried to take revenge. David Jones was charged with, and acquitted of, writing the illegal correspondence to the auctioneer.[59]

James Thomas, another auctioneer living a few miles to the south of Llanwrda, received several public warnings about his behaviour. Early in 1843 a midnight attack was made on one of the cottages which he owned at Talley, and windows were broken. Some months later his namesake, of Llandysul, was involved in the plight of Mrs Jones of Blaenbedw, a widow who was imprisoned for a small debt. She had borrowed most of the money from Thomas, and Rebecca demanded her release on pain of setting fire to property. Whether Thomas was affected by this advice remains doubtful. According to Edward Lloyd Hall the old auctioneer gave as good as he received when a large party visited his home in October 1843. They had many grievances and one of them was a familiar complaint that, like other auctioneers, Thomas did not always dispose of all the property of debtors at public aution.[60]

The reasons for this can be surmised. All those connected with the execution of distraint orders and auctions of debtors' goods faced special problems in the Rebecca period. Amongst the most popular resolutions at the nightly meetings were those to resist sheriff's officers, and to boycott all sales. Court bailiffs already had, as we have seen, a bad public image, and during these years they needed the help of Metropolitan police, parish constables, and other watchmen. In anticipation of their visits to farms and cottages, Rebecca often arrived just before the bailiffs and removed the goods about to be distrained. At Narberth the pigs of a farmer who could not pay his rent were lowered by ropes down a coalpit, while in Llanfyrnach parish about a hundred people, faces spotted with black, pleasantly surprised farmer Richards by threshing his corn and removing it, and his furniture, before another distraint for rent could be executed. Stock even disappeared as bailiffs were asleep in possession of a farm. When farmers, constables, and pound keepers agreed to take charge temporarily of debtors' goods on behalf of creditors, they, too, became a target of large Rebeccaite mobs. In the parish of Llanegwad, where a number of incidents were reported, these people were compelled to return the distrained property to its former owners.[61]

Right across the three counties the descent of bailiffs on a farm was signalled by the sound of a horn which, in turn, heralded the arrival of armed men. This happened to John Evans and John Lewis when they called on 10 October 1843 at Tyrypound, a couple of miles to the east of Carmarthen, to execute a warrant for non-payment of rent against William Phillip. Phillip Phillip, William's 56-year-old father, had warned Evans on the previous day that 'the mother and her children' would be there to greet them, and so it proved. People in disguise appeared from every corner of the yard, and caught the two men as they attempted to escape. A tall man, wrongly identified as William Harris the blacksmith, ordered their legal papers to be burnt and then Evans had to swear on the Bible not to attempt another distraint. Lewis, his helper, was beaten until his clothes were soaked in blood.[62]

There were at least twenty other attacks on bailiffs about the same time, and, unlike those recorded before and after the Rebecca period, they were usually accompanied by elaborate ritual. On 8 September 1843 people in the Llangunllo area were summoned together by a solitary crier traversing the countryside; he was disguised in black and carried a pitchfork. The object of his attention was the bailiffs then in possession of property at Penbeili-mawr in the Teifi valley. Soon a crowd gathered, and four armed men, dressed in women's clothes, were sent to the house; they entered the back kitchen and walked around the bailiffs in silent protest.[63] It was an effective gesture. A month later one bailiff would have appreciated the warning. He was sleeping at Towy Castle, south of Carmarthen, where he was in possession under a sheriff's warrant for £130. On hearing the loud knock on the door, he spun under the bed, but he soon realized that Rebecca would not be denied. She was a forbidding figure, with a mane of horse hair down her back and a large feather in her cap, but Davies's thoughts were on the guns and swords which gleamed in the moonlight. 'We have enough of your sort in the country', was the tone of the evening. While Rebecca negotiated with the female tenant of the farm, Davies was sent on his way to Carmarthen, skipping above the shots fired at his feet.[64]

When bailiffs remained in the homes of debtors until the very day of the sale of property, their presence was a constant irritation. In the attacks on them, especially those in the daytime, women played a prominent part. At Maesgwenllian, up the Gwendraeth Fach estuary, Sophia Evans and Frances Howells badgered two bailiffs who had taken residence in the farm of Phillip Howells, mayor of nearby Kidwelly. Phillip owed rent to John Colby, and the females negotiated with the bailiffs for the purchase, by Howells's relatives, of all the distrained goods for £35. The money was never paid, and on the night of 10 September, the eve of the auction, Rebecca arrived with many daughters and took the furniture. The bailiffs were forced to swear not to execute another distress and sent away on an old horse. In other incidents women enjoyed the spectacle of bailiffs undergoing the ordeal by water, while at Brechfa there was the chance to mock a man tied up inside a common pound. This bailiff, who had been removed by Rebeccaites from a Llanfynydd farm, was released from his overnight prison when the pound keeper received the ancient discharge fee of $4d$.[65]

Pound keepers were, like bailiffs and gamekeepers, regarded as accomplices of the people's enemies. There were several types of pounds, but their general purpose was the same, to hold the animals which had trespassed on private and common land. Attempts were made in this period to rescue animals being driven to these enclosures, keepers were subjected to abuse and violence, and the parish pounds at Newcastle Emlyn and Llandyfriog were demolished. In the spring of 1843 a gang attacked the house and two stacks of hay belonging to a pound keeper living near Preseli Top. His wife and baby fled to neighbours, and he escaped through a hole in the wall. The precise reasons for his unpopularity are unknown, although his pound could well have been used to hold the animals taken or sold because of debts. This was the explanation for the destruction of part of the enclosure at Slebech in August 1843. Baron de Rutzen had distrained for arrears of rent on a tenant at High Toch farm, and impounded cattle and colts. Although William Merryman, the pound keeper, was supported by one or two friends, he was unable to hold the very large crowd at bay. They arrived about midnight in military fashion, and Rebecca gave orders that the animals were to be released. One of de Rutzen's keepers, James Rowe, was hit senseless but recovered just in time to see the underside of a horse passing over his body.[66]

People were mistaken when they concluded that the Rebeccaites were able to prevent all legal distraints for debts. A number of sales of property took place, though a large proportion of the goods remained in the auctioneers' hands. 'No purchasers were forthcoming', admitted one angry gentleman. The depression no doubt contributed to this situation, but so, he was certain, did the threat of violence against possible buyers. At the Cardiganshire Quarter Sessions in October 1843 the chairman 'adverted to the rumour that property in cases of distress for rent would not be purchased; but he desired it to be understood that even government itself would be a buyer in the event of distresses not being bought. It must be understood that law is strong enough to yield protection and will protect.' It was a fine speech, but a little optimistic. Local inhabitants were wary of making themselves unpopular, and when, on the Orielton estate in September, a stranger bid for a horse in a farm sale already once postponed by Rebecca's antics he was immediately set upon by the wife of the debtor.[67]

In spite of assistance from relatives and neighbours some of the occupiers of property had to face the immediate prospect of loss of their homes. It appears that the great landowners were too sensible at this time to remove too many rent defaulters, but there were lesser landlords, living in both town and country, who were more desperate or willing to court unpopularity. John Williams, the Carmarthen solicitor with a country seat outside Bettws, gave one of his tenants notice to quit, and by return he received a Rebecca message, of which this is a part: 'the neighbourhood has at last risen its head, and ... there is justice and peace on her banners ... this is the cause that determine us to defend from the great and the gentleman, if thou art a gentleman ... The element of fire is an instrument to bring people to order. Does thou know how to work a day on barley bread and cheese like us? ... Remember thou about William Francis of the Penybanwen; do thou justice by him, or else we will do justice by thee.' To reinforce the message a crowd marched to Williams's country home on 28 September 1843, terrified the servants, and broke the windows.[68]

As we saw in earlier chapters, the removal of tenants was sometimes traumatic, not least when people resisted the notice to quit, but now they had the promise of Rebecca's support. At Laugharne in September 1843 Mrs Jane Morris and her daughter Mary screamed loudly when the sheriff's officer arrived to claim their home, and soon forty people came racing to their aid. They pushed poor Thomas Jones into the pond, but all in vain, for within a short time, troops were on their way to the farm from St Clears.[69] Three months later a more interesting, and salacious, drama unfolded at Treleddyn in northern Pembrokeshire. The recipient of this ejectment order was Mrs Devonald, a widow who had once enjoyed a close personal relationship with farmer and coach-builder John G. Partridge, who now had rights to the property. Once he had secured her removal, the threats of arson began. Mrs Devonald was determined to return to the farmhouse again for some of the fittings and her clothes. The authorities, anticipating trouble, agreed to guard the untenanted residence with two armed Metropolitan policemen. A short time afterwards, Mrs Devonald arrived with 100 chanting Rebeccaites, and a ferocious battle ensued.[70]

Changes in the occupancy of holdings were always important in the countryside, and Rebecca naturally took an interest in them. Wise landowners took account of the wishes of people in the district concerned, and tried to balance rights, custom, and requests. To ignore tradition was to invite trouble. Thus in October 1838, not long before the appearance of Rebecca, John Jones, a farmer of Llangoedmore parish, was startled by the sight of people with torches moving swiftly through the night towards his buildings. It was, said magistrate Herbert Vaughan, an extension of 'ceffyl pren', intended 'in the present instance to deter a coming in tenant from taking possession of a farm'. About the same time, at Llanegwad, a new tenant of John Jones MP underwent a similar experience, although he took none of the Llangoedmore farmer's subsequent precautions and lost a barn and stable to the arsonists.[71] Rebecca, in her attempt to 'make the world plain', drew on this long tradition of community feeling and intimidation.

Because of the economic conditions and Rebecca's command to tenants not to take land on estates where rents were kept high, certain landlords were in the unusual position of being able to contemplate wholesale changes in tenure. John Rees of Llanarth said in November 1843 that many farms in the county were vacant, and the same information came from the Slebech estate in Pembrokeshire and that of Middleton Hall in Carmarthenshire. A few tenants sat tight, not paying the rent and afraid, or reluctant, to consider another yearly agreement. Thomas Cooke, agent at Middleton Hall, reported on 27 November that he had almost 1,000 acres on his hands, and some of this could not be let, in spite of the presence of soldiers and police, because Rebecca willed it.[72]

Those landlords who proceeded to replace tenants were guaranteed to incite the Lady's wrath. Threatening letters were received promising that notices to quit would be followed by arson, and so it proved. Buildings and hay ricks went up in flames in Llanddarog, Meidrim, and a dozen other Carmarthenshire and Cardiganshire villages. The careful incendiarists took the precaution of burning the property of non-resident gentlemen and woman, at the very moment when their farms were untenanted. Amongst the sufferers were Mrs Nicholls, sister-in-law of Captain Bowen Davies of Maesycrugiau, who had moved to England. She had, according to a report in the *Welshman*, a reputation for encouraging potential tenants to bid against one another for her farms, a practice which the

Times correspondent repeatedly condemned. The Reverend Thomas Rees was in a slightly different category. He had not lived at Aberduar, near Llanybyther, for some time past, but had let his extensive holding in parcels to a number of people. Although he later denied the details given to the press, it seems that Rees wished to replace one of his tenants with another, and Rebecca letters to both the landlord and the newcomer expressed disapproval of the change. In the event, outhouses were set on fire.[73]

In the case of Howell Davies of Cynwyl the punishment was somewhat unusual. Davies, who was outspoken in his hostility to Rebecca, asked the wife of one of his tenants if a prospective occupier could see over the farm. She gave permission, but remarked that 'my aunt has told us that we cannot be compelled to quit the farm'. On the night of 29 August 1843 part of a fir plantation and some ornamental trees were cut down, as well as a couple of gates on another of Davies's farms. 'We hope that sufficient evidence will soon be obtained to bring the guilty parties to receive the punishment they deserve', said the *Carmarthen Journal*. 'Unless a check be put to their proceedings, the rights of property will be disrespected, if not entirely cease in this part of the country.'[74]

Thomas Cooke believed, from the evidence of his own eyes, that a large proportion of Rebecca's threatening letters were sent not to landlords but to tenants. A few were received by farmers about to renew their tenure at unfavourable rents, but most were warnings to people not to take vacant farms. This message appeared on a building on the Cwmgwili estate: 'The notice is given to any persons that takes any farm, before asking, and getting leave from the present holder, that his life and property will be in danger.'[75] Those who still felt an overwhelming desire to feed their land hunger were hounded by torch-carrying mobs and voices in the dark. 'I give you warning,' boomed a voice at a Brechfa window one midnight, 'that you must not go to the farm you have taken, or we will burn your hay and corn, as we have done to your neighbours.'[76] From Cardigan in the west to Llandovery in the east, farmhouses, outbuildings, and other property were set ablaze to convince unpopular incoming tenants of the error of their ways. As we shall see in the next chapter, Rebecca reserved to herself, and the old tenant, the right to decide who should occupy a deserted farm or fields, after consultation with her farmers' committee. Two reports of the destruction of buildings in the Llangadog area were said to have been a form of punishment for tenants who had taken land without the consent of the local 'Union'.[77]

Rebecca's views on the occupancy of empty holdings were well known; the first preference was to be given to the old tenant, if he or she were prepared to take the farm at a lower rent, and when that was not possible the land was to be offered to someone without a place. 'Woe unto them that join house to house, that lay field to field' (Isaiah 5: 8), was the biblical slogan on several of her notices. One of these was sent to farmer Thomas Thomas on 5 August 1843. He was about to take a farm in the parish of Llanegwad at a price which his predecessor could not afford. In the letter Rebecca described her anger at learning how, after demanding a lower rent, the old tenant had been supplanted by this 'shameless devil'. 'We have been informed that you are guilty of that self-same transgression, which is virtually prohibited in the Bible . . . In consequence of your coveting the farm . . . now occupied by Rachel Jones, we deem it advisable to inform you that we do not allow you or any other individual to . . . make any proposal or offer to the landlord for the said tenement, and thereby peremptorily cast out the said person . . .' For a while Thomas was too scared to enter the farm, but eventually, in November, he sent four loads of corn ahead of him to the premises. A few days later it was set on fire.[78]

Rebecca's determination to prevent a decline in the number of tenant farmers caused much comment, and drew upon itself the obloquy of 'dangerous agrarian ideas'. Amongst those who protested vigorously against such a proposal were Miss Yelverton, her agent George Wood, 'a hard gentleman to be amongst us Welshmen', and 'high farmer' Thomas William, all of Whitland. The lady, of Irish extraction and brother of a Whig politician, was told that 'you doust not good to others nor yourself you living here and your God in Dublin . . . I mean to rent Talybont to some person that are without a place to do him good . . . Rebecca.' Wood was criticized for the manner in which farms on the estate were let, and for his lack of charity to the poor, and Thomas William had broken 'the law' by 'retaining three farms in your hand'. T. Jones, of Henfynyw near Aberaeron, a 'gape-mouthed Devil' and possibly another agent, was also accused of breaking the 'very Command of God' by engrossing farms at the expense of existing tenants.[79] Such people received threatening letters and a few were attacked by gun-firing Rebeccaites. When there was no sign of property being disgorged, the arsonist set to work. Several fires were reported in the north-eastern district of Carmarthenshire only because, said the local press with muted contempt, gentlemen had exercised their natural rights to rent or purchase a few extra fields. No wonder James R. Lewes Lloyd of Dolhaidd, on his brief stay in the Principality in the autumn of 1843, felt obliged to deny in public the rumour that he had appropriated another man's property.[80]

Colonel Love, writing as late as October 1844, claimed that because of these pressures tenants had, as in Ireland, given up their farms, and others still chose not to enter new properties. Love gave the example of the farm in Cilrhedyn belonging to the Carmarthen banker William Morris. The existing tenant had been irregular with rent payments, and had generally neglected the property, and so Morris refused to renew the lease until he received sureties guaranteeing a change in behaviour. When these were not forthcoming, the banker advertised for a new tenant, and someone on the Cawdor estate agreed to pay a much higher rent. The person was immediately subjected to all manner of abuse, 'every one in the neighbourhood appearing leagued against him, and he was forcibly ejected from his meeting house and not permitted to enter it'. In the mean time the old tenant produced the required sureties, but Morris refused to close the agreement with him until apologies had been extended to the man who had almost replaced him.[81]

Some idea of the extent of Rebecca's influence can be obtained from her public conversations with Joseph Jones of Llanllawddog. This farmer had been directed by the Lady to quit the holding which he had taken. Unable to get a new place by 30 October 1843, he wrote a note to Rebecca asking for six months' grace. On 7 November he had his reply, a comforting address granting his wish, but advising him not to move subsequently to Glangwili mill. 'Your house is as safe as if it were dressed with humility and bound with brass', so long as he obeyed these instructions. 'Becca cannot injure any humble man . . . we cannot injure a man who is so easily handled.' A similar message was received by John Hugh of Felinfoel, Llanelli, who considered taking a mill at Kidwelly and then thought better of it.[82]

If land tenure was a major concern of Rebecca and her daughters, there were nevertheless some aspects of land ownership and land use which also held their attention. A number of private gates and fences were pulled down during these years probably, one suspects, for a prank or out of spite, but occasionally, too, by mobs who claimed that the barriers blocked old rights of way. There were other types of encroachment of which Rebecca disapproved, and she expressed her feelings on the matter to David Prothero, the vicar of Eglwyswrw. She resented sheep and horses being allowed to graze in the churchyard, and contested the Church's claim to land and a Methodist chapel which had long been regarded as the property of the Lloyds of Bronwydd and neighbouring families. 'I fear there is nothing short of a visit from Rebecca will bring him to his proper senses', said one of the letters posted in 1843, and there were similar warnings to other clergymen and farmers to be satisfied with what their 'predecessors had quietly and peaceably enjoyed'.[83]

Although enclosure by Act of Parliament was not an important issue at this time, feelings remained strong over any threat to common rights. When John Hughes wanted to fence a piece of land on the common at Newport, Pembrokeshire, he found that the two women who opposed him now had the support of Rebecca. Her children west of Llanelli discussed the recent enclosure at Pinged Marsh and Pembrey mountain, and the manner in which the interests of the poor had been neglected.[84] Some wanted to reopen the land, and one or two hedges were torn up. In other parts of the region, too, fences and corrals on common land were removed. The houses of a squatter and a poor woman near Cardigan and Cross Hands were razed to the ground. At Tre-lech a crowd of 200 Rebeccaites, led by men on horseback, tore down the hedges set by farmers years before on Crugebolion common.[85]

Rebecca the Redresser

The outstanding example of retrospective justice was at Llandybie where encroachments, by both the rich and the poor, had limited the privileges of the commoners. Over ten years previously the green adjoining the common, where the Whitsun fair was held, had been enclosed with a wall and put into the hands of David Lewis, a tenant of Lord Cawdor. On 31 July and 23 August 1843 two large Rebecca mobs ceremoniously removed the wall of this and other encroachments. The last squatter in the area to suffer was Mary Rees of Plasybont, whose *ty unnos* had gone up two years before. This old widow claimed that Thomas Mainwaring, the butcher, whom she sometimes supplied with meat, had been against the new property from the beginning, but she was a little surprised when he burst through the door on 28 September with a hatchet in his hand. He was accompanied by five men, with flour or chalk on their faces, and the old woman was made to swear on a gun that she did not know them. Then, after rescuing a few of her belongings, Mary Rees watched the Rebeccaites dismantle her home. No one came to her assistance, either then or at the Assize three months later.[86]

In view of the conflicts over Game and Fishing Laws which we described in Chapter 4, and the complaints by farmers of the damage done to their corn, it came as no surprise when Walter Morgan of Llanarthne, amongst others, tried to encourage 'his Mother' to act upon them. The *Times* correspondent reported that notices were sent to the gentry of southern Carmarthenshire in the autumn of 1843 warning them not to shoot any game in the approaching season, as it all belonged to Rebecca.[87] A few of the letters, and speakers at the nightly meetings, stated that there was no property in wild birds and fish, and magistrates and witnesses in poaching cases were subjected to intimidation. Most of the animosity, however, was directed towards the bailiffs, keepers, and underkeepers, 'idle fellows . . . no good to the rich man nor poor', who patrolled the woods, rivers, and fields of the gentry's estates. At Dolhaidd the garden of James R. Lewes Lloyd's bailiff was pulled up and his beehives destroyed, and several gamekeepers went in fear of their lives. William Bassett, who lived in a cottage close to Gelliwernen, Llannon, fled into the woods when the mob approached on that night in August 1843. They destroyed most of his goods, and took away his gun and ammunition. Acres of game preserves were set alight, and Rebecca letters called on non-Welsh conservators to return from whence they came. To protect such estates, soldiers were rushed to Newport, Whitland, Llanarthne, and Castle Malgwyn.

Castle Malgwyn, the home of Abel L. Gower, played a major role in the saga over fishing rights. The people of the three counties had some of the best rivers in Britain, and hundreds of fishermen lived, legally and illegally, from their waters. Their main concern was that the Teifi, with its salmon, the Tywi, and all the other rivers should be free of both physical and legal restrictions. Along the Pembrokeshire coast during this period attempts were made to destroy the huge Scotch weir nets which prevented hundreds of fish swimming up the Nevern. At the same time, despite threats and a conciliatory address from Edward Lloyd Hall, there were successful assaults on a small weir of the Bowens of Llwyngwair, and on Felingigfran weir further up the river. The people had complained for years that the keeper of this last obstruction took far more fish than he claimed. Three months later, in mid-October 1843 in a different part of the county, a smaller party of Rebeccaites from Llawhaden re-enacted the attack of 1830 on the de Rutzens' Blackpool weir. This stopped salmon swimming up the Cleddau, and had always been regarded as illegal. It took the men five hours to destroy the impressive structure, and the baron, who had already received two Rebecca notices, was furious with the authorities for ignoring his request for police protection.[88]

Other fishing traps were smashed, including one belonging to William Chambers near Pontarddulais, but undoubtedly the most famous incident of its kind was the pulling down of the old Llechryd salmon weir, a few miles up the Teifi from Cardigan. It was the property of Thomas Lloyd of Coedmore and had been leased to Abel L. Gower. The unpopularity of this weir, especially with the men of Cilgerran and Cenarth, was long-standing and efforts had been made years before to have it removed. In 1843 Gower received groups of protesters at Castle Malgwyn, but insisted that he would sack his workmen if they proceeded illegally. He was supported by soldiers at Cardigan and Newcastle, extra watchmen, and a primitive communication system. After the threatening letters began to arrive he persuaded the authorities to give him twenty-five of the soldiers, and thirteen of the Metropolitan policemen, all of whom he paid. In the event the Rebeccaites set one of their false trails, and when, on the night of 13 September 1843, the soldiers rushed off into the countryside, the disguised men waded into the water and began to saw and hack. A parish constable was forced to watch the proceedings.[89] Gower reacted in typical hot-blooded fashion, sacking men and tearing up one threatening letter in public, and he had to be rescued by Edward Lloyd Williams of Gwernant, David Saunders Davies MP, Augustus Brigstocke, and other friends. At a public meeting at Llechryd on 9 October the gentry promised to raise a subscription to buy the weir with a view to its permanent removal. According to the most detailed reports, the population were lyrical in their thanks, and cheering crowds swore that the Rebecca movement was now finished.[90]

Behind this prognosis was the hope that some at least of the farmers of the area were genuinely concerned about the direction of Rebeccaism. For the first time, in October 1843, one or two of the larger farmers were coming forward to act as special constables, no doubt emboldened by the recent arrivals of professional policemen and soldiers, and their dispersal across much of the three counties. Edward Lloyd Hall, who looked for these things, reported about this time on disagreements at Rebeccaite meetings and on acts of near rebellion by farmers as they watched another building go up in flames.[91] Yet, as Table 5 shows, such activities were by no means extinct in October, and the following weeks witnessed the climax of a form of Rebeccaism that has never been properly studied, the attacks on individuals. On 22 December 1843 the editorial in the *Carmarthen Journal* exclaimed against one murder and one attempted murder by the Lady's supporters, and insisted that it was only the military and civil forces which kept such hostility in check. In the opinion of Lancelot Allen of Cilrhew, the continuation of these barbaric outrages during the next few months was the best argument for a Rural Police force in his native Pembrokeshire. 'Surely,' he told the Quarter Sessions in April 1844, 'they could not for one moment say that a country in which such things as these took place, was in a state of perfect tranquillity.'[92]

Altogether during the years 1839–44 there were reports of over a hundred assaults by Rebeccaites on their enemies, both inside and outside their homes. The most serious cases occurred in the two triangles of land bounded by Llangrannog, Eglwyswrw, and Brechfa in the north-west, and by Carmarthen, Llandeilo, and Llanelli in the south-east. In addition there were several unpleasant incidents close to the border of southern Pembrokeshire and Carmarthenshire. We have already noted a few of the worst attacks, like that on John Edwards of Gelliwernen, but there were many others, not all of which were willingly disclosed. Frequently, the precise reason for the hostility was known only to the victim, the newspapers simply stating that the man or woman had become 'obnoxious to Rebecca'. What alarmed reporters was the 'class feeling' behind such attacks, for the mobs threatened both Liberals and Tories, self-confessed friends and opponents. A bemused Thomas Cooke, agent of reforming landowner Edward Adams, gave up trying to understand the rationale behind the ferocity: 'I think the whole country is gone quite mad', he declared on 16 September 1843.[93]

In fact, there was some order to the madness, for one can identify certain groups of people who were subjected to most of the assaults. These included, as we saw in the last chapter, farmers who would not join the Rebeccaites nor contribute financially to the cause. One farmer in the Meidrim area, braver than a neighbouring magistrate and a clergyman, told a collecting mob that he would never give them money, and promptly lost about £25 worth of wheat when his cattle were turned into a field.[94] Damage to growing corn and, later, the burning of ricks were the most common form of punishments for refusing to support the Lady's activities, though truly stubborn characters were manhandled and lost all their farm buildings. On the night of 30 August 1843 the house of a small farmer openly hostile to Rebecca was pulled down about his head. Perhaps the most frightening vengeance on such people was the violent armed robbery, when small gangs of disguised men fired into houses and forcibly levied money for the cause. The reporter for the *Morning Herald* heard a rumour that three or four people had called on a farmer near Mansel's Arms, put a gun to his head, and walked away with £80. Amongst the other victims were Mrs Williams of Manordeifi, and Jonah Williams of the Llangadog area, who reluctantly parted with 18s. 2d. for his 'brothers [on trial] in Cardiff and Brecon gaols', convinced that the three men in women's clothes and red cravats would soon spend his contribution on drink.[95]

Landlords of public houses were expected to be sympathetic to the popular movement, and provide the Rebeccaites with drink, a meeting place, and news of the movements of the soldiers. A few, at St Clears, Pembrey, and Llanelli, acted as treasurers for Rebecca. Those who hesitated in their allegiance were threatened, and several, including the occupant of the Farmers' Arms in Llangendeirne parish, needed a military guard to save them from attack. They were accused of acting as double agents, taking money for information from both sides, and altogether doing rather well out of the troubles. On 4 September 1843 Rebecca mounted her horse and led her daughters to the Bear in Felinfoel, north of Llanelli. The landlord, David Williams, had already received a letter, and was now called before the Lady. He was told of the rumour that he was about to split on the daughters, and swore that it was not true. Rebecca declared that she was in a generous mood, and the guns fired only into the night sky. Not far away, Francis McKiernin, the publican who had been carried home in triumph after the Sandy gate court case, lost his popularity and a hay rick when he provided accommodation for the soldiers. To be scrupulously fair to these publicans, one should remember that it was hard to refuse lodgings to soldiers and policemen, though an innkeeper at New Inn in Cardiganshire threatened to 'throw up his licence' rather than risk Rebecca's displeasure.[96]

Those who showed no solidarity and actually took up the fight against Rebecca knew the consequences. It was said that, until mid-September 1843, her greatest enemies were not the professional soldiers, but the parish and police constables, and those who assisted them. Energetic natives like superintendent George Jones of Llandovery, Metropolitan policemen such as John Daniel of Cardigan, and a Porthyrhyd blacksmith on the side of authority were abused, beaten, and had their homes attacked by angry mobs. As part of the war waged against George Jones a deputation of Llandovery inhabitants said that they would act as special constables only if he were dismissed. One of the first English recruits into the Carmarthenshire Rural Police was told by Rebecca to leave the country within two weeks, and to encourage him a mob fired into the door and windows of his house. Others, including the inspector of police at Rhayader, received threatening letters, open taunts in the street, and a gun barrel in the chest. One can well understand why they chose to be missing on certain days, and why angry burgesses demanded investigations into their inactivity. For their part, John Daniel and the other constables complained that the attacks on them and their families were rarely turned into court cases; they suspected, perhaps with good reason, that some magistrates had neither the courage nor the sympathy to give them support.[97]

This was one explanation for the reluctance of people to act as special constables. At a meeting of magistrates held at the Guildhall, Carmarthen, at the beginning of August 1843, five people appeared to be sworn in as specials. Three of these said that they were afraid to take the oath as they had received warnings the night before that their homes would be destroyed and their lives were in danger.[98] Several hundred more of their friends in the parishes of Abergwili, Newchurch, Llanpumpsaint, and Merthyr ignored the summons, but, in an effort to defuse tension, the magistrates decided not to impose fines on them. To be a special required a certain type of hero; those caught by the Rebeccaites were whipped and otherwise treated very harshly. At Llandeilo, Llangeler, and Porthyrhyd they returned their staves of office, along with the threatening letters and claims for fire damage. The English workmen at William Chambers's Llanelli pottery, and the pilots and coastguards of the area, had no choice; unless they acted as specials they feared the loss of their jobs. Rebecca understood, but took selective revenge. Sir John Walsh, in a bid to show that Radnorshire had not fallen to such depths, promised Major-General Brown in November 1843 that all the principal farmers of his disturbed district would volunteer to act as specials, but for a time none would. Brown was aggrieved; one of his main objectives in south Wales was to force people to do their duty. Only late in the day, when their safety was more assured, did tradesmen and farmers come forward in numbers to aid the magistrates.[99]

Rebecca earned her nickname of 'Redresser' partly because of the punishment meted out to those willing to give evidence against her children. In a number of instances prosecutors were so frightened that they did not appear in court to support their cases. 'The man that informed', said John Lewis, the farmers' leader of Llangadog, 'deserved to be turned out of the country.' For this man, and his friends, the Royal Proclamation and the many offers of rewards and pardons were just a way of 'getting people to commit perjury'.[100] Informers and witnesses were denounced by their neighbours and beaten. Miss Evans, the daughter of an Eglwyswrw builder, who gave evidence against two brothers for threatening the police, was followed into the farmhouse one night and fired at as she lay on the floor. This was in January 1844, and Colonel Love, who had just removed soldiers and police from the village, had to ask them to return. Rowland Daniel, curate of Llandysiliogogo, received similar treatment. One night a disguised man came to his home and threatened to set it alight. The curate recognized him, and a court case followed. On his home from the magistrates at Aberaeron a crowd pulled Daniel from his horse and, when he refused to say a prayer for Rebecca, beat him until his ribs cracked.[101]

As intended, the plight of those who gave evidence hardly encouraged others to do likewise. Although the rewards of £50, £100, and £200 were tempting, informers needed police protection before and after court cases. At Llandeilo in September 1843 a 'very questionable man', drawn by the £200 prize for a successful conviction in the case of the destruction of Lord Dynevor's wheat mows, gave the magistrates the names of a 'respectable farmer' and his son. The evidence soon disintegrated and the men were carried home in triumph, but the informer had to be rescued from a gang of women and placed in the lock-up for several days. Evan Thomas, 'the lion of Porthyrhyd', promised information against seven or eight people if he were paid 14–15s. a week for two years in another place. A year later the police failed to stop a *ceffyl pren* being carried against a man who had secretly informed on the

Rebeccaites. Perhaps, like some of the other pathetic examples that recur in the Home Office correspondence over the next decade, he ultimately found refuge outside the region.[102]

Thomas Thomas of Pantycerrig, Brechfa, paid a different price. On 29 September 1843 a gang of 15–20 armed men told him that the people were displeased by the conflicts between him and his neighbours, and a tall Rebecca dragged him to Byrgwm, the home of farmer David Evans. There it was agreed that he should have nine days to settle the damages caused by his cattle straying into the corn of Byrgwm. The sum requested was £2, reduced from £5, but it was still four times as much as Thomas was prepared to pay. Evans, who played a major role in organizing the attack, refused Thomas's lower offer, saying that the matter 'is now in the People's hands'.

The crotchety 71-year-old Pantycerrig farmer unwisely disclosed the names of his assailants to magistrates in Carmarthen, and two were arrested. Within hours of his returning home from giving evidence against them, his farm was engulfed in flames. In spite of pleading from his wife and servants, the large crowd of jeering Rebeccaites prevented anyone dousing the fire. Nor was this all, for Thomas subsequently lost several sheep, and named two brothers, Benjamin and Evan Jones, as the probable criminals. Thereafter, there were two versions of the story. According to one account, Thomas decided on 18 December to accede to the request of magistrate John Lloyd Price of Glangwili and visit the home of Thomas Jones to discuss his sheep-stealing sons, and accidentally fell off a makeshift footbridge into a brook on his return from the errand. Being an old man it was possible that he collapsed because of a stroke or heart attack. The other version is that the Jones family, and perhaps others, always intended that harm should befall Thomas, and sometime in the late afternoon he was knocked unconscious and pushed into the shallow stream. A coroner's court was carefully chosen, but after a long deliberation returned the verdict 'Found Dead'. Colonel Love told the government that the medical evidence of a bruise on the temple and concussion allowed no other verdict. Yet within months Love, and public opinion, were strongly convinced that murder had been committed. To add insult to injury the three men charged with the arson attack on

Thomas's farm were acquitted at the Carmarthenshire Assize in the spring of 1845. It was only one of a number of judicial decisions that were regarded, in legal circles at least, as the outcome of sympathy or, more likely, of fear.[103]

Amongst those also intimidated by Rebecca were magistrates, clergymen, and landowners. Of the first of these, Alban L. Gwynne of Mynachty wrote: 'threats of incendiarism and even murder are constantly received by those in a higher class of life who exert themselves in repressing any rebellious proceedings.'[104] At Llandovery in October 1843 magistrates' homes were guarded every night by soldiers, and those who examined suspected Rebeccaites needed police protection. 'Manly', 'uncompromising', and 'active' John Lloyd Price of Glangwili was saved several times from the mob by the close proximity of two military stations.[105] More isolated colleagues like George Bowen of Llwyngwair, James Bowen of Troedyraur, and Thomas Lloyd of Bronwydd either withdrew from their country homes every evening or left the district after threats were received. George Griffiths of Newcastle Emlyn and borough magistrates Jenkins and Nugent of Cardigan also decided to keep on the right side of their poorer neighbours. 'A good many . . . showed the white flag', said the grandson of Lieutenant-Colonel Herbert Vaughan many years later, '. . . and held aloof from the movement altogether.'[106] The Home Secretary could not hide his annoyance: 'the execution of the law is not aided by those who hold the Royal Commission . . . The effort must be made by the resident gentry, who have everything at stake.'[107]

Part of the problem was highlighted by the experience of Rees Goring Thomas of Llysnewydd, Charles A. Pritchard not far away at Tyllwyd, John Henry Philipps of Williamston, and John H. Rees of Cilymaenllwyd, Pembrey. They expected immunity from Rebecca's vengeance by publicly expressing sympathy with some of the people's grievances, and yet all four required military protection because of threatened or actual midnight visits. In fact, there was a certain satisfaction in some quarters that Liberals such as Thomas C. Morris, Edward Adams, and William Chambers suffered quite as much as Tories. 'What fools we have to deal with', said Colonel Love to Major-General Brown on 21 December 1843, as he tried to cope with the rapidly changing moods of both Whig and Tory reformers on the Bench.[108]

Rebecca was unrelenting in her pursuit of those who persistently sought her demise. Her *bête noire* was George Rice Trevor, son of Lord Dynevor, and the man who co-ordinated the local response to the outrages and received the government's praise. He was a moving target, constantly in the company of soldiers, but his father's property was set on fire, and in September 1843 men dug a grave for him within sight of Dynevor House. Rebecca promised that Rice Trevor would occupy the plot by 10 October, and news of the threat produced a 'sensation' at Carmarthen, where he was in temporary residence.[109]

Trevor was able to ignore all the threats, but more than twenty magistrates were publicly humiliated or attacked by Rebeccaites. In the spring of 1843 Timothy Powell, of Pencoed near St Clears, was unpopular for several reasons, one of which was his work as a magistrate. He was a quiet, if determined, anti-Rebeccaite, who had already received a threatening letter. On at least two occasions, one coinciding with a church service, six acres of his woodland was put to the torch, and only the actions of the police and helpers saved his house.[110] His nearest colleague, Rice P. Beynon, evaded two attempts on his life; on one evening, as we have seen, he was away from home when shots were fired into his bedroom. A more oustpoken opponent of the Lady, and one who criticized the concessions made to the turnpike rioters at St Clears, was John E. Saunders of Glanrhydw. Having defied his enemies to do their worst, he could not have been too surprised when the lawn gate in front of his mansion was set on fire.[111] The same could be said of David Davies and Bankes Davies of Carmarthen, and of fellow magistrates William Morris, a banker of the town, and the mayor Edmund H. Stacey, who signed distraint orders against prominent rebels. Shots were fired about their homes and property destroyed, and someone broke the hind leg of the banker's valuable brood mare with a pickaxe.[112]

Three magistrates who mounted a serious challenge to Rebecca were Dr Walter D. Jones of Lancych, near Newcastle Emlyn, William Chambers junior of Llanelli, and the Reverend James W. James of Robeston Wathen. Jones and his family were threatened several times, mainly, it seems, because he was adamant that even unpopular laws have to be obeyed. Unlike some magistrates he insisted that people should pay tolls even where gates were down, and he let it be known that he was ready to supply information on people suspected of being Rebeccaites. He was a Tory in politics and deeply suspicious of all Dissenters. In his capacity as a Poor Law Guardian he agreed that the rates were too high, but blamed this partly on labouring men who spent too much of their income on drink. On the night of 4 August 1843 a volley of shots shattered the lower windows of his seaside farm at Pennar in the parish of Aberporth. The physician grabbed his loaded gun but, before he could fire, another shot smacked into the bedroom window. In all eight bullets entered the bedroom, narrowly missing Jones and his wife. After this fright the family returned to Lancych, and filled every room with firearms, but still the threats came. Six weeks later he was contemplating taking his wife to Cheltenham or London for her nerves. His parting words to the Welsh peasantry have the tartness of the offended paternalist: 'you were not formerly unthankful to those who had been kind to you, but at present, to me, some evil spirit has taken hold of you . . . Pray to God for forgiveness against whom you so shamefully transgressed.'[113]

William Chambers junior was very different from Dr Jones, although the family were also prominent in the local Board of Guardians. The young industrialist and landowner stated publicly that he had done more than any man to assist with the gates, rents, and tithes, which was partly true. Yet he was always more hostile to Rebeccaism, and more ruthless as an employer, than perhaps his Liberal politics indicated. The local population had several reasons for disliking the man, though frustratingly we will never know all the details. The correspondent of the *Swansea Journal* felt that certain matters were 'outside the province of a provincial reporter, [and] I altogether suppress them'.[114]

Chambers's major crime, in the eyes of Rebecca, was his actions on the night of 6 September 1843, when his secret information allowed the Llanelli and Swansea authorities to ambush the men attacking Pontarddulais gate. Chambers was accused of shooting some of the wounded, and such was the animosity towards him that money was offered for his death. For a week after the affray John Jones, David Davies, and other Rebeccaites pursued him unceasingly. Despite promises of military help and half-hearted assistance from the magistrate's servants, ricks, outbuildings, and farmhouses on five of the Chambers's properties were set on fire. One of the horses which escaped from a burning stable was later stabbed, and the men who were intent on doing the same to the magistrate washed their hands in its blood. This was an orgy of destruction, more frenzied and costly than anything which had gone before, and compared immediately to the worst Irish excesses. Chambers junior issued a defiant address, but his relatives were already packing. Two months later his family had gone, having sacked employees and sold off the harvest produce. Only, so it was said, a begging requisition from the tradesmen of the town encouraged the industrialist to stay.[115]

Before the year closed the Rebeccaites carried out their most desperate attack on a magistrate, and one which removed all doubts about their 'murderous intent'. The Reverend James W. James of Robeston House, a clergyman with a large family and a small living, was chairman of the Narberth Board of Guardians. He had a deserved reputation as a firm upholder of the laws, and, it was widely believed, was the man behind the attempt in December 1843 to re-erect gates on the road between Robeston Wathen and St Clears. The clash between him and Rebecca was unavoidable; amongst the threats held out to him was that he would be forced to join in one of the Lady's raids. Unlike many of his colleagues on the Bench James willingly examined people suspected of being Rebeccaites, and on 19 December 1843 a person whom he had committed for breaking open a pound was tried at the Assize. On the previous night James was in the dressing-room at the back of his house. At about 10.30 p.m. he opened the shutters to check the weather, when two shots rang out. One pierced the muscle in his arm. While he was reeling backwards Mrs James, in the bedroom on the other side of the house, was taking cover from more shots. No one knew how many people were outside, but James had no doubt that they were connected with the Rebecca prisoner. Despite a government reward of £200 and extensive police enquiries not one of the assailants was detected, and a few months later the clerical magistrate announced his intention of withdrawing from the Bench.[116]

The Reverend James was only one of fifteen known cases of Anglican clergymen who were intimidated by Rebecca. Dissenting ministers were more fortunate, although one or two of them invited attention. Thomas Rees, who had been a minister at Capel Iwan before moving to a chapel in Glamorgan, returned in September 1843 to give a sermon before his old congregation. He lectured them on sin and the illegality of nocturnal outrages. On the following night he was woken by the sound of guns near the house, and rightly took it as a warning to be silent on the subject of Rebecca.[117] Other clergymen, including those in the parishes of Cenarth and Llangunllo, who condemned criminal forms of protest received more direct messages, and their terrified servants had to clean up the debris left by gunshots and stones. Like the magistrates whom Colonel Love so despised, these vicars and curates expressed a determination to carry on regardless, but then considered moving because of their wives' ill-health. The Reverend Henry Nathan was a tougher character. He bombarded the ministers of the Crown with complaints about the breakdown of law and order in Fishguard, and was held responsible for the stationing of troops in the town. After they departed, in January 1845, an attack was made on his home, and suspicion immediately fell on three persons, 'notorious characters' who had once been committed for destroying a turnpike gate.[118]

The English press typically gave no explanation for the hostility towards Anglican clergymen, saying only that these men were well loved and considerate towards the poor. Some were, but scrawled jottings which have survived list the moral failings and criminal activities of the Reverend Ebenezer Morris, Thomas B. Gwynn of St Ishmael's, and other 'surpliced and Tory tyrants'. Charles A. Pritchard of Tyllwyd declared that it was all part of a concerted campaign against the Church itself, and the *Carmarthen Journal* proffered this theory: 'the more active and zealous a clergyman is, the more obnoxious he is to the Rebeccaites'.[119] They were referring specifically to the Reverend J. Jones of Llansadwrn, who was forced to flee with his family to lodgings in Llandovery. On 29 September 1843, in the early hours, he was awoken by a volley of gunfire and obliged to take a message from Charlotte, one of Rebecca's family. He was given the choice of having his home burnt down within ten minutes, or of agreeing not to take into his possession fields recently bought 'as it was contrary to Becca's law that he, as a clergyman, should hold any lands'. He took the second option, and, in characteristic fashion, a gunman returned the next day and left a reminder in the stable door. Like another property-seeker, the Reverend David Prothero, whom we have already met, and unlike 'the pious example' of the Reverend John Jones of Nevern, the Llansadwrn clergyman had 'fallen astray far and wide' from the 'great principles of Christian virtue'.[120]

Several of the attacks which were regarded as pointless may well have been concerned with the tithes or church rates. Two men who received overtures about the former were the Reverend John Williams of Llandybie and the Reverend David Lewis of Cilwen in the parish of Abernant. Eventually both proved conciliatory, although smoke from their ricks and the sight of white-clothed gunmen around the house must have been a reminder of their vulnerability. Another who wanted the military close by was the Reverend Benjamin Lewes of Dyffryn, vicar of Cilrhedyn, and one of the architects of the new Poor Law in the district. He was reviled in poetry and letters for this work. He was, as we noted in Chapter 3, already suffering from an anxiety neurosis, but in the case of the Reverend Isaac Hughes, vicar of Llandyfriog, on the other side of the Teifi, the fears for himself and his property were justified. The Rebeccaites pulled down one of his cottages, stripped a field of barley, and maimed a valuable horse.[121]

Hughes was a strong critic of the popular movement, and some of the local population later got their own back on his family by accusing his son of participating in Rebecca-like activities. This vicar seems to have been a target for the anti-tithe feeling along the border of south Cardiganshire, and in this he was at one with the vicar of Llangrannog, Eleazar Evans. Threatening letters condemned the latter for his arrogance, for his 'large' tithe claims, and for taking the people's money on false pretences. Contributions had been made to a building in the upper part of the parish, which, it was believed, was to be a 'free school'. Evans got the bishop to license it for divine service, and was immediately denounced by the Dissenters. Rebecca added her pennyworth: 'I with 500 or 600 of my daughters will come and visit you . . .' So many threats were made against him that every night his home was barricaded, and help was sought from the magistrates. His wife became seriously ill, and the farm and stock had to be sold. On the day of the sale, his main opponent—'delegated . . . by Rebecca'—played a satisfying part.[122]

No one knows how many landowners suffered similar humiliation. The *Times* reporter, in August 1843, spliced his account of the turnpike riots by a note that a dozen gentlemen's homes had been attacked, and this number was to rise sharply before the year was out.[123] Those gentlemen with the proudest reputations were often the last to admit that they had been visited by Rebecca, or dismissed her actions as insignificant acts of private malice. The poisoning of the large fishponds at Lord Cawdor's Stackpole Court might have been an example of the latter, but the firing of wheat mows on Lord Dynevor's home farm was not.[124] Nor was the burning of the corn and hay at Dolhaidd, the mansion of James R. Lewes Lloyd. He was the first chairman of the Newcastle Emlyn Poor Law union, and was noted for his direct language. In the spring of 1842 he went to the south of France, where the weather suited the health of a young member of the family, and, when he returned briefly for a sale of property, he immediately angered Rebecca. According to one of the threatening letters which he received, Lloyd's failings were his non-residence, his keen eye for land-acquisition, and his forthright comments on the troubles. At a public meeting he said that he was appalled to see troops in his homeland and called on farmers to unite against the evil in their midst. When he was struck on the head by a stone one Sunday, he told the church-goers that he was not afraid to use the pistols which he carried.[125]

During 1843 there was constant conflict between the richer members of Welsh society, who felt that they had a right to personal protection, and the government, which did not want its troops becoming private bodyguards. 'They must not expect', wrote the Home Secretary to the duke of Wellington on 24 September 1843, 'every stackyard, every tollhouse and every plantation to be guarded by soldiers.' The exit of so many landowners during the troubles was a tribute to the government's determination. 'Those who have bullied them [the people],' exclaimed Graham, 'are the first to run away.' Not all did, for Baron de Rutzen, and others like him, stayed and fought, and threatened to sue the authorities for injuries done to themselves and their property.[126]

Some requests for assistance were taken seriously, and small parties of soldiers and policemen were billeted in mansions for a brief while. Amongst the beneficiaries were Saunders of Glanrhydw, Vaughan of Llangoedmore, Owen of Cwmgloyne, and Gower of Castle Malgwyn. After a visit from a mob, and an obligatory donation to their cause, Gower managed to turn Castle Malgwyn into a military fortress. Of the rest of the frightened men, Howell Davies of Cynwyl and Edward Adams of Middleton Hall had perhaps the most justifiable complaints of neglect against the magistrates. The former was rather careless in his dealings with tenants and female servants, and his views of Rebecca were common knowledge. He was a determined prosecutor of criminals and urged his neighbours to be the same. In May 1843 one of his cottages was set on fire, and several more were saved, but two months later his large stacks and ricks succumbed to the arsonist. This was followed by at least two further attacks on his property, one of which we have already noted.[127]

Edward Adams was a more formidable opponent of Rebecca. His radical politics and his support for Dissenters' causes had apparently endeared him to the population of northern Carmarthenshire, and he fought harder than any one for concessions from the Kidwelly turnpike trust. He was, therefore, as he kept telling everyone, the people's natural ally, but his Achilles' heel was his firm belief in the rights of landowners. Despite petitions and threatening letters he was not ready to reduce rents which were termed 'shamefully high' by Thomas Cooke, his own agent. Nor was he willing to discharge his agent, steward, and other unpopular members of staff. On 18 July 1843 a mob visited the hall, and four days later smashed every window in the bailiff's house. He had been helping with distraints for rent arrears. Other incidents followed, and late in August Thomas Cooke received a notice to quit the country.

As he kept three loaded pistols, two swords, and a dagger within reach, he relished the prospect of a fight, and when nothing happened he wrote to his mother in a tone of confidence. On 12 September the attack came, when his master was away at the Quarter Sessions. The woods were suddenly full of armed men, and two huge fires were started close to the hall. Even the plugs in the large fishponds were removed to prevent water being used to fight the blaze. An express was sent to Carmarthen, and, after a cruel delay, soldiers arrived and were accommodated in an empty house near the mansion. Adams, having gathered up the valuables, left the place, and Cooke's family, too, had an extended stay in Ferryside. Before returning to France the landowner called all his tenants together, and in a voice which could be heard a quarter of a mile away, condemned Rebecca, supported his agent, and offered to take their farms back into his possession.[128]

Although Adams was as defiant of Rebeccaism as anyone in the three counties, his flight from Middleton Hall suggested that he, too, was apprehensive about the immediate outcome of the outrages. His agent, a cool character, writing on 23 July said that 'the country is in a state of all but open rebellion'. As the 'war against private property' became more serious in the weeks that followed, Cooke's words were repeated by many observers. Dr Walter D. Jones of Lancych warned that, because of the attacks on himself and others, Rebecca now appreciated the power which ordinary people could wield. The gates had almost been forgotten, he said, and, like Edward Lloyd Hall, he believed that the issue of the turnpikes was only a cover for more insurrectionary designs. Chartists and other 'bad people' had infiltrated the movement, and if they were not stopped 'all the soldiering you have at your disposal will be insufficient to prevent the occurrence of the scenes of the French Revolution'.[129]

Not everyone shared this bleak outlook, because all analyses of the seriousness of the troubles were highly subjective. For the optimists, though there were not many of them, the incidents described in this chapter were not part of a confident and well-planned Rebeccaite conspiracy, but deviant behaviour by small sections of a dying popular movement or just individual acts of malice. The argument continued along these lines: the respectable mob, unwilling and unable to function once the soldiers had arrived, had been replaced by the arsonist and the hired gunman. There were, it was alleged, 'new Beccas' about who had no connection with the turnpike rioters, and did not have their mass support.[130] According to Edward Lloyd Hall, who wanted to play a leading role in the peaceful removal of the Llechryd weir, many of those present at its destruction were reluctant Rebeccaites. Some of them had been dragged from screaming wives who were convinced that the soldiers would kill them. These divisions within the movement were, in the opinion of the Newcastle Emlyn attorney, signs of internal tension and weakness, and were to be encouraged by sound policing, conciliatory landowners, and peaceful meetings. The newspapers certainly did their bit, suggesting that those responsible for impressment, fire-raising, and murderous attacks were not even 'distant relations to "Rebecca"', and could not have been Welsh people.[131]

The activities of John Jones (Shoni Sgubor Fawr) and David Davies (Dai Cantwr) were used to illustrate this theme, and some people forgot, or chose not to recall, that from the earliest days Rebecca's daughters had collected recruits and money in the countryside, and indulged in acts of arson, vandalism, animal maiming, and brutality. Shoni and Dai did not therefore, as several writers state, introduce these things into the movement; rather they brought them to a fine art. We know a good deal about these two single men in their early thirties because they were, unlike most Rebeccaites, extremely talkative. They were both natives of Glamorgan, had worked in a variety of jobs, and might have played a minor role in the industrial troubles and the Newport rising. Shoni was notorious for his drinking achievements and his formidable fists, but Dai was a quieter person, less violent than his enemies claimed. When exactly the two men arrived in Carmarthenshire is not known, but they were first identified as Rebeccaites in August 1843. One of their favourite meeting places was the Stag and Pheasant at Five Roads. There they discussed and planned activities in the company of a 'committee' of Thomas Phillips, farmer of Topsail, Pembrey, David Thomas, the 25-year-old son of the tenant of Cilferi-uchaf, 'the reverend' David Jones, a shady and shadowy figure, and others from the surrounding farming and mining communities. These men joined the large Rebecca mobs that destroyed turnpike property at Porthyrhyd and attacked John Edwards of Gelliwernen, and they also formed part of smaller groups that intimidated John P. Luckcraft, the harbour-master of Llanelli, Mrs Slocombe, the wife of the manager of the Gwendraeth ironworks, John Evans of Gellyglyd, Llannon, and James Banning, William Chambers's steward.[132]

For Shoni the Rebecca movement provided him with an opportunity to indulge his passion for violence. William Chambers said that he 'ran riot over that part of the country... He was the most despotic governor there they ever had.'[133] His activities were financed by farmers and others about the district, who later claimed that they had been too terrified to refuse him anything. Apparently Shoni and his closest friends could earn 2s. 6d. or even 5s. a night for intimidating constables or burning William Chambers's property. They were accused of adding an unwelcome dash of self-interest and private enterprise to the people's movement. But they were not, as some implied, the first to receive such payments, nor were they at all unusual in gathering up farmers and labourers on their expeditions or in drinking into the late hours after an evening's work.

Forms of impressment and levying were common in other districts of the three counties. One gang in mid-August 1843 called on John James, the 57-year-old tenant of John Lloyd Davies of Alltyrodyn, pointed guns at his chest, and gave him a quarter of an hour to raise £10. Others, too, mixed business with pleasurable protest. David Thomas, who had most to gain from squeezing legacy money out of Daniel Harris of Llanfihangel-ar-arth on the night of 11 August 1843, promised, said the prosecuting counsel, a share to members of the gang. We know that money collected by Rebeccaites from door to door, and by armed raids, was used to buy guns, and to pay labourers and innkeepers, but it seems likely that some of it just disappeared. In 1849 the authorities were still trying to catch Henry Jones and William Harries, farm workers, who were accused of taking money 'on behalf of Becca' from a box of Thomas Pugh in Llangadog parish during the riots.[134]

The Rebecca troubles offered criminals, vagrants, and the

needy a chance to advance themselves in the people's cause. After a small wave of determined begging by gangs of vagrants in northern Cardiganshire, a correspondent in the *Welshman* declared that 'a new danger arises from the number of trampers that are prowling about the neighbourhood who call themselves the children of Rebecca'. These 'sham Rebeccaites' included John Francis of Merthyr and his vagrant friends, whose burglaries in the winter of 1843–4 probably had nothing to do with the popular movement.[135] There were others, too, who made the unlikely claim that their activities were approved by Rebecca. One of these was David Howell, who had been committed for a felony about seventeen years previously. He had been a miner at Ebbw Vale, and played a leading role in the Newport rising of 1839, but he was now back in his native Carmarthenshire. Together with at least four other men he was accused of forcing his way into the Bwlchytomlyd home of David Thomas, a farmer and tax-gatherer. While two of them held swords over Thomas and his wife, the others found the money bag, and recovered land-tax payments. A comparable case occurred on Christmas Eve at Llanwrda, when two men with blackened faces and worsted caps on their heads called, with others, at the house of William Williams and refused bread, cheese, and a warm fire in favour of £5, a sum that was often requested during the Rebecca period. Incidentally, those committed for these two burglary offences were acquitted, which indicated perhaps that the victims of these crimes were themselves not above settling old scores and otherwise profiting from the troubles.[136]

Inevitably, there were cases of malice or vengeance which also had no connection with the Rebecca movement. Threatening letters were sent, as they always had been, by individuals who had their own particular grievances. Thomas Richards, who had an unsavoury reputation, was found guilty of this crime in December 1843; for some undisclosed reason he harassed John Irwin, proprietor of a steam packet which plied between Llanelli and Bristol. Others found guilty of a similar offence were Martha John, the angry mother of a bastard child, and Bridget Williams, the dispossessed pauper whom we have met before. Although the juries refused to believe them, it is interesting that these two poor women claimed that the letters, with their Rebeccaite language, had been brought to their doors with instructions about their delivery.[137] This was common practice. Perhaps there was a small and lucrative trade in the drafting of literary threats on behalf of aggrieved parties.

Detached observers insisted that arson and murder were by their very nature individualist crimes, and formed no part of the popular movement. It was true that the figure of Rebecca was not always present at the incendiary attacks on farms, and the first suspect was usually the tenant with a grievance. Fire-raising, and attempts to kill landowners and magistrates, required few actors, and there were people in authority during the last period of rioting who saw advantage in absolving Rebecca from the responsibility for one or two of the worst incidents. Thus a petition congratulating the Reverend James W. James on his recovery from his arm wound assured him that it was the work of one of two assassins and not genuine Rebeccaites.[138] The clergyman thought otherwise, and it would be naïve to assume that all the activities described in this chapter were random and unorganized.

The most detailed reports and confessions confirm that many of the attacks on property and people were the result of decisions taken either at small committee or larger open meetings. There was evidence that the countryside was divided into 'unions' of parishes, where prominent Rebeccaites kept an eye on 'every injustice'. Their committees discussed questions of rents, tolls, rates, land-holding, breaches of public morality, and other matters. They almost certainly called the night meetings, the noise of which terrified the respectable along the Teifi valley and near Aberaeron and Pontyberem. Meetings on the hills around Capel Iwan passed resolutions concerning changes in property ownership in the area and about popular resistance to bailiffs who came to execute warrants. Not everything was agreed; calls by individuals for support in private battles over land and money were sometimes ruled out of order, and at a Llangendeirne assembly a demand for direct action was postponed while shopkeepers and publicans were given time to lower their prices. When it was decided to proceed, the noise of horns and gunfire was heard through the night, and people were deputed to carry out the attack. The communal and planned nature of much of the action described in this chapter needs to be stressed; even the incendiarism was not the same as the secretive individual revenge of a sacked East Anglian labourer or the instinctive spiteful gesture of a rejected vagrant.[139]

The parties which carried out these raids usually comprised about forty people, but there were spectacular exceptions. The attack on Gelliwernen was made by three large units, which met near the scene of the crime. On this occasion the people had horns, rockets, and a variety of other weapons, but the smaller mobs had only guns and torches. When they arrived at the target, the gunmen often placed themselves in a circle around the person or the property under threat. Given the amount of firing at such times it was a miracle that more people were not injured, but some of the men had powder only and Rebecca herself did not want deaths on her hands. Half a dozen bailiffs, keepers, and farmers claimed that they were saved from possible death by the intervention of the leading figure in the mob. This was no consolation to Benjamin Jones's victims, nor to the Reverend James James, but it showed that Rebeccaism had its limits.[140] On rare occasions the Redresser even preferred compromise and conversation to direct action, and there were debtors, unmarried mothers, and other harassed individuals who persuaded her not to take up their grievances.

Customs and ritual gave Rebeccaism a degree of legitimacy. Certain forms of action were an extension of the *ceffyl pren*, and nowhere was this more apparent than on the southern Cardiganshire coast. Before attacks on tithe collectors and other unpopular characters, their merits and demerits were discussed, and the actor who played their part was sold to 'the Devil'. In visits to wife-beaters and bailiffs there was, too, a public account of their wrongs before humiliation, and the removal of clothes and ordeal by water. Even on the smallest errands a form of disguise and ritual was adopted. Gunmen had hoods over the faces or tied handkerchiefs across their noses. Not all the members of a group were disguised, but the leaders had some of the regalia associated with the gate-breakers. The visits to the wealthiest of the farmers, agents, and fathers of bastards were splendid affairs; the lustrous carriage, the bedecked Lady, and the black footman brought an irreverent sense of aristocratic decorum to the proceedings. The colours of the people's costumes were varied; the traditional white being intermixed with a few greys, browns, blacks, blues, and reds. Even the horses ridden by the Redresser were black and brown as well as white. Red was popular in certain areas; red coats, red cravats, and the red of real blood. David Davies, forgetting the Merthyr insurrection of 1831, told some of his colleagues one evening near Llanelli that he was unfamiliar with the ritual of washing hands in the gore of sacrificed animals.[141]

The Redresser was always anxious to give an air of authority to her work. Each mother had to swear formally to the identity of her illegitimate child, and each father had to kiss it as a sign that he accepted his responsibility. Other victims had to make sacred promises before the act of vengeance was complete, with hands on the best bayonet, or preferably on a Welsh Bible. John Evans, sheriff's officer at Tyrypound, was forced to face the fireplace and say, with the Bible in his hands: 'As the Lord liveth, and my soul liveth, I will never come here to make any distress again.' The religious tone was appropriate, for one or two midnight adventures began with 'a sort of prayer calling upon God to assist them', and behind this Rebeccaism were mental images of the Old Testament God and the Great Reaper. Threatening letters reminded people of the fate of God's enemies. The vicar of Llangrannog was referred to Judges 6: 27 and 28, where there is an account of a grove set on fire and bullocks sacrificed.[142]

This hint of religious fundamentalism—'the spirit of Cromwell'—combined with a perceived threat to property rights, convinced many people that Rebeccaism was a serious danger to the established order. This was reinforced by a growing belief that the movement was 'being taken up by a different set of men, and for different objects'. 'The farmers in the first instance waged a war of extermination against turnpikes and toll houses, to relieve of the tolls, and this they could not do without the assistance and co-operation of their labourers and farm servants', said the *Carmarthen Journal* on 1 September 1843. 'It now appears that the labourers and farm servants have discovered that they have class interests distinct from their employers . . .' The evidence for this was the resolutions in favour of better rates of pay, more food, and other allowances passed at a handful of August and September meetings in the parishes about Bettws Evan, Newcastle Emlyn, Abernant, and Llannon.[143] Colonel Love, taking up a well-worn theme, told the Home Secretary that such class divisions in the movement would become more obvious during Michaelmas hiring time and the difficult winter months. To encourage this, a small number of landlords and churchmen pointed out to the labourers that the greatest protesters over tithes and rents were inclined to be the worst masters. It was a dangerous game to play, and not very successful. One is struck by the degree of collaboration within the Rebecca movement; it provided something for everyone, and even when farm workers and colliers held their own meetings in south-east Carmarthenshire late in 1843, their agenda was not that different from the farmers. At one Pembrey meeting they resolved to meet the latter to draft a joint petition.[144]

Some observers, anticipating the break-up of the movement about this time, argued that industrial workers were gradually withdrawing their support. Now, it was said, if they used Rebecca's methods it was only for their own sectional ends. There is a degree of truth in this, but not the whole truth. The colliers were able to serve Rebecca and themselves. The context was the growth in unemployment and reductions in wages, which led to a storm of protest against unpopular managers, foremen, and their truck shops. In July 1843 a large mob, some in the usual disguise, fired into the imposing home of John Thomas of Cwmmawr, one of the oldest industrialists in the area, and left a note commanding him to dismiss an employee and promising vengeance on his son, who had sent the dragoons after Rebecca. Two months later they reappeared at the Gwendraeth ironworks of Charles Newman, terrified the armed guards, and warned his lodger, and managing clerk, to leave the country. James Slocombe, the clerk, had become unpopular partly because he had acquired secret information about the Rebeccaites. On 28 October there was another attack at the same place, after soldiers and policemen had been attracted away from Pontyberem. Many of these incidents were therefore not simply industrial disputes: the 'elegant Lady' did not involve herself greatly in these.[145]

It was perhaps logical to suggest, as many did, that fewer farmers were involved in the forms of Rebeccaism described in this chapter than in the destruction of turnpike gates. Yet farmers undoubtedly organized some of these activities, and had much to gain by them. In fact, the occupational breakdown of a sample of 83 persons arrested for these crimes is very similar to that of 86 arrested for offences against the gates. About a quarter of the former were farmers, another quarter were farm labourers, and almost a tenth were farm servants. The main difference between the two groups were these: the people taken into custody for Rebeccaism were on average a few years older than the people in the last chapter, more of them were female, and there were more outsiders and known criminals in the list. Although the samples are small, it seems that fewer men with large holdings were involved in assaults and the destruction of property than in the attacks on the turnpike gates. It is tempting to make too much of all this. Those who thought that there was a complete separation between these Rebeccaites and the turnpike mobs were wrong; at Cilrhedyn, Cilycwm, Porthyrhyd, and many other places, crowds walked for miles behind their Becca, and in the same evening destroyed trust property, knocked up unpopular clergymen, and fired ricks.

If there were tensions and divisions within the Rebecca movement, these were not always along occupational or class lines. Some of the leading militants and some of the greatest doubters were farmers, both freehold and tenant. There was resentment amongst the waverers when direct action pre-empted the possibility of successful peaceful bargaining, and when arson was used to remove private grievances which few of them shared. Edward Lloyd Hall, predictably, gave one such example, at Gilfach Dafydd, to the north of his Newcastle Emlyn home. During the evening which he described two farmers' sons, armed and disguised, ordered neighbours from their beds until, when they reached the appointed spot, the crowd was 200 strong and completely ignorant of the purpose of the visit. When they were told that they had to fire a haggard, some of the farmers present refused and the meeting broke up in chaos.[146] Other attacks were aborted, possibly for the same reason.

As the destruction of private property gathered momentum the readers of newspapers all over Britain were assured that the most respectable supporters of Rebecca now deeply regretted their actions. The Reverend Henry Lewis Davies, curate of Troedyraur, claimed that as soon as the farmers felt threatened by events in the movement they decided to bring it to an end.[147] As Christmas drew nearer, so more ratepayers came forward to pledge their loyalty to the Queen and their willingness to be special constables. There was, one suspects, a growing fear of the effects of further nightly armed gatherings, but there was an awareness, too, of the limitations imposed upon all of them by the new policing methods. It was the latter, more than anything, maintained Hugh Williams the Chartist, which stopped Rebecca.[148]

The emergence of an effective response to Rebeccaism took months to achieve. Initially there had been an element of panic. Within hours of an attack by gunmen, fearful landowners and clergymen were packing their bags for a stay at a nearby garrisoned town, or a visit to England and the continent. Those left behind displayed the usual mixture of courage and terrified inactivity. 'The Welsh justices . . . rely too much on the soldiery', wrote the Home Secretary, 'and will do nothing without them and succeed in doing little with them.'[149] Wagner of Manordeifi wanted those in authority to give him hand grenades to lob at would-be assassins. Rees Goring Thomas wondered if the government would approve of an armed association amongst the landowners, and Edward Lloyd Hall endlessly repeated his suggestion of a volunteer cavalry force, comprising the gentry and wealthy tenant farmers. Virtually everybody was agreed that the existing machinery of law and order was incapable of dealing with Rebeccaism. Much of the anxiety to obtain professional policemen and soldiers sprang from this conviction. 'The affair hangs fire too long', Sir James Graham told the duke of Wellington on 24 September 1843. 'It must be put down speedily, or it will become dangerous and a very bad example.'[150] The blanket covering of soldiers and policemen across the three counties after October 1843 was reminiscent of the Irish scene, and was intended to stop more than gate-breaking. It was not popular; the new arrivals had to face threats and insults from the retreating Rebeccaites.

Even these well-trained men were not enough to bring stability to the region. The more informed observers, writing in the winter of 1843–4, detected 'a wicked and malevolent spirit' still at work, and were grateful that a good harvest and rising prices reduced the likelihood of 'outright rebellion'. In the spring of the following year Colonel Love felt confident enough to withdraw some detachments, but six months later noted the continuance of 'a very hostile feeling against the upper classes' in the Teifi valley.[151] The *Carmarthen Journal*, speaking for the conservative forces in society, believed that it was now important for every respectable person to declare his or her unqualified hostility to the Rebecca cause, and for the legal arm of government to support them by making an example of the committed offenders. The attempt to detach certain social groups from the popular movement had begun. Newspapers kept before their readers the horrifying images of the mob on the rampage at Carmarthen and of the lifeless body of Thomas Thomas of Pantycerrig, and they welcomed the holding of the large open meetings which we shall be discussing in the next chapter. The county leaders who attended these meetings gave this consistent message: those who wanted grievances redressed should first make a commitment to peaceful forms of action. For their part the government stressed the urgency of catching 'some of the principal offenders before long . . . [for] some examples of severe punishment are necessary'.[152]

The treatment of those taken into custody for these offences met with the approval of the *Carmarthen Journal*. There were over 100 such people, and a half of them were convicted. With a few exceptions, those found guilty of arson, burglary, and wounding with intent at the higher courts were punished more severely than the gate-breakers. The government prosecutor was prepared to make deals with many of the latter, and they received more sympathy from the media and philanthropic gentlemen. Shoni Sgubor Fawr, convicted of shooting at a female publican, deserved, and got, a worse reception, and 28-year-old Benjamin Jones, who attacked the home of Thomas Thomas of Pantycerrig, and then battered the main witness, was chased and prosecuted with unrelenting zeal. Both men were transported for life. Other offenders, found guilty of similar but lesser crimes, were surprised by the judges' verdicts. When the seven men accused of demanding money from Daniel Harris got ten years' transportation, and their organizer twenty years, there was pandemonium in the court and the men in the dock had to be forcibly restrained. Only the people accused of sending threatening letters could afford a smile. The comparative leniency of their sentences was, it was said, an admission by juries and judges that they were ignorant rustics, confused and manipulated perhaps by others. Even the normally unforgiving *Carmarthen Journal* felt that some unnamed persons bore a heavy responsibility for debauching 'the simple and primitive character of [such] an industrious, quiet and professingly religious people'.[153]

7
A Slow Death

ALTHOUGH a few people accepted that the Rebecca riots would have been impossible without a pre-history of radical ideas and leadership, there were many more who proclaimed that the movement itself transformed the innocent Welsh peasantry into violent criminals and seditious rebels. The *Carmarthen Journal* warned its readers in June 1843 that dangerous notions were spreading across the countryside, and there was worse to follow. Before the end of the year it was said that nowhere else in Britain were farmers and servants so critical of their social betters and the church establishment, and so politically advanced. *Yr Haul*, for the clerical conservatives, reminded people of the days of the French Revolution. Only, some argued, the lack of leadership of the type given by Daniel O'Connell in Ireland kept the people within bounds. Thomas Lloyd, of Bronwydd, a member of one of the oldest gentry families, said that he found it almost impossible to believe the changes that were taking place in the Welsh psychology. William E. Powell, MP and lord-lieutenant of Cardiganshire, told the education commissioners in 1846 that 'the people are much disimproved since the Rebecca riots, which have tended to engender a spirit of dissatisfaction'.[1]

In the next quarter of a century, as the impact of Rebecca ran its course, so the process continued. Outwardly little changed over much of the region. In south-east Carmarthenshire there was spectacular growth in the coal industry, and Llanelli doubled in size between 1841 and 1871, matching its rival and larger neighbour Swansea. Elsewhere, the coal, iron, lead, copper, and other industries stagnated or grew only slowly. In Cardiganshire and Pembrokeshire no great commercial or industrial developments affected the balance of wealth and power; if anything, the hold of the large landowning families was tightened.[2] As the agricultural economy entered another golden age, the physical and ideological threats to landlordism seemed to disappear completely. It was many years before the big estates succumbed to the economics of depression, and before electoral politics passed out of their control. At first glance, paternalism, dependence, and deference were as strong as ever. This region, like every other, was said to have settled down to a mid-Victorian stability and respectability. Yet at the lowest level, there was movement and change throughout this later period; thousands fled the land, Rebecca still presided fitfully over illegal protests, and, of most importance, political education never stopped. The election of 1868, with all its triumphs and failures, was an indication that in this isolated corner of Britain social and political attitudes were different from those described by travellers two generations before.

IDEAS

The epithets most used about the 'lower orders' of south-west Wales before the outbreak of the troubles were 'apathetic', 'backward', 'simple', and 'religious'. It was customary to contrast these 'simple agricultural people' with the 'violent Chartist politicians' of the south Wales coalfield. The *Welshman* was saddened by their lack of interest in secular issues, and some radical leaders bemoaned their apparent indifference to reform. Hugh Williams declared that the people of the country were 'peculiarly inoffensive and quiet—even bordering on apathy'.[3] Prime Minister Peel and James Graham told the nation that the agrarian disturbances had no connection with politics, and Thomas Campbell Foster, as befitted a *Times* correspondent, obtained an assurance that the people's grievances were economic and social.[4] He preferred Rebecca, he said, to the female Chartist Mary Ann Walker.

This analysis fitted in well with the claims of landowners that they had been utterly astonished by the 'vindictive and dangerous feelings' of the local population, and that only 'outsiders' and 'demagogues' could have produced such a change. Here were the origins of the indoctrination theory that was to be proffered at various times during the nineteenth century. William Ferrand, the Young Englander, suggested that the peasantry of west Wales had been manipulated by Chartists or Anti-Corn Law Leaguers, and Colonel Love, echoing these sentiments, claimed that on their own the Rebeccaites were incapable of widening a narrow debate about tolls. Others placed the responsibility on the *Times* correspondent himself, or on Tory populists, men like Lewis Evans of Pantycendy, who called on farmers to unite, stirred up feelings against the Poor Law, and invoked God's wrath on new breeds of landowners, employers, centralizing bureaucrats, and 'cursed machinery'. But perhaps the greatest blame was reserved for the Dissenting ministry. Anglican propagandists documented the power which these rival preachers seemed to have over their congregations.

Of all the explanations for the troubles of this period, the manipulation theory is the least convincing. Radical papers in London regretted the lack of interest which their own agitators had taken in south-west Wales, and accusations that certain landowners and ministers of religion were behind the movement were never proved. What was outstanding about the disaffection was its local origins, its insularity, and the way in which ordinary people played their part. The last aspect was especially interesting because reporters, like education commissioners some years later, sometimes mistook the taciturnity of Welsh-speaking farmers for a want of independence and ideas. If some of them were quiet, there were many reasons why people chose to remain silent in Rebecca's country, but when they did open their mouths in public they were impressively articulate. Farmers Michael Thomas and Thomas Evans, Thomas Williams, the blacksmith, Thomas Davies, the currier, and William Thomas of Rhosfawr, Llannon, received prolonged applause at protest meetings. 'He had no talent to speak, neither did he profess much understanding, but he had experience to guide him,' the last of these told a large crowd at Llyn Llech Owen, near Gorslas, Carmarthenshire. 'He was willing to be reduced to poverty by Providence, but not through the iniquity of man . . . He had determined not to vote again to please his landlord or any other person.'[5]

The violence that we have chronicled was accompanied by an outburst of popular emotion and ideas. Many people, including Caleb Morris, a prominent London minister, and writers in the *Nonconformist* and *Y Diwygiwr*, welcomed this 'popular effervescence', the end to 'numbness and inactivity in civil matters'.[6] The reporter for the *Swansea Journal*, who had both the language and ear of the rioters, insisted that the ideological development of ordinary people in the second quarter of the century was the most exciting revelation of the Rebecca movement.

> Up to this period we have been accustomed to consider the Welsh people as being politically apathetic. These meetings convince us not only that such is the state of things no longer, but that, judging from the sentiments embodied in the speeches ... this [growing strength of] feeling, though called forth more immediately by recent occurrences, is not altogether of yesterday's date. Many of the speeches evinced a degree of intelligence and singleness of purpose as to gratify us; and we cannot but regret that means are not more generally adopted by ... the 'higher classes' to conciliate and acquire the confidence of those hardy sons of toil.[7]

As the reporter noted, feelings of independence had been present before the riots. The parish was one level of education and conflict; throughout the second quarter of the century one issue after another had divided ratepayers. In the 1830s, as we saw in Chapters 2 and 3, matters came to a head. The culmination was the assault on parish autonomy, and parish leaders, by the centralizing governments of the 1830s; the intention of the Whig ministers was to pass control over policing, the Poor Law, and other important areas of life to a different élite, with novel English bureaucratic support. This only confirmed the doubts which some people had always entertained about the Whigs, but for other reformers it was a great disappointment. Few regions had been more exhilarated by the Reform crisis than south-west Wales. Farmers about St Clears and Llansadwrn had petitioned for a wider franchise, and Dissenters pressed for a change in Church–State relations. Some of the people whom we have met during the Rebecca story attended the large public meetings in 1830–2, and participated in the violent hostility shown towards the opponents of the Reform Bill, men like Sir John Owen, George Rice Trevor, John Edward Saunders, and Rees Goring Thomas.[8]

Nor did it end there. The split in the gentry over Reform, and the subsequent elections of 1835 and 1837 in Carmarthenshire, prolonged the debate over taxes, tithes, and church rates. Meanwhile, within months of the passing of the Reform Act, there were also calls in the region for the ballot and universal suffrage. Five years later Hugh Williams established the first Working Men's Association in Wales at Carmarthen. With his colleague William Jenkins he toured the south-west, holding meetings and taking signatures. By the summer of 1839 hundreds were enrolled as Chartists at Aberaeron, Cardigan, Llandysul, Newcastle Emlyn, Fishguard, Haverfordwest, and Narberth, and also in some of the smaller village communities. The movement, which shocked James Lewes Lloyd of Dolhaidd and his friends, soon collapsed, but it drew a number of tradesmen, craftsmen, and labourers into political discussions, and their voice was heard again in 1843.[9]

Not everyone was interested in national politics; some were preoccupied with local religious and educational battles, others had strong ideas on the changing nature and cost of land-holding, and many were probably indifferent to anything but making a living. Of the very poorest members of this society we know all too little. One should not be surprised therefore at the variety of views within the Rebecca movement; David Rees, who tried to claim the movement for the advanced Whigs, or Lewis Evans, who did the same for the Tory radicals, retired defeated. One popular analysis was the division of the Rebeccaites into the Chartists, who led, and the non-political, who followed. It was an interpretation that appealed to the conspiratorial, and had an attraction for early labour historians, but it was wide of the mark.

The secrecy that surrounded the Rebecca movement makes it difficult to know precisely what the participants thought and said. The threatening letters provide some clues, as do the lists of grievances and petitions that accompanied the riots. Petitions were sent to landowners, titheowners, magistrates, turnpike trusts, government commissioners, Parliament, and the Queen. The most famous petitions were adopted at the mass meetings which have been such a neglected feature of the movement. These were not as large as the great Irish demonstrations of 1843, but they were impressive enough. 'It is so large an assemblage,' said John Lloyd Davies of an October meeting of 2,000 near Lampeter, 'that it is alarming, I think, in this country, where I never saw 200 assembled in my life for any such purpose.'[10] There were more than seventy-five such gatherings in the second half of 1843, and their geography closely resembles that of the non-tollgate outrages. Many were held within a few miles of Newcastle Emlyn, Cynwyl Elfed, Llangadog, Llannon, and Narberth. They were a feature of the late summer and early autumn months, though it is likely, because of the lack of reporting, that we have underestimated the number of meetings held before that time. From the first there were vague accounts of night meetings which both planned action and gave vent to popular complaints, and one or two of these, and especially the large daytime assemblies, could be compared with rallies in other parts of Britain. Although parish meetings were not uncommon in the Welsh countryside, these particular gatherings were special, and were labelled, in some instances, 'the first political meeting' of the district.[11]

The first gathering about which we have detailed knowledge was that of 300 farmers and servants at Cwmifor, near Llandeilo, on 20 July 1843. One speaker called it a 'Convention . . . held in the first year of Rebecca's exploits', and the presence of Thomas Campbell Foster at this meeting ensured it maximum publicity.[12] During the next months there were four main types of public meetings: the first to protest about tolls and related issues, the second to form farmers' committees and unions to negotiate with landowners and titheowners, the third to draw up petitions to Parliament and the Queen, and the fourth to appoint deputations to wait on government commissioners. Many were comparatively minor affairs, with 150–300 in attendance, held out of the rain in a barn or chapel schoolroom. The meetings began at all hours, partly because people were so cautiously slow in arriving, but midday and the evening were the most popular times.

Reporters were annoyingly vague about who attended these smaller assemblies, but a typical description was 'respectable freeholders, farmers and ratepayers'. Gentlemen, farm servants, and labourers, as well as artisans and industrial workers, were also present on occasions, and even dominated meetings. Much to the delight of political reformers, women and children were spotted amongst the crowds, especially at the largest demonstrations. There were a dozen meetings where audiences were estimated at 500–1,000 strong, a handful of even larger free-trade meetings up the Swansea valley, and, most famous of all, the huge hillside gatherings. The assemblies at Mynydd Sylen, Llyn Llech Owen, Allt Cunedda, Pencrwcybalog, Cefn Coed yr Arllwyd, Pen Das Eithin, and Llanarth were district conventions, usually held in the daytime and attended by parish delegations and several thousand people. They were planned weeks in advance; notices were posted up, petitions prepared, and speakers, gentlemen, and newspaper reporters invited.

Certain meetings, notably those called to discuss specific issues like tithe reductions or turnpike trust affairs, were dominated by landowners and clergymen. William Edwardes of Sealyham, Walter R. H. Powell of Maesgwynne, Rees Goring Thomas of Llysnewydd, Captains Lloyd, Child, and Evans, and John H. Rees of Cilymaenllwyd, all thought that it was worth attending public debates on grievances in their localities. While others of the gentry class pleaded with the government to ban all such meetings, these individuals believed that they were an effective safety valve. Altogether the gentry appeared at about a third of the meetings, even though a few gentlemen, like Lewis Lewis of Gwynfe, disappeared quickly once they had heard the tone of the speeches. Edward Lloyd Williams of Gwernant chaired the great Pencrwcybalog meeting near Newcastle Emlyn, and several others; so did William Lewis of Clynfiew at Penrhiwfach in the parish of Cilrhedyn, and Williams Chambers junior at Mynydd Sylen. In this capacity, they made valiant attempts to limit the topics of discussion. When the gentry were absent, speakers frequently condemned them for their lack of interest, and respectable farmers and Dissenting ministers took the chair. These people were invariably identified as community leaders, already prominent in parish and Poor Law Board meetings. Amongst them was the Baptist John Morgan, who presided at two of the largest meetings.

Many of the speakers were also from this social category. Men like John Williams, another Baptist minister, from Pontarddulais, spoke at several meetings, as did Griffith Thomas of Pontyberem, while 'substantial farmers' John Lewis and John Rees prepared their own petitions for public approval in the vicinity of Llangadog and Lampeter.[13] Yet it would be wrong to assume from this, as the conservative journals did,

4. *Great Meeting on Mynydd Sylen* (*Illustrated London News*, 1843)

A Slow Death 327

that the tyranny of the 'parish gentry' had simply replaced that of the honourable variety, because important contributions were made at the meetings by small farmers, shopkeepers, schoolteachers, carpenters, colliers, labourers, and others. They spoke in Welsh, and the English correspondents who heard them had interpreters by their side. Sadly the most poorly reported meetings were those where farm servants, labourers, and colliers discussed their own particular problems, but it seems likely that their spokesmen also contributed to the more open assemblies. The largest gatherings were representative of a cross-section of public opinion. When William Edwardes of Sealyham said at a Wolfscastle meeting that only the principal farmers should be present on such occasions, he was received with laughter.[14] 'Everyone is here' said one delighted contributor, and no convention was complete without its loud youth section, and its jeering Rebecca supporters.

The range of subjects discussed at these meetings was very wide, the matter of tolls rarely dominating the debates. After two or four hours, the assemblies agreed lists of resolutions, along the lines of those passed at Tre-lech on 25 September. Changes were demanded there in the turnpike tolls, the new Poor Law, the Tithe Commutation Act, the county stock, magistrates and legal fees, land tenure and rents, trade restrictions, and parliamentary representation. In some places, especially if magistrates or landowners were present, there were attempts to restrict the discussions, and proposals on land tenure, the regulation of rents, the relationship between Church and State, vote by ballot, and extension of the suffrage were sometimes ruled out of order. Such measures, said Edward Lloyd Williams to a large Llechryd meeting, 'would create confusion, and . . . were against the sense of the country'.[15] Naturally, the order of priorities varied from district to district; in Pembrokeshire, and in parts of southern Cardiganshire and western Carmarthenshire, many of the speeches were concerned with the tithe and Poor Law, while the Corn Laws were the focus of attention along the eastern border of the last county.

Behind the speeches was not one but several ideologies of protest. Conservative voices, even amongst Dissenters, could be heard at some gatherings, reminding people of their place in the order of things and of the wealth and influence in the hands of great men. Society was, they insisted, organic and interdependent, and gentry, farmers, and labourers must 'pull together in the shafts'. 'He freely admitted,' said John Rees of Llanarth, 'that it was ordained by Divine Providence that there should be gentlemen and poor men ... He called on them to join in petitioning for a reduction of rents, and to leave the Llechryd weir, vote by ballot, and all that sort of thing ...' Lewis Evans, who had been in Ireland during its worst troubles, warned people of the dangers of setting tenants and landlords 'at variance'.[16]

For Rees, Lewis Evans, the Gwynfe farmer Michael Thomas, John Williams, a prosperous freehold farmer of the Llangyfelach area, Mr James, a Unitarian minister of Gellionnen, Pontardawe, and other traditionalists, 'Reform' was a dreadful word. The first Reform Act, and the Whig Poor Law Amendment Act, Tithe Commutation Act, and County Police Act had brought them nothing but sorrow. The Rural Police was a favourite topic at public meetings; 'if the great folks feel desirous of having them, let them pay for them', was the spirit of the speeches.[17] The support which the gentry had given to the reform of the Poor Law was, for these Tory radicals, the hardest pill to swallow. At the Llechryd meeting, where the gentry were at their most numerous and conciliatory, expressions of distaste for aspects of the Act, and promises to remove the weir and take on more labourers, had the required result. Nocturnal outrages were condemned, and three gentlemen, Thomas Lloyd, Abel L. Gower, and Edward Lloyd Williams, were carried from the meeting by a grateful population.

Despite the prodigious efforts of Lewis Evans and others, most meetings were infected by the 'disease of Liberalism'. It had been incubating for some time. Attacks on the politics and efficiency of aristocratic government, and on class legislation like the Corn Laws, had been the staple diet of Dissenting periodicals and reforming speeches for well over a decade. William Chambers, James Child, and a few of the other chairmen at the meetings of 1843 were themselves political economists, and many of the leading speakers and petition writers were prominent radicals. The *Morning Herald* claimed that the series of meetings was little more than an ego-trip for the Reverend Thomas Emlyn Thomas of Cribyn, the reformer, scholar, and Unitarian, William Anthony of Carmarthen, the socialist shoemaker, and Hugh Williams, the Kidwelly Chartist.[18] There were other reformers, too, who made their mark at half a dozen meetings, men like farmers Stephen Evans of Cilcarw and David Gravelle of Cwmfelin, a 'leader of the dissenters', and John Jenkins, lecturer of the Anti-Corn Law League.

Of them all Hugh Williams was the most important, even though he spoke in English.[19] He made his first public appearance at the Cynwyl meeting of 16 August 1843, when he accepted an invitation to address the *Times* correspondent on behalf of the farmers. A short time later he was asked to prepare a petition for the great Mynydd Sylen meeting, and, with Evans and Gravelle's help, variations of this were presented to other gatherings. In direct contrast to the Tory radical manifesto, this programme highlighted the differences between the interests of landowners and those of farmers, and the close ties between the latter and the labourers. 'Reform' was the answer to aristocratic tyranny, reform which left the landowners with less authority, the Church with less money, the farmers with more security, and Parliament with a broader class representation.

Although petitions of this kind were passed with overwhelming majorities, it was difficult to estimate the depth of commitment to change. Landowners who were confident of their tenants' deference and their lack of unity dismissed the public debates as 'ravings of outsiders and scoundrels'. Abstract 'words ... do not frighten us', but what did concern them was the combination of ideas and action. They resented anything which interfered with the natural relationship between landlord and tenant. In the high summer of 1843 reports circulated of strange 'agrarian notions' and secret farmers' committees in all three counties, and the *Carmarthen Journal* warned of delegates interfering with landowners' rights. On the very eve of the great march on Carmarthen, meetings of Rebeccaites took up the matter of farmers' rents and taxes.[20]

LEX, an anonymous lawyer who badgered newspapers with his ideas, had long advocated a 'Farmers' Union'. This letter writer, who gave his address as Haverfordwest, believed that farmers should turn to co-operative agriculture, and argued that they had the collective power to conclude satisfactory deals with merchants, financiers, and lawyers. He wanted them to collect information on land prices and to limit the over-priced competition for land. There was a farmers' union of some kind in Pembrokeshire by the autumn of 1843, but its relationship with a later and similar body is not clear. The second, the Pembrokeshire Farmers Philanthropic and Benevolent Institution, met at Llawhaden in January 1844. Its membership was restricted to certain groups of freehold and tenant farmers, and in its sober and moral tone it had something of the character of a friendly society. Its rules obliged members to provide information on their own property and rents, and forbade them to make offers 'clandestinely' for other people's farms.[21]

The structure of this union, and some of its rules, were significantly similar to that of a group of farmers who met in a barn at Penlan, near Cwmifor, on 3 August 1843. A fortnight earlier, at the Cwmifor convention, resolutions had been passed calling for an equitable adjustment of rents and for committees to be appointed in each county to sanction farm changes. Other resolutions condemned the unbiblical coveting of a neighbour's land, and protested about the cost of borrowing and the appointment of Englishmen as estate stewards. Within a short time 200 people had joined a farmers' union in the neighbourhood, and another branch was being formed at Llansadwrn. The precise connection between these bodies and Rebecca is not clear, though she was offered honorary membership, at least one of her letters was sent from 'The Farmer's Union', and she punished people who broke its rules.

The *Times* correspondent was welcomed suspiciously into one of the union meetings, for members of the audience had doubts over the legality of the whole enterprise. Others thought that it was an excellent idea, which should have been put into practice years before: 'when they elect members of Parliament they do just as they please, and we have no voice, but here we have.' According to government sources the guiding union spirits were a freehold farmer, John Lewis, possibly the same man who spoke at the open assemblies, and a small group of militant farmers, who established contacts with well-known Chartists. An informer confirmed that Merthyr radicals were in the area at the time of Foster's visit, and were talking about common property in land, and the need to take united action in defence of farmers' rights. The latter was a prominent theme at public meetings, as one speaker after another criticized land-hungry tenants for playing the landowner's game.[22]

As we have seen, Rebecca tried to impose solidarity across the farming community, and the public meetings increasingly took an interest in the question of land tenure and the cost of holding land. Many speakers, especially in Cardiganshire, supported the idea of landlords letting only one farm to each tenant. In the opinion of farmer Evan Thomas, agents along the Teifi valley had not dealt 'rightly' with this matter. John Lewis, at the Cefn Coed yr Arllwyd meeting on 10 October 1843, also won overwhelming backing for his notions of guaranteed tenure and fair rents, while Hugh Williams proposed a land court, the regulation of rents, and the removal of taxes from those occupying land.[23] Such interference in the landlord–tenant relationship was advocated in one of Rebecca's notices, and it is a reminder that the Land Question in Wales was not completely dormant before the 1880s. The Irish *Nation* wondered if the British government responded quickly and leniently to the riots because concepts of rights to land, fixity of tenure, and moderate rents had entered the heads of people on the mainland.[24]

Another sign of the growing independence of the farming population was in the sensitive area of Church–State relations. The secular and religious voices of the establishment later claimed that even before the census of 1851 they were awakened to the weakness of the Church's position by the Rebeccaite attacks on clergymen and by the Cromwellian tenor of the speeches of 1843. When the Reverend Thomas Emlyn Thomas put forward a list of grievances at the Pen Das Eithin meeting in October 1843, he excluded reference to the difficulties of supporting a state Church, but John Rees of Llanarth immediately offered his own petition, which called for the abolition of compulsory church rates, a much reduced level of tithes, and greater rights for Dissenters. It was a moderate request, in line with other proposals at the first public meetings, but there were members of his audience who disliked the very notion of a state Church. Griffith Evans, a shopkeeper, declared at the Llanarth meeting on 20 November that he was sure that everyone was in favour of Disestablishment, and the applause for such sentiments here and elsewhere gave him hope.[25]

As we have seen, such sentiments had been heard in the reform and church-rate controversies of the 1830s, and they were strengthened by Graham's Factory Education Bill of 1843. 'I hold, sir, that Government has nothing whatever to do either with the religion or the education of the people', said one Baptist minister, at a Haverfordwest meeting against the Bill in April 1843. '. . . I am a parent, and sooner than I would allow my children to be taken from me and taught the mummeries of an apostate Church, I would burst the bonds of mortality.'[26] *Y Diwygiwr*, edited by David Rees, and Edward Miall's *Nonconformist* were only two of the more important publications demanding a separation of Church and State at this time, a prospect brought nearer by the foundation of a national Anti-State Church Association, later renamed the Liberation. In 1844 it appointed a Council for Wales, and amongst its members were ministers H. W. Jones of Carmarthen, J. Griffiths of St Davids, J. Saunders of Aberystwyth, and David Rees of Llanelli. The early impact of this association is hard to evaluate, but landowners and clergymen were worried by the prospect of co-ordination between their own Dissenting protesters and those elsewhere in Britain.

The attendance of David Rees at the Mynydd Sylen meeting, the contribution of Henry Davies of Narberth, Daniel Davies, and other ministers at similar assemblies, and their appearance on committees appointed by the same fuelled speculation that religious disaffection was behind the troubles of this period. Colonel Love was told that their object was 'to upset the church and have a Republican Government', and there was talk of scriptural warrants being given for the destruction of gates and other resistance to injustice—'the Devil's flesh'. At Mynydd Sylen one farmer simply directed the attention of the rulers of the country to the first part of Judges and the twelfth chapter of the first book of Kings, which was an account of rebellion against oppression, and Foster did not allow his readers to forget that in Genesis 24 God had blessed Rebecca. The *Times* correspondent, who was not the only commentator to be impressed by the religious flavour of the public meetings, accused David Rees of fomenting the Rebecca discontent in his monthly publication, and claimed that sedition found its way into chapel sermons.[27] Dr Walter D. Jones of Lancych, who had been attacked by the Lady, and *Yr Haul* went further, implying that the whole Rebecca affair had been organized in the Baptist and Independent chapels.

The evidence for the latter charge was extremely thin; in their speeches at services and public meetings the ministers almost always condemned violence to persons and property. Their periodicals expressed concern about the workhouse mob at Carmarthen and the subsequent shootings and murder. Meetings of Cardiganshire Independents in June and August 1843 passed resolutions condemning such behaviour, and in October the Calvinistic Methodists at Newport in Pembrokeshire recommended expulsion of Rebeccaites. At Llanboidy Anglican and Dissenting ministers worked closely together to influence public opinion against nocturnal meetings, and Henry Richard, for all his later nonchalance about the movement, helped to organize the distribution of leaflets urging people to renounce the use of the gun and torch. The voice of religion was also heard in the Oddfellows, True Ivorite, and Rechabite lodges, as, one after another, they condemned those members who had given support to Rebecca.

This was reminiscent of the response to the Newport rising. Ministers of the Baptist and Independent persuasions were prepared to admit that with 'so much oppression . . . protest cannot be stopped', and occasionally expressed solidarity with the victims of police and military brutality, but they asked their followers to petition openly and peacefully. John Lewis, the Independent minister of Henllan, Carmarthenshire, was horrified by the whole business, and regarded it as a diversion from the true path of moral and temperance reform. William Chambers, who had no reason to love Rebecca, admitted that David Rees had consistently denounced violence. At a Wolfscastle meeting Daniel Davies of Hook asked 'what would the state of the country have been had not the Dissenting

Ministers neutralised [the] spirit of dissension . . . Because perhaps one or two had been foolish, was it fair to charge a large body of men with inciting the people . . .?'[28]

The answer was 'surely not', but it was nevertheless true that the autumn meetings of 1843 bore testimony to the aggressive Nonconformity of some of Rebecca's sympathizers. Speakers denounced the power and costs of the established Church, declared their determination never to pay the church rate, and demanded liberty in religion as in other things. At a Llandyfaelog gathering on 6 September 1843 a third of the proposals submitted by David Gravelle had a religious dimension. Apart from the separation of Church and State and the removal of bishops from the House of Lords, these resolutions called for greater local control over clergymen, the ending of obligatory services in workhouses, radical changes in the church rates and tithes, and the use of the latter for their original purpose. In the opinion of Gravelle, whose sentences were sprinkled with apocalyptic images, there were comparisons between the turmoil of 1843 and the state of Germany at the time of the peasants' rebellion in the early sixteenth century.[29]

Gravelle, like so many other speakers, included a few political resolutions in his programme. It was said that these meetings were the first independent political demonstrations by the ordinary people of rural Wales. This was not strictly true; some of the matters discussed had been sources of discontent for years. The malt tax, the Game Laws, and the income tax had inspired petitions before, and to these were added more recent anxiety about the effect of Peel's commercial legislation on the landed interest and the iniquities of certain by-laws and the county and poor rates. For many people, especially Michael Thomas of Gwynfe, the key political question of the day was, and had always been, the Poor Law. Speaking in Welsh at a meeting near Llangadog on 10 October 1843, he said that because of their experience over the Poor Law Amendment Act the people would in future ask more of the gentry and their parliamentary representatives.[30]

This Act was near the top of the political agenda during the Rebecca years, not least because it symbolized the loss of political power at the local level. In southern Cardiganshire and Pembrokeshire the violent events masked the peaceful attempts by vestries, public meetings, and individuals to reverse this 'Ungodly measure'. One campaign was organized by Robert Waters of Penally Court, Richard Llewellyn of Nash, James Child, and a host of yeoman families. It helped to convince the government that everybody except the greatest men was against them. These people held protest meetings throughout the autumn of 1843, fearing that unless they organized themselves, the government commission would deliberately direct attention towards the tollgates and ignore their main grievances. At a meeting in Pembroke town hall on 11 November they agreed to form a county Anti-Tithe-Commutation and Anti-Poor Law Association. This was a short-lived body, independent of similar institutions elsewhere in Britain; its main remit was to collect information and act as a local pressure group, presenting petitions to commissioners and Parliament.[31]

Although Walter R. H. Powell and other gentlemen regarded such activity as a justified extension of local politics, the support of free trade was a different matter. This was an attack on the very heart of landlordism. 'As might naturally be expected', said John Jones of Brynamman, 'they [the landlords] made laws to suit themselves. They made a corn law to keep the price of corn above its natural value. They made the law of distress in order that they might be able to seize the produce of the land if the tenant failed to pay . . . The laws of this day are contrary to the laws of God.'[32] Great efforts were made to show that the people responsible for the economic crisis, for the low prices, high rents, and discontent were the monopolistic and restrictive landowners. Amongst some of the Welsh religious leaders the campaign against the Corn Laws was a moral crusade, and the outpourings of *Seren Gomer* and *Y Diwygiwr* and the provocative speeches at the public meetings of 1843 are a reminder that the Rebecca riots took place in the context of a parliamentary campaign over the issue of free trade. In the spring of 1843 John Jenkins, the Anti-Corn Law League organizer in south Wales, gave his first lectures in Carmarthen.

As a non-Welsh speaker his impact was greatest in the industrial districts along the Carmarthenshire–Glamorgan border. Jenkins, Hugh Williams, David Gravelle, and Dissenting ministers convinced people at Llanedi, Llandyfaelog, Allt

Cunedda, Bryn Cwmllynfell, Cwmtwrch, Pentre, and Pontardawe that at least a temporary experiment with repeal of the Corn Laws was justified. At Pembrey a meeting of colliers and labourers argued that they could not be worse off, and called for an end to all restrictions on trade.[33] In other parts of west Wales the issue was often missing from their published lists of grievances, but even in the most unpromising territory that outspoken minister David Hughes of Tre-lech, and a handful of committed tradesmen and farmers, made their opinions heard. It is difficult to estimate their success, though the *Carmarthen Journal*, the main protectionist journal of the region, was obviously saddened when, on the eve of repeal, so few of the tenant farmers came forward 'to defend the agricultural interest'.

One fear, which the *Carmarthen Journal* shared in 1843, was that these political discussions woud turn into requests for changes in the British constitution. Frustration with their parliamentary representatives, both Tory and Whig, was evident at these meetings, and there were admissions that people had not used their votes wisely. Had they ignored their landlords' political directions in the past, said William Thomas of Llanon, 'Mr Goring Thomas would not be taking all the farmers' cream, and leaving their children to live on whey.'[34] Of course, independent action had its consequences; there had been evictions in the past 'for voting against the church rate, and . . . their landlords', and there would be more in the future without protective legislation. One farmer of Newcastle Emlyn was ruled out of order when he tried to document intimidation at elections.[35]

Even so, the ballot was a popular theme in both the meetings and the periodical literature of the time, and so, to a lesser extent, was an extension of the suffrage. The Mynydd Sylen, Llyn Llech Owen, and other meetings addressed by Hugh Williams called on the Queen to dismiss her ministers and appoint a Parliament representing a wider spectrum of public opinion. Resolutions in favour of the ballot, a dissolution, and franchise reform were also passed further to the west, at Pen Das Eithin, Llanarth, Cilrhedyn, and other places in the Teifi valley where the toast was 'the Queen, Wales, and our Liberties'. The precise nature of the ideal electorate was not always clear, though there were a number of the peasantry who demanded universal male suffrage. 'Notwithstanding their poverty,' declared Thomas Thomas at Llanarth on 20 November 1843, 'they ought to have a vote in electing a representative to Parliament.'[36]

These views convinced a few observers that Chartism lay at the bottom of all the violent and peaceful protests of the period. The movement for the vote had begun in Wales outside the main industrial areas, and it reflected the interests and priorities of small-town and village radicals, and of those many farmers who were not enfranchised. Even the Chartist press of Merthyr was full of articles attacking landowners, aristocrats, tithes, church rates, and the like. A considerable number of the leading agitators in industrial south Wales had grown up in places such as Haverfordwest, St Clears, and Llanboidy, and returned there just before the Newport rising seeking assistance, and afterwards, seeking asylum.[37] As we have seen, Hugh Williams and William Jenkins had done their best to bring the new political gospel to countryfolk. One report stated that two Birmingham Chartists were in the vicinity of Efailwen not long before the first attack on tollgates in 1839. At the time local proprietors William B. Swann and John H. Allen were uncertain about the significance of all this; the general conclusion was that the population were concerned about the issues of gates, county rates, and the Poor law, while the political visitors had their sights fixed on universal suffrage.

By 1843 the Chartist clubs in the region had been disbanded, and Hugh Williams had ceased to play a prominent role in the political movement, but the ideas were not forgotten as Holloway, the reporter of the *Northern Star*, discovered when he travelled around south-west Wales in the autumn of the year.[38] Whether the return of industrial workers from the south Wales coalfield to their Carmarthenshire and Cardiganshire home parishes in the spring and summer stimulated this interest in radicalism was an open question, but there were reports of unnamed Chartists attending meetings of Rebeccaites in both counties. Perhaps, as their enemies alleged, they provided organization skills, and promoted seditious ideas on land ownership and natural rights.

At the open meetings in the autumn of 1843 several speakers

espoused democratic principles, and had strong views on class legislation, royal and aristocratic corruption, and the labour theory of value. They were, understandably, from the ranks of the poorer farmers and those without land. 'The working classes gathered the wealth of the nation, and it was but fair play that the working classes wanted,' said Thomas Thomas the carpenter at one mass meeting near Lampeter. 'All those who had to obey the laws should have a voice in making them.' Significantly, Thomas, like William Anthony, the Carmarthen rationalist and socialist, and Benjamin Grey, the Llangyfelach collier, was a local radical and not the travelling agitator so beloved of conservative magistrates and landowners.[39] The promised visit of Feargus O'Connor to Tenby and district in the autumn of the year never materialized. The Irishman, like Bronterre O'Brien and William Hill, kept his distance from this kind of rural protest, fearful of its violence and uncertain about its working-class credentials.[40]

The correspondence in the Home Office files indicates that some at least of the Rebeccaites felt rather neglected by the Chartist body. In June, July, and August 1843 messages and people were sent from Carmarthenshire to the Chartists at Merthyr, seeking contacts between the rural and industrial malcontents. The Chartist association there was the largest in Wales, and its secretary David Morgan and colleagues David Thomas and David Rees were keen on an alliance, if only the right political priorities could be agreed. Some of the Merthyr men, like David John and his sons, had strong personal ties with south-west Wales, and shared the same pre-industrial Painite vision as their country cousins. The Johns, together with David Thomas, had a sneaking admiration for the physical-force tendencies of the Rebeccaites, but other leading Welsh and English Chartists counselled restraint. The Merthyr association, which had established good communications across Britain, was anxious to maximize the effect of strikes and protests in south Wales. It asked the Dowlais ironworkers, the Swansea coppermen, and all other rebels to delay their actions until everyone could come out for the Charter. The Chartists believed, or affected to believe, that simultaneous Irish, English, and Welsh troubles would overstretch the forces of law and order and precipitate a political revolution.[41]

In the event the greatest threat was ideological. The Merthyr Chartists prided themselves on their missionary work, and responded positively to requests for people to lecture on their democratic programme. For example, in the late summer and autumn of 1843 Henry Thomas, Benjamin Havard, and two other lecturers, possibly David and Rees Rees, arrived in the vicinities of Llangadog, Llanarthne, and Carmarthen. There they encouraged Chartist sympathizers, attended open meetings, and debated with smaller groups of Rebeccaites in private. Their message was consistent; now that much had been achieved by the destruction of gates, it was time to turn to other, and more important, matters. Universal suffrage offered the prospect of lower taxation, a repeal of the Poor Law Act, free trade, and the recovery of land appropriated centuries before. In the mean time, Henry Thomas and his Chartist friends welcomed the local interest in the ballot, the boycotts of unpopular taxes and of the sales of distrained goods, and the unions of small farmers. They urged the rural population to form defence clubs and to watch out for spies. Altogether, the Merthyr Chartists found that the peasantry were rather more advanced than they had expected. They proceeded to give them moral and financial assistance during the riots and later at the special commission in Cardiff.[42]

The national leaders of Chartism were perplexed by the composite nature of the support for Rebecca. On the whole, as the public meetings illustrated, the various groups in west Wales worked together on a common platform, though just occasionally clashes of interest did emerge. As we have seen, the farmers held their own assemblies from time to time, and established their own unions. The agricultural workers also had half a dozen separate meetings, to discuss wages, perks, conditions of service, and prices.[43] It seems that they were trying to establish a sound bargaining position before the hiring time. There were claims, not always substantiated, that farm workers remained discontented, and joined night meetings about Troedyraur, Newcastle Emlyn, and Llandysul in the following year. Yet, to the surprise of Feargus O'Connor and Bronterre O'Brien, farm servants and labourers never went further than this; neither man could understand why they should have sacrificed their independence in 'a farmers' movement about tolls'. The explanation was complex, but it was to be found both in the nature of the society, and in the character of the movement; common feelings were strong, common enemies abounded, and Rebeccaism was wider than just a farmers' crusade. It contained the anger and hopes of a surprising number of people.

The greatest dissatisfaction within the movement arose, we are told, from those colliers and ironminers who were for a time keen Rebeccaites. They became restless over the lists of grievances discussed at the autumn meetings. On 10 August 1843, at Pontyberem, they demanded lower food and beer prices as their reward for helping the farmers with the reduction of rents and tolls, and threatened shopkeepers and publicans who refused to comply. A month later, at a Pembrey meeting, the colliers withdrew from the gathering after the chairman, John H. Rees of Cilymaenllwyd, agreed to listen to their particular grievances on another day. About 300 colliers, and some of the poorest farmers and labourers of the area, met on 20 September, and discussed the Poor Law, the Game Laws, commons' enclosure, free trade, and the stoppage of the works nearby.[44]

The last was the one matter that was to weaken their loyalty to Rebecca. The colliers turned eastwards for guidance and help to their industrial colleagues. Communications were established with the discontented miners in the Swansea valley, where unionists from England were at work, and with the men about Merthyr and beyond. Union officials, representing the Miners' Association of Britain, arrived in Carmarthenshire sometime in 1844, and for a while they operated, as they had done at Merthyr, with prominent local radicals. Hugh Williams was a useful contact, and he seems to have taken them on a short guided tour around his beloved Kidwelly. Reports coming to Colonel Love in the late summer of 1844 even suggested that these union agitators were stirring up both the industrial and rural population, and a few of their meetings were conducted with the ritual and noise of a Rebecca or Scotch Cattle assembly. It was a frightening omen, which Colonel Love watched closely.[45]

Colonel Love and others in authority during these difficult years were intrigued by two questions: what connections were there between the Rebecca riots and these peaceful forms of protest, and what communications were there between agitators in this region and in other parts of Britain. The simplest answer to the former question was to distinguish clearly between the Rebeccaites—'primitive rebels' motivated only by cattle prices and tollgate dues—and the people described in this chapter, with their reforming but rational zeal. Thomas Emlyn Thomas, Stephen Evans of Cilcarw, David Rees, Hugh Williams, and lesser-known figures at the meetings were anxious to dissociate their campaigns from those of Rebecca.[46] Williams claimed that he was a moral-force Chartist who had disowned violence in 1839, and none of the spies set on him could prove otherwise. No doubt some of the more respectable members of his audience hoped that public debate would reduce the incidence of arson and personal violence. Resolutions condemning nocturnal outrages were presented to the meetings, often by landowners or their professional hangers-on, and supported, increasingly, by everyone present. Edward Lloyd Hall, who had once talked wildly of revolution, now looked forward to an era of peaceful and sustained political action.

The temptation to see violence and peaceful agitation as two mutually exclusive types of, or stages in, protest meant that the connections between the two were ignored or distorted. It was no accident that complaints at Llanfynydd or Llannon meetings against, for example, an unpopular gate or tithe payment were immediately followed by direct action; in fact it was implied in the very discussion of them. There were reports of small groups leaving these meetings and committing illegal acts. Stephen Evans, who was identified as Rebecca on one occasion, Michael Thomas, and several other prominent speech-makers admitted that they had met the Redresser, and William Thomas of Llannon offered public thanks to her.[47]

In the audience at the large gatherings were many who had ridden with Rebecca, and these mocked the Royal Proclamation and rewards for informers. The stone-cutting Jones family of Llanddarog, and vociferous Phillip Phillip of Pound, who were singled out for their contributions to the debates at Llyn Llech Owen and Allt Cunedda, were tried for being involved in illegal protests. Philip Howells, who chaired the Pembrey assembly, was the insolvent mayor whose family had called on Rebecca to resist the bailiffs.[48] An exasperated William Edwardes of Sealyham, chairing a Wolfscastle meeting, was greeted with laughter when he accused the farmers present of seeking a discussion of grievances after they had physically removed them. A few of the Rebeccaites claimed that talk in or out of Parliament would not answer their immediate problems, but many of their friends saw both ideas and action as part of the same process. One can see why Thomas Lloyd of Bronwydd felt that an important shift was taking place in the rural psychology.

There was a suspicion amongst both the Whig and Tory gentry that the organizers of popular unrest in this corner of Wales were in contact with agitators elsewhere. The newspapers of the time were full of the religious crisis in Scotland and the demonstrations against the government and Church in Ireland. Government opponents mocked the ministers for giving a boost to Celtic nationalism: 'In Scotland the Church has been rent in Twain under their sway; in Ireland the people are clamourous for repeal; and even in secluded Wales we now have an insurrection.'[49] There were, in and about the Rebecca movement, elements of national feeling. In her letters 'the Lady' sometimes refers to the oppression suffered at the hands of Saxons, and voices were raised against the appointment of outsiders as agents, clerks, and tenants, and, at the mass meetings, against the lack of respect for the Welsh language. One or two of the travelling reporters looked for, and found, 'bitter distrust of or hatred . . . towards Englishmen', and some anger at the centralizing tendencies of the latter and the weak-kneed resistance of Welsh MPs.[50] But the politics of Rebeccaism did not include nationalism. This was not too surprising, for the very conception of modern Welshness was in the process of being formed, and the main targets of this movement were indigenous. It was the government's reaction to the riots, as much as anything, which stimulated national feeling.

Everyone was agreed that the Rebecca riots were not a separatist movement, but there were similiarities with the Irish situation. The duke of Wellington thought that the Welsh were consciously imitating the tactics of the Irish Repealers, and one or two of the latter were travelling across south Wales at the time.[51] The existence of arson, violence, and murder in both countries caused some to label Rebecca 'the Captain Rock of Wales'. When land tenure and rent, tithe, and rate boycotts were discussed by Rebecca sympathizers, and when mass meetings were being held on either side of the Irish sea, the natural reaction was to say that 'the Celts are disaffected', and to call on the government to extend their new repressive Irish legislation to this part of west Wales.

There was a report of letters being received in the region from friends in Ireland, and general support for the Welsh Lady from Daniel O'Connell and the *Nation*. The latter was impressed by the spirit of 'Localism' if not nationalism behind the rural protest, and suggested that if the Welsh, Irish, and Scottish MPs got together, the electorate could look forward to a federal structure of government in Britain.[52] This was the only way to safeguard Wales's identity and resources. So far as one knows, there were few Irish nationalists present at the evening meetings to present this case, and even fewer references in the speeches to the turbulent land from which they came. Most people, including David Gravelle, presumed that, though there were similarities in the Irish and Welsh experience, the popular movements were at different stages of development and intensity. Rebecca ignored the contemporary protests over the Irish sea, and she was said to have shared the common Nonconformist dread of popery. The latter was illustrated in 1850 by the huge orchestrated outcry in the three counties over 'papal aggression'.[53]

There were, more mysteriously, hardly any comments in the Welsh meetings of 1843 about the contemporary problems of Scottish Dissenters and crofters and on the plight of the English agricultural labourers. People were not ignorant of these matters; they were accounts in the local press of the first demonstrations of hungry labourers and of the waves of arson that swept across the East Anglian and Home Counties in the early 1840s. However, as in the days of Captain Swing, there was no attempt to establish contacts between the two regions. Rural protest has always been remarkably self-contained. Even the parallel attacks on tollgates outside the Principality were largely ignored, including the incidents in Scotland and nearer home, about Bristol and north Somerset. In the last district the destruction of tollgates by men in women's clothing had been a feature of the early eighteenth century, and the most recent comparisons were sufficiently interesting to warrant attention in the *Welshman*.[54]

There were even closer parallels with events in east and north Wales. At least a dozen gates were destroyed across the industrial counties of the south-east, while the roads of Caernarvonshire, Merioneth, and Denbighshire were the scene of protests said to have been inspired by Rebecca. Threatening notices were sent to a number of workhouses in the Principality, and poachers on the Wye, and people resisting bailiffs in Anglesey and Pontypool, were accused of 'playing Rebecca'.[55] Yet no one was able to prove that this unrest had been instigated by people from west Wales. The same was true of the incendiarism that disturbed parts of north Wales at this time. In the neighbourhood of St Asaph, Cerrigydruidion, and Wrexham in 1843 ricks and outbuildings went up in flames, but much of this was a manifestation of private malice and the work of vagrants.[56] Unlike the Chartists, who sent David Ellis from Merthyr to Montgomeryshire in 1843, the peasantry of the south-west seemed indifferent to the struggles elsewhere in rural society.

The same, to a lesser degree, was true of their attitude to industrial discontent. Colonel Love's first concern throughout the period was to prevent, or if not to monitor, contacts between leaders of the agrarian movement and those of the striking miners, colliers, and copper workers in neighbouring counties. Communications were established, but little came of them. 'With Rebecca & her followers they [the copper workers of Swansea] have nothing in common,' J. H. Vivian, their employer, declared in October 1843, 'nor have they taken any part in the struggle or mixed themselves up in any way.'[57] Meanwhile, Colonel Love and Captain Napier busied themselves with the promised co-ordinated revolt of northern radicals, Welsh rebels, and Irish nationalists, but by the winter of 1843–4 this threat, too, was receding.

The lack of communication between the Rebeccaites and other movements together with the absence of 'right leaders' were, in the opinion of Love, the reasons why the 'doers' and the 'thinkers' in west Wales never ultimately posed a serious threat to the State. 'The Welsh agitation will spend itself within its own mountains and valleys', concluded the *Times* correspondent. In general it was true; the Rebecca movement had a very specific regional character and it quickly came to an end once its immediate objectives had been achieved. Its strength, as the government realized, was its community nature, and the range of its support and ideas; it could be seen as the last or first of its type. Hugh Williams always believed, at least privately, that 'the Rebecca affair' lacked the class and ideological polarization necessary to be a lasting reform movement. The *Sun* newspaper agreed, though it anticipated that the knowledge gained through the riots and meetings would produce its effect in time.

Very soon, however, it was hard even to recall the more outrageous notions discussed in 1843, a fate shared by the Chartist and Owenite campaigns for a new moral world. In the mid-Victorian era the actions and programmes of all these early protesters seemed strangely inappropriate. Not all the ideas were rejected, but the vision and the methods of change were. It was as if one tape or language of popular politics had come to an end. History served to reinforce the superiority of the new ways of thinking. In these later years the accounts of the old mass movements were exercises in exorcism; what was left out was as important as what was kept in. Chartism, the Newport rising, and the Rebecca riots were slimmed down, and interpreted in a way which suited the new agenda of politics and the current definitions of rational behaviour and the Welsh character. Henry Richard, the apostle of reforming Dissent, said in 1866 that the riots had been 'greatly misunderstood'; they were 'just about tollgates', 'that was the whole origin and meaning of what has been called the Rebecca riots'. A shroud was drawn over the other aspects of Rebeccaism described in this book and over many of the discussions in the unions and the mountain conventions. 'The Rebecca disturbances of 1843 . . . had no political significance whatsoever,' declared Richard.[58] He insisted that the political life of the ordinary people of west Wales began only with the Liberal crusade, and it became harder, in the years of peace and achievement, to disagree.

REACTIONS

The reactions to the Rebecca riots form part of the explanation for the change in climate between the turbulent 1840s and the mid-Victorian years. The unique nature of the troubles forced a response. Whatever their later claims, no one had predicted the riots and the accompanying ideological ferment. The Newport rising of 1839 had removed the complacency about industrial Wales, and troops were stationed there well into the mid-century. What surprised everyone was a rebellion amongst the rural population; the Swing riots had warned people of the tension in the English countryside, but south Wales had been largely free of such outrages. The Rebecca riots raised the spectre that the rural Welsh were as dangerous and violent as their more notorious industrial cousins.

When the riots reached unacceptable proportions, the bewildered Tory government faced a two-pronged attack. Alarmists joined the Queen and Prince Albert in their private annoyance at Prime Minister Peel for his slow response to the crisis, and warned that the success of the attack on tollgates invited further trouble, not just in Wales. Others on the Whig and radical side mocked the government for taking the riots too seriously, and poured scorn on the number of soldiers, the policemen quartered in the region, and the talk of martial law. What united all of these critics was an agreement that a quick response to Rebecca, whether conciliatory or repressive, would have ended the affair almost before it had begun.

After an initial delay by the government, the response to the events in the three counties was considered and comprehensive. In fact, some politicians hoped that it would become a model for dealing with protests elsewhere, chiefly in Ireland.[59] The Home Secretary sought a political solution to the troubles, demanding enquiries, suggesting concessions, and offering the prospect of long-term reforms. The approach undermined protest and stifled criticism, and was undoubtedly a contribution to a more civilized dialogue between government and people. It was a calculated policy. Both Peel and Graham were concerned by the depth of feeling over the grievances in Wales, and by the ideas being canvassed at the public meetings. At the same time, the Home Secretary was determined to strengthen the machinery of law and order in the region, convinced that lasting changes in this area were also necessary to prevent further Welsh rebellions over other issues. Like the Irish troubles, the Rebecca riots proved to be an important stimulus to new forms of control and policing.

After seeking immediate explanations of the Welsh outrages from those in authority, the Home Secretary sent Thomas J. Hall, the stipendiary magistrate of Bow Street police station, to south Wales on 28 July 1843 for over two weeks. He was assisted by solicitor George H. Ellis, and it was he who produced a critical report on the operation of the trusts. Hall's role was something of a mystery; the Chartists believed that his real task was to investigate the riots rather than their causes. The speed with which Hall operated, and the silence about his report, proved a source of confusion and guaranteed prolonged resistance in the region. The commission of Thomas Frankland Lewis, Robert Clive, and William Cripps, which began its work at Carmarthen late in October, was a more satisfactory enquiry, and, in its tour of south Wales, was praised because it took evidence from a wide selection of people. It did not escape criticism altogether, but the very existence of the commission was assumed, rather naïvely, to be the cause of Rebecca's demise.

The report of this commission was made public on 6 March 1844, somewhat later than had been hoped. The government was accused of dragging its feet, but, having accepted the main recommendation to consolidate the trusts, another enquiry was needed to determine the cost of this change. The liabilities of the various trusts amounted to almost a quarter of a million pounds.[60] Once a method had been devised of paying off these debts, the South Wales Turnpike Trust Amendment Bill was introduced in Parliament. The Act, which received the royal assent on 9 August 1844, placed the management of the turnpike roads in the hands of County and District' Roads Boards, and insisted on a uniform system of tolls. At the beginning people claimed that the County Boards, nominated by magistrates in Quarter Sessions, simply ignored the condition of the roads, being preoccupied with paying off government loans and balancing budgets. For a while complaints of illegal and excessive tolls continued. Colonel Love, still at his

Carmarthen base in 1846, warned that Rebecca would strike again, and in December a gate was destroyed. When the Castell Rhingyll tollgate was re-erected in 1848 it, too, was quickly demolished, and one or two serious incidents were reported in 1850–1, but in general large-scale destruction became a rare occurrence.[61] Magistrates like Saunders of Glanrhydw and Lloyd of Dolhaidd made sure that the costs of repairing bridges and roads were kept to a minimum, and within two decades the management of the roads in south Wales became a matter of self-congratulation. Of course, grievances did not completely disappear; late in the nineteenth century feeling ran strongly against the level of highway and county rates.

The political influence of the riots was considerable if hard to define. The events in south-west Wales gave a particular resonance to the debates over certain Bills then, and in the immediate future, and the riots themselves were partly responsible for changes in government thinking. Discussions at Westminster on the punishment for malicious injury to property, on the recovery of small debts and on imprisonment for debtors, on the setting of salaries for magistres' clerks, on the abolition of church rates, and on policing and other matters were all slightly different because of Rebecca. The same was even more true, as we shall see, of parliamentary debates over education both in the 1840s and beyond. Nor was Rebecca totally forgotten in the appraisals of free trade, the malt tax, and other legislation affecting the agricultural interest, and her name was heard again during clashes over new Game, Fishing, and Enclosure Acts in the years 1844–65.

Probably the most obvious example of Home Secretary Graham's sophisticated political response to the Welsh crisis can be seen in the amendments made in 1844 to the Poor Law Act. He told the Commons that the unpopularity of the bastardy clause in Wales, and to a lesser degree in the north of England, obliged him to reconsider the position of the single mother. He had become aware, he said, of the wisdom of separating the question of bastardy from that of the Poor Law. Under the new proposal, the unmarried mother could, within the first six months of her child's birth, herself summon the father before magistrates at Petty Sessions, and, once paternity had been proved, he was liable on pain of loss of goods and liberty to maintenance payments of 2s. 6d. a week. This was less generous than it seemed at first reading, but in south-west Wales the immediate outcome of the change was a rash of summonses; a troop of unhappy men passed through the courts to the cheers of crowds of women.[62]

There were other signs, too, that Poor Law administrators in London were at last beginning to appreciate the special character of the region. In their report of 1844 they admitted that the poverty in the three counties had increased the instinctive resentment against high salaries, especially for its medical officers.[63] Moderate reductions in salaries were approved, not least in the Cardigan union where, for years after the riots, difficulties were found in collecting poor rates. Resistance to a full workhouse policy remained a problem in the area, and the Commissioners had to deal with thinly disguised attempts to continue paying relief to distressed able-bodied paupers. There were also in south-west Wales special difficulties in implementing the new settlement law and the increasingly strict directives from London on policy towards vagrants. The Welsh authorities, who were badly affected by the problem of non-resident poor in the mid-century, were anxious to keep expenditure low yet wished to reduce the dangers of popular rebellion. In the event, they found the Poor Law Commissioners more accommodating than they had been at the time of Rebecca.

If the political response to the troubles was careful and judicious, and on a par with the succession of Factory, Mines, and Health Acts that followed the industrial strife in northern England, the same can be said of developments in the field of law and order. The pace of change was faster in Wales than in many parts of Britain, for the government displayed here a determination that characterized its policy in Ireland. The Whig Home Secretary Lord John Russell expected magistrates during the first Rebecca scare in 1839 to manage with parish and special constables, and the assistance of the yeomanry and Metropolitan policemen, but in 1842–3 a greater force was needed. Sir James Graham, the new Home Secretary, was blamed for not responding more quickly to Welsh requests for help, but he had other priorities at the time, and was anxious to give military aid only in return for promises of improvement in local policing.

The collapse of authority in the three counties forced his hand, and by the end of 1843 more than 1,000 soldiers had been sent to the region. What astonished Graham was the ineffectiveness of these men, and, along with the duke of Wellington, he placed the blame on Colonel Love. Major-General George Brown, who was sent to rectify the situation, decided that Love was too sensitive to the wishes of the native landed gentlemen and too cautious about placing soldiers in small numbers where they were most needed.[64] As the map shows Brown located them right across the region, and placed Metropolitan police alongside them. They were initially praised as a preventive force, but the cost of the additional constables proved too much for the authorities and by the end of 1844 all the London police had returned to the capital.

The most fearful of the propertied residents, even in Aberystwyth which was hardly affected by the Rebecca troubles, wanted the military forces to remain after the riots had subsided. The main problems were expense and accommodation; the central government, committed to extending the Brecon barracks, was reluctant to sanction another headquarters in south Wales. This caused difficulties both during and after the disturbances. Workhouses, public, and private houses were not the ideal places for soldiers and, as soon as the threat passed, Colonel Love removed his men from the smallest villages. To his embarrassment, they sometimes had to be returned. As the months passed in 1844 the commander became more and more irritated by the vacillation and ingratitude of magistrates, and their claims for financial compensation over billeting.

Only Carmarthen took the long-term view; in the winter of 1843–4 Colonel Trevor and his friends raised over £2,000 towards improving the barracks in the town. The government was asked to make the town into 'a second Brecon', but there were interminable delays in choosing and purchasing a suitable site.[65] By 1851 Pembroke Dock had become the government's favoured base and headquarters for this half of the south Wales military district. The site had obvious advantages, and during the Crimean War hundreds of men were temporarily stationed there. Not that Carmarthen was forgotten; for ten years after the Rebecca riots regiments came and went, and for much of that time soldiers also gave security to the people of Cardigan and Newcastle Emlyn. In September 1851 men of the 82nd Regiment staged a mock fight with Cardiganshire labourers, a friendly reminder that the latter were as much on trial as the Merthyr miners.[66]

Two of the changes that were contemplated as a result of the riots were in the nature of the magistracy and of the police. Colonel Love and Major-General Brown told the government that one major source of discontent in south-west Wales was the quality of the JPs, and many of the public meetings passed resolutions calling for stipendiary magistrates. The temporary appointment of Colonel Hankey as a roving justice was a nod in this direction, but the only significant developments were a reduction in the number of Petty Sessional divisions and an increase in that of magistrates. The latter occurred at the end of the riots and again in 1850. There were still places, such as Narberth, where the scarcity of active magistrates hindered legal business, but in general judges were increasingly complimentary about the amount of work performed by justices at the lower courts.

Complaints of magisterial tyranny and their lack of Welsh language and sympathy did not disappear. Henry Leach, the chairman of the Pembrokeshire Quarter Sessions, and Rees Goring Thomas took their whips to vagrants, and others were accused of being vindictive towards those who breached the Game Laws. Yet there were signs over the next decades that lessons had been learned, especially in their administrative work. There was, for instance, greater care over the level and equalization of county rates, over the cost of building bridges, county halls, and gaols, and over the administration of the turnpike roads through the County and District Roads Boards. There was a determination, personified by Rees Goring Thomas and those two self-appointed spokesmen of the farmers John E. Saunders and John H. Rees, to get value for money and more accountability. Appointments of clerks, coroners, surveyors, inspectors, and constables were closely watched, and these men found it harder than their predecessors to make a little on the side.

Amongst these magistrates there was a growing appreciation of the value of professional policemen. Discussion over the nature of policing had pre-dated the Rebecca troubles. The Pembrokeshire magistrates, amongst others, had responded to the Home Secretary's circular of 1839 over the Shropshire experiment with a county constabulary. In Pembrokeshire, despite Henry Leach's enthusiasm, there was a reluctance to move far from the traditional reliance on parish and special constables. In December 1843 and January 1844 there were meetings all over the county, and petitions signed by hundreds to keep the status quo.[67] But the events of 1839–44 made this position virtually untenable and people in authority recognized that changes were inevitable. In Pembrokeshire a form of paid constabulary was initiated in 1844, with superintending constables based at lock-up houses. It was retained until the County and Borough Police Act of 1856 and, although there were sometimes grumbles about its efficiency, it was widely praised. In the bad economic years of 1849–50 parishes in Carmarthenshire and Cardiganshire petitioned for a similar system in their counties.

Carmarthenshire experienced the worst of the Rebecca troubles, and from an early date George Rice Trevor and most of the county leaders supported the idea of a county police in preference to the embarrassingly expensive London men. The introduction of a rural constabulary was agreed at the Quarter Sessions in July 1843, when Rice Trevor, Earl Cawdor, and Rees Goring Thomas expressed the characteristic sadness of law-and-order pioneers, but there were a few hitches and a paper mountain of protest to be overcome. One of the main complaints was the number of men in the new force, some 57 at its foundation. It was estimated in 1848 that there was one policeman for every 2,180 inhabitants in the county, and Edward P. Lloyd of Glansefin and Sir Erasmus Williams declared them to be incompetent and unnecessary. In 1850, when farmers submitted memorials against police costs, the force was reduced by 13 men.[68]

This made the task of Captain Scott, the first chief constable, even more difficult. There were parishes in the north of the county, such as Llangeler, where policemen and stations were few or non-existent, and he had to turn down requests for assistance from angry ratepayers. At the same time, well-policed rural neighbourhoods, benefiting from a reduction in crime and arson, were reluctant to see officers depart for other areas. During the 1850s more police were needed in the industrial district about Llanelli, and along the turbulent path of the new railway lines. To obtain them, and, more importantly, to keep them, Scott had to offer good wages, for too many policemen looked eastwards, where other forces and other employment offered better prospects.

The fiercest battle over the reorganization of policing took place in Cardiganshire. In this, the least populated county, there had been no strident call for salaried constables, not even in Aberystwyth, Cardigan, or the smaller towns. As the northern districts were comparatively free of the Rebecca troubles, it was presumed that any scheme for a Rural Police would encompass the southern half of the county only. They had reckoned without police reformers like Dr Llewellin of Lampeter College, Edward Lloyd Williams, and Lieutenant-Colonel Herbert Vaughan, and the support of Colonel Love, Colonel Hankey, and Home Secretary Graham. While the northern opponents of a Rural Police, led by Pryse Pryse of Gogerddan, were busy drafting their resolutions and counting the petitions of support, a poorly attended Quarter Session at Cardigan in January 1844 gave Graham all he wanted.[69]

Although the three-year effort to disband the force was unsuccessful, the anger over the establishment of the Cardiganshire force curtailed the growth of the new police and the schemes of Captain Freeman, the first chief constable. In the northern parishes the old parish constables carried on much as before, and there was a general belief that the new uniformed men were just the personal servants of unpopular landowners, a charge that was to be repeated many times until the 1880s. Yet there was, Edward Lloyd Williams had predicted, a subtle change of views as the years passed. Those places which had a police presence, like Llanbadarn Fawr and Llwyndafydd, claimed that traditional tolerance of stealing and riotous behaviour rapidly vanished, and during the mid-century there were more requests for county policemen from small mining communites and from parishes intimidated by vagrants. Landowners, 'principal inhabitants', clergymen, and some farmers and Dissenting ministers in Cardiganshire, and indeed in the neighbouring counties, soon dropped their opposition to the new force. Even some of the peasantry possibly welcomed the new standards, though the clashes between them and the police, and the abuse heaped on the latter during the early days, suggests that acceptance was not always easy. Where the professional peace-keepers were the only English-speakers in a Welsh community, communications must have been particularly difficult.[70]

The County and Borough Police Act of 1856, which established a uniform system of policing across Britain, resurrected old debates about local power and self-government, but there were fewer protests in south-west Wales than in many other areas. People had become accustomed to full-time paid policemen, and parish constables had been made redundant. Moreover, committals to gaol and the numbers for trial had fallen by the late 1850s, and many attributed this to better policing. Then there was the memory of the Rebecca riots, and the change that had come over the region since those days; even the Powells of Maesgwynne and the Lewises of Gwynfe were now reconciled to professional policing. 'The country', said Earl Cawdor in October 1853, 'was never in a more quiet, peaceable and prosperous condition than at present . . .' 'In Carmarthenshire there was dissatisfaction with it [the Rural Police] once—at its introduction', he added four years later during a discussion of the new Act, 'but if they were now told that the police were to leave, the whole of Carmarthenshire would rise up in arms against their going.'[71]

Although the importance of better policing was well understood, the first reaction of many people to the unrest of this period was to call for more religion and better education. Already, by the autumn of 1843, Quaker missionaries were at work in the region, and at Penygroes in north Pembrokeshire a large religious meeting called the misguided people back to the true spirit of Christianity. Meanwhile, the judges who presided over the trials of Rebeccaites regaled their listeners with horror stories of the ignorance of farmers and labourers. John S. Harford of Peterwell and the William Edwardes of Sealyham, both of whom had the taste of political rejection in their mouths, chose this moment to tell the Welsh-speaking peasantry that they were 150 years behind the more civilized members of the British community.[72] In contrast to the industrial population, which was responsible for other risings, these people could not, however, be accused of irreligion. What they lacked was 'proper religion' and 'the right schoolmasters'.

The riots gave added impetus to the campaigns for more religious buildings, more and better clergymen, more intelligible sermons, and a sound religious education. As early as April 1844 the *Nation* had predicted that the rebellious Welsh would be subjected to the classic imperialist remedy of 'English language and more churches'.[73] Almost before it realized, the region was in the middle of an Anglican revival; during the 1840s and 1850s more churches were built in the diocese of St Davids than for many years past, and a greater effort was put into the repair and restoration of old church buildings. Llanegwad, Crunwear, Llanfihangel Abercowin, St Dogmaels, and St Clears parish churches were restored in the midcentury, and new ones appeared at Pembroke Dock, Llandeilo, and Pontarddulais. The Cawdors and de Rutzens, appalled by the events of 1839–43, made a special contribution to the churches in their neighbourhoods.[74]

According to the census of 1851 there were almost a thousand places of worship in the region, and about a third of these belonged to the established Church. In many districts of Pembrokeshire, and about Lampeter and Cardigan, there was at this time a good ratio of church buildings to population. Anglican clergymen in the two counties worked hard with clerical meetings, Welsh services, agricultural and charitable endeavour, and the occasional wash from a revival, to recapture old loyalties. The Reverend Herbert Williams of Llanarthne and friends went a little too far; in 1845 they were amongst the hundreds in Carmarthenshire who were captivated by 'Prince', an ex-clergyman, and his vision of an imminent second coming.[75] In the mid- and late 1850s, and again in the early 1860s, some churchmen were also swept along in the evangelical fervour that was gripping Dissenters across much of the northern half of the region. The keenest clergymen and their landowning patrons used gentle pressure to persuade tenants and labourers to attend church services, and Bible societies did their best, but there were still glorious buildings with long empty pews.

The census of 1851 provided both comfort and anxiety for the Dissenters. Having been berated for their part in the Rebecca troubles, and having felt the aggression of a revitalized Church, it was pleasing to bask in some statistical warmth. Two-thirds of the faithful still attended chapel rather than church on Sundays. Yet the census had a threatening message for both church and chapel in the three counties. Although the attendance figures were generally good, there were districts where only a third of the population bothered to enter God's house. The era of almost continuous revivalism was coming to an end, and the number of new chapels had recently begun to decline.[76] The revivals of the next decade, especially that of 1859, temporarily filled empty seats, and the political outcry over 'papal aggression' and Anglican taxation kept their ministers in the public eye, but in the mean time the Mormons were working their way quietly along the coast, and there were hundreds of people who were soon to lose interest in the fate of the rebuilt chapels. The mid-Victorian battles over education, Church–State relations, and the wisdom of the new English Home Missions contained evidence that things were not well. This made the role of Sunday schools even more important, and here the success of the Dissenters, especially the Independents and Calvinistic Methodists, was overwhelming.

These Sunday schools were at the centre of the debate about education, for they provided religious knowledge and disputation in the Welsh language. Prominent members of Welsh society argued that behind the rural malaise of the years 1839–44 was a special kind of ignorance, an inability to come to terms with the modern secular world and the English language. 'It grieves the heart of every true Briton', said the Reverend D. Lloyd Isaac at Llangathen, 'to find most places of importance and trust, whether a "pwll glo" [coalpit] or a railway station, occupied by strangers; and the Welshman dull, vacant, and clumsy, when you talk to him about figures, history, etc.'[77] The inability of the monoglot peasantry to rise in the social scale was a common theme of the public meetings of the mid-century. The Welsh, said David Pugh, chairman of the Carmarthenshire

Quarter Sessions, lacked the educational facilities and the social mobility of the Scots, and he believed that whichever government took up 'the sacred cause of public education' would reap a bounteous harvest in Wales.[78]

Never far from education was the question of language. The Rebecca affair, and even more the attention which it received from outsiders, initiated a long discussion on the merits and demerits of spoken Welsh. John Lloyd Price of Glangwili, and Joseph Downes, who made his home in the Principality, claimed that the riots had proceeded from 'no justified cause', but solely from the differences in culture and language. The only guarantee of security for the future was 'the extirpation of Welsh'.[79] The commission on the turnpikes, which omitted so much, felt that it, too, had to comment on the ignorance of the English language, and the damaging effects of this on progress, education, and the established Church. The *Times* correspondent took a relaxed approach to the subect, but there were others calling for the next generation of Welsh people to be denied the use of their native tongue, certainly at school and perhaps by compulsory and prolonged visits to England.

The Reverend John Jones ('Ioan Tegid') of Nevern denounced this response, and some Dissenters went further and condemned English law, English agents, and English railways. The Welsh should, said Ieuan Gwynedd, editor of the *Principality*, 'take care of ourselves as we have hitherto done'.[80] One fear, often stated in the post-Rebecca analyses, was that the people themselves might widen their educational horizons via the Sunday and day schools which were increasing so rapidly in the 1830s and 1840s. In the first few years after the riots there was a great flowering of new schools, especially in Pembrokeshire and Carmarthenshire; some were started by adventurers, and even, as one observer put it, 'by Chartists and Rebeccaites'.

It was concern over the weaknesses of private and Welsh educational endeavour generally, and an obsession with the virtues of the English language, which prompted William Williams, the Coventry MP and a native of Carmarthenshire, to press for a government enquiry. He had been more affected than most at Westminster by the Rebecca troubles and, like David Morris MP, saw unhappy comparisons with the Newport rising. The government, impressed with the recent educational reforms in Ireland, reacted positively; three commissioners and ten assistants were appointed and for months they took detailed evidence across the Principality, largely from Anglicans. Although the report has been absolved from the worst charges against it, in the south-western counties the findings had a considerable bias. The Reverend Henry L. Davies, curate of Troedyraur, and other clergymen who had been the victims of Rebecca's tongue now took their revenge, not just on the poor quality of education but also on the character of the Welsh peasantry. The lengthy report, which appeared in 1847, recommended a thorough overhaul of elementary schooling, partly financed by government and with trained teachers and English-language classes. The implicit hope, which several witnesses articulated, was that the people could be guided back to the true Church.

The storm that broke over the publication of the report had important results. It increased the sensitivity and unity that had been growing amongst Dissenting ministers since 1843, and ensured that most Independents and Baptists would stoutly resist further Church–State interference in education matters. Their concern for the education of the poor in Wales, which was expressed several times at the height of the Rebecca troubles, and again at the Llandovery conference in the spring of 1845, was balanced by a keen suspicion of the intentions of Sir James Graham and his Anglican friends. The work of the bishop of St Davids and the Welsh Education Committee, an early response to the fears engendered by rural violence, was carefully watched by the voluntaryists. In the late 1840s and 1850s new Church of England schools swamped those of other denominations. Some were National Schools, but there were also a good number of Madam Bevan's Circulating Schools and others promoted by landowners like the Cawdors, the de Rutzens, the Peels of Taliaris, the Gowers of Castle Malgwyn, and the Philipps of Williamston. By the late 1850s all three counties were covered with district committees of Anglican educationalists. They were matched by groups of Dissenting ministers and professional people, who, with the aid of money from small farmers and labourers, established a few British schools in the late 1840s and early 1850s, as well as the training college at Brecon.

By their own admission David Rees and the voluntaryists found it hard to meet the popular demand for education. A delighted *Carmarthen Journal* chronicled the opening of each new Anglican schoolroom. 'Better late than never' was how the vicar of Llangathen described the laying of the foundation stone of their National school in 1859, and he gave due thanks to Earl Cawdor and Lord Dynevor. 'Gratitude to our Benefactors' was the motto of such schools, and, in the years of the Crimean War, it was decorated with flags of the Allied Powers. By 1864 there was a considerable number of schools in the three counties receiving education grants, and visits by government inspectors, but there remained poor isolated communities where schooling was minimal.[81] This state of affairs did not escape the notice of Dissenters, and some voluntaryists, seduced by the Liberal victory in 1868 and Forster's educational compromise two years later, finally accepted the need for a complete and compulsory system.

The education question was ultimately part of a wider debate about morality and social relationships. George Rice Trevor, speaking to one of the many agricultural societies of the mid-century, called on his audience to make individual and collective efforts to improve social harmony in their localities, and thus remove forever the spectre of Rebeccaism. The argument used so often in the mid-century could be summarized thus: the interests of landowners and tenant farmers were one, and therefore they should work more closely together, not least to improve the welfare, morals, and behaviour of their servants and labourers. Agricultural societies and the smaller farmers' clubs, many of which were formed in the 1850s, could do much in this direction.[82]

The irony was that, outside Pembrokeshire, there were constant complaints about the absence at such meetings either of the great landowners or, alternatively, of the smaller tenant farmers. There was another problem, too, namely the intrusion of 'tenants' rights' and 'political matters' into agricultural discussions during the depression of 1849–53 and beyond. The Carmarthen and Pembroke Farmers' Clubs were fairly radical in their demands for greater security of tenure and rewards for improvements, as was the Vale of Towy Agricultural Society, which flourished in the neighbourhood of the Farmers' Union of 1843 and had some of its members.[83] However, most agricultural societies were preoccupied with social activities, and sponsored the ploughing matches which were so popular at this time. In addition, there were the shows and dinners held by the county and district societies, where prizes were given to successful farmers, and where financial rewards were offered to employees with long-service records and to those large families who had ignored the temptations of crime and pauperism. There were even plans to give certificates to servants without sexual blemishes.[84]

According to David Saunders Davies, the Carmarthenshire MP, the main virtue of these competitions, meetings, and dinners was the contacts made between people of different classes.[85] This was also an attribute of charity, when properly conceived and given. In 1852 a judge at one of the county Assizes praised the greater willingness of the upper class to help the lower. It was hardly spontaneous. The Rebecca riots brought an awareness of the dangers which sometimes accompanied poverty; during the worst winters and depressions of the next twenty years there were constant reminders in the press of the wisdom of acting generously. In the spring of 1847 cargoes of barley were brought in for the hungry, and subscriptions and soup-kitchens were a feature of the coldest months. At Llangadog, Llangeler, and other old Rebecca strongholds, the cup of charity was said to have been 'overflowing' as the Misses Lloyd of Danyrallt, Mrs Thomas of Llysnewydd, and their friends came to the rescue.[86]

There was an interesting debate in these years over the comparative benefits of money spent on charity and public health. There was certainly a growing concern for the health of the poor, encouraged no doubt by the horrors of cholera in 1849. There had been an interest in the dumb and the deaf well before the mid-century, and in the welfare of the mentally ill, and it seems likely that the obsession with the riots delayed action on this front. The Asylum Act of 1853 required the counties to provide a mental hospital, but it was another ten years before one was opened at Carmarthen. Almost immediately, there were fears that it was being used as a place to hide unmarried mothers and the unruly elements of rural society. For years the very name of the institution was a warning to women against becoming an embarrassment to their families. The campaign to improve physical health had a faster momentum. Earl Cawdor, the bishop of St Davids, and William Chambers gave their public support to the moves which brought infirmaries and dispensaries to Carmarthen, Aberystwyth, Llanelli, Tenby, and other places during the mid-Victorian years. The Boards of Health, the Common Lodging Houses Act, as well as compulsory vaccination, made some improvement to the disease-ridden quarters of these towns.[87]

Such charitable activity was, as the Carmarthen Baptist minister H. W. Jones declared in October 1846, good in itself and good for class relations. Yet it had to be controlled. Dr Bowen at Carmarthen and Henry Leach reminded the givers in the worst winters that others, including the small farmers, suffered as much as 'the poor', and warned that 'charity itself, the best of our virtues, must be exercised with discretion, without which we may pervert its purposes, and make even benevolence mischievous'.[88] One detects more calculation in the hand-outs of the 1850s and 1860s, and in the clothing and penny clubs established with the aid of Lady Williams of Rhydodyn, the Powells of Maesgwynne, the Lloyds of Bronwydd, and others. For those who wished to benefit there was a price to pay. Because of such charity the 'deserving village poor' were said to have recovered 'proper feelings' of respect for their betters in the years after Rebecca. It is an interesting claim, but one which should be viewed with scepticism. The reception given in the cottages to lady bountiful, the elaborate rituals of welcome for the returning squire and his new-born son, and the tokens of gratitude by tenants at the rent dinner are impossible to quantify.

Social contacts and improvements were also to be encouraged by voluntary service, sport, and recreation. The region had a poor reputation in the early nineteenth century for the size and quality of its county militia, and there were later expressions of shame by Lord Dynevor for the anti-militarism of Dissenters and for Wales's contribution to the Crimean War.[89] In fact, the Volunteer movement of the mid-Victorian years was reasonably popular in the south-west, and most towns and districts had their rifle corps. These corps, which had officers from the sons of the lesser gentry and the middle class, offered the attraction of a good social life and competitive sport. Sport was taken seriously in the mid-century, as if the young farm labourers and urban apprentices could not be trusted with the small amount of free time at their disposal. The attempt to change old habits gathered a little speed during the 1850s: in Carmarthen, Aberystwyth, Cardigan, Haverfordwest, Pembroke, and St Clears the new police were drawn, with some success, into the long battles against the burning torch-lit celebrations on Guy Fawkes day, Old New Year's Day, and Christmas Eve, and against Shrove Tuesday football, juvenile gambling, and nude bathing.[90]

In their place the people were offered athletic competitions and the 'good Old English rustic pastimes' of vicarage tea-parties, donkey, pony, and sack races, the greasy pole, and fireworks. Cricket matches were especially common in the immediate aftermath of the Rebecca riots, with villages and towns playing their nearest neighbours, or visiting teams of Oxbridge scholars and military men, but the popularity of these matches was not sustained. For people like Charles Bishop of Llandovery they bore too close a resemblance to real life, with working-class fast bowlers bearing down on gentlemen with only a bat in their hands, and it was not until the late 1860s that organized cricket regained widespread support.[91] In the mean time there were quoits and other hobbies; brass bands were established in villages and mining communities in these decades, and choirs and glee clubs abounded.

Gradually the new, and to us familiar, pattern, of Welsh yearly and weekly recreation was established, though we need to remember that most agricultural workers still had only two or three days' holiday a year. At Haverfordwest, for example, Christmas and New Year's Day were transformed by Rechabite perambulations, missionary thanksgivings, junior songsters, and tea and muffins, while Easter and Whit Mondays were occasions for excursions and temperance festivals. Having declined somewhat in the later 1840s, the temperance and teetotal movements quickly recovered their strength in the three counties. Temperance festivals were a failure of the 1850s, and Whitsun holidays became synonymous with Sunday school and Band of Hope processions. At Capel Als, Llanelli, David Rees insisted on total abstinence. Like other ministers he gave a special Welsh dimension to the growing Puritan campaign, and large petitions reached Parliament asking for a ban on Sunday drinking in the Principality.

This mission was complemented by other efforts to influence the minds of the people when at work and at rest. Even the loudest defenders of the Welsh during the 1840s admitted privately the need to reform some aspects of peasant and urban culture. At Redberth Lodge, Pembrokeshire, Miss Thomas—'the chief civilizer of the neighbourhood'—conducted a personal crusade against the 'loose songs' and drinking of wedding feasts, and the Reverend Henry Lewis Davies of Troedyraur became secretary in 1850 of the Society for the Promotion of Morality among the Lower Classes.[92] The Rebecca riots concentrated the minds of people like the Reverend John Evans of St Clears, John Saunders of Aberystwyth, and David Rees of Llanelli. 'The Elevation of the Working Classes' was now a prominent theme of their religious and lay sermons. Evans, a clerical magistrate, together with Dr Llewellin of Lampeter, tried to give friendly societies an even more respectable image. Other elevators helped to rejuvenate the Carmarthen Mechanics Institute in the later 1840s, providing it, and the Young Men's Christian Society ten years later, with uplifting talks on 'The Worm that teaches Man' and 'The Building of the Pyramids'. Within a short time of the Rebecca riots similar institutes had been launched at Llandeilo, Haverfordwest, Llanelli, Pembroke Dock, Llandysul, and Aberystwyth. These bodies sometimes merged with the literary and scientific societies of the time, but their membership figures, like those of the so-called working men's clubs and co-operative societies of the 1860s, were disappointing.[93]

Perhaps the most characteristic and natural cultural developments of the 1850s and 1860s were lectures, penny readings, and eisteddfodau. The Llanelli Atheneum, opened in 1858 to the collective delight of David Rees, the Reverend Ebenezer Morris, and Rees Goring Thomas, lesser assembly rooms, village pubs, chapels, and schoolrooms were the venue for talks by the Reverend T. Price on Bunyan, Michael D. Jones on America, Henry Vincent on British history, and Dr Stone on Mesmerism. Even more popular in the 1860s were readings from Shakespeare and the Welsh bards, interspersed with hymns and words of wisdom, and concluded by the national anthem. Edward Lloyd Hall (now Fitzwilliams) and other veterans of the Rebecca campaign presided benignly over these proceedings, as they did over a number of eisteddfodau. The popularity of the latter in the mid-century owed much to the Cymmrodorion, the True Ivorites, the Oddfellows, and the chapels. They produced a rich diet of local, district, and national competitions, some in the Welsh language and others bilingual. The assembly at Cynwyl Caio in August 1858, the first of its kind in the neighbourhood, was typical. John Johnes of Dolaucothi presided, and the marquee was crowded to capacity.[94]

Already at this eisteddfod, the familiar note of half-apology was sounded; the president did not want to push the Welsh language at the expense of English, for Wales needed England and vice-versa. Over the years most of the opening and closing addresses turned on this theme, but by the time of the national eisteddfod at Aberystwyth in 1865 and at Carmarthen two years later there was a change in emphasis and a falling away of gentry and clerical support. 'The tastes of the Welsh were different from those of the English', was now the message, and perhaps the survival of their ancient language accounted for this and should be protected. Despite the frenzied criticisms of *The Times* and *John Bull*, nothing more seditious than that was intended, and the social science sections and the essays of eisteddfodau reaffirmed the glories of religion and nature, the importance of 'forming a good character in early life', and the value of self-improvement, hard-work, and harmonious relationships between master and servant. For some romantics, the literature and music of these festivals, and the fostering of native talent, did as much as anything to sooth the savage breast of the peasantry.

No one doubted, however, that economic change was a major contributor to the comparative social peace of the decades after Rebecca. Those with simplistic analyses of social tension studied the movement of prices with a special interest. The agricultural world took a little time to recover from the depression of the early 1840s, and the permanent reduction of rents, which Thomas Campbell Foster advocated as the answer to rural poverty, never came. During much of 1846 and 1847, to the delight of free traders, conditions improved, and there were even signs of a labour scarcity in parts of the countryside.[95] The vagrants and the poorest members of urban society were the casualties; within the workhouses the years 1844–8 were described as 'one long rebellion', as scores of inmates refused to perform their appointed tasks, while outside the walls subscriptions were raised for the starving poor. In 1849, as the agricultural economy worsened, farmers once again could not pay their rents and rates, and meetings were held to consider the distress. Indeed, Parliament and the press at this time were preoccupied with 'the state of agriculture' and 'the condition of the working classes', but the Young Englanders and the Cobbettites who wanted allotments and smaller farms were reminded that something of the kind already existed in Wales.

The low prices of the early 1850s brought fears of a further outbreak of rioting, but the hardship was not as general as in 1843, the targets of hostility less obvious, and soldiers and police were still on patrol. Soon both tenant farmers and farm labourers were enjoying the long mid-Victorian summer of agricultural prosperity. Some tenants experienced a doubling of rents in the period 1844–70, but there were profits to be made now, and when land came on to the market they mortgaged themselves to the hilt to buy it. Although a number of farming families, especially the young ones, looked outwards to the richer pastures of the border counties and the milk trade of London, most occupiers remained, like old sheep, close to the land of their birth. In the 1860s there were favourable comments on the recent advances in methods, livestock, and productivity on Welsh farms, and the hopes for the new monthly cattle markets at Carmarthen, Llandysul, and Llangadog proved fully justified.

The change in economic fortunes owed something to the railway boom, which brought markets closer and gave a boost to urban development and to the incomes of the working class. The prospect of the civilizing and commercial advantages of the railways was the one beacon of light in the post-Rebecca gloom. The South Wales Railway Bill was introduced into Parliament at the beginning of 1845, but such were the legal obstacles that the line between Swansea and Carmarthen was not opened until the autumn of 1852. In just over a year it was extended to Haverfordwest, and then more lines were added in rapid succession until, by 1864, much of Carmarthenshire and Pembrokeshire was connected by rail. The enterprise was carried out by a dozen railway companies, the directors of which were members of Parliament, local landowners, and the prominent businessmen whom we have met before. As the stations were built each town had its own reason to cheer; Milford's position as a port and communication point was secured, Llandysul's and Llandeilo's trade multiplied, and Tenby and Ferryside benefited from the revolution in holiday habits.

Not everyone bowed to 'King Rail', or welcomed his horde of alcoholic navvies, but, as isolated New Quay later realized, the coming of the trains could not be ignored. It was sometimes the catalyst which shook corporations out of their lethargy. Urban improvement was not new, but it had been held back by the depression and the Rebecca troubles. As economic conditions changed, so pressure mounted for spacious markets, better town and shire halls, assembly and reading rooms, libraries and institutes, savings banks and parks. Even Cardigan, Llandeilo, and Fishguard, famous for their lack of innovation, were moving forward rapidly in the late 1850s and early 1860s, and by the end of this period every major town had its gas, lighting, and water works. As we have seen, the presence of diseased Irish immigrants and the regular cholera scares kept the Boards of Health alert, and slowly, but ever so slowly, sanitation improved. Yet, in spite of everything, there was by the 1860s a rather sad aspect to this story of urban change. One or two of the sharpest commentators had already noticed that Carmarthen and several other towns in south-west Wales were showing 'signs of decay'; it was proving increasingly difficult to match the growth, entrepreneurial zeal, and architectural elegance of the wealthier towns of Britain.[96]

The point was made by the movement of population. There were parishes in the three counties which experienced a decline in population in the 1830s and especially in the 1840s, well before the coming of the railways, but in the next two decades the loss of people took on more serious proportions. The modern contraction and stagnation of the Welsh countryside had begun; well over 2,000 people every year left the three counties, and some villages were transported across sea and land.[97] Farmers and craftsmen from communities to the west of Carmarthen were packing their bags before the Rebecca riots were over, and during the next depression hundreds more arrived at Aberystwyth, Milford, and Swansea to board the ships bound for America, Australia, and Canada.[98] There were many reasons for going, but gold fever and Salt Lake City were prominent attractions. In later years more young dissatisfied miners and railway workers left for better rewards abroad, but always there was the steady haemorrhage of the farming population, drawn by Welsh cousins, the promises of free land, and the romance of an egalitarian and Godly homeland.

Most people, not least the farm servants and labourers, travelled much shorter distances, to the booming industrial hinterland of Llanelli, to Swansea, Merthyr, and the mines opening in the Rhondda. Twelve months after the ending of the riots the roads of Pembrokeshire and southern Carmarthenshire were said to have been crowded with these migrants. There was also work to be found on the railway lines being built across south and mid-Wales, in the building trade, and in the lead mines and ship-building industries which were enjoying a late revival. One result was growing expectations amongst those left behind. Employers in the three counties faced a succession of strikes in these years, notably in 1846, 1853, and 1856. Masons, carpenters, shipwrights, painters, weavers, tailors, and shoemakers demanded wage rises to match the advances being given elsewhere in Wales. There were colliers' strikes, too, and protests by navvies and lime-burners over their truck system. A few employers, notably R. J. Nevill and the owners of the Dafen tinworks at Llanelli, saw the importance of conditions of work and play, and provided reading rooms, concerts, and excursions for their workforce in the 1850s. Along with their Merthyr brethren, these workers were given an annual day-trip to the very districts from whence so many of them had come.

In the summer of 1850 the employees on the estate of Clynfiew were taken by carts and wagons for a day's holiday at Gwbert sands, near Cardigan.[99] It was an event deserving of comment in the local press, and a sign of the change that was to come in the position and treatment of this large body of workers. Between 1843 and 1867 agricultural labourers obtained on average a 50 per cent increase in their money wages, and there were parishes in Carmarthenshire where both indoor and outdoor servants obtained half as much again. Inevitably, the reduced agricultural workforce showed new signs of independence. In the courtroom there were complaints about conditions of employment and the quality of the food given by the farmers. For their part, farmers spoke of extravagant demands at hiring fairs, and of people disobeying orders and deserting their work. Hundreds of male and female servants chose these years to cut and run, secure in the knowledge that there was better-paid work not far away, and the chance to be master of their own social and political fate. When the Llandeilo and Carmarthenshire agricultural societies asked in 1867–8 whether better cottages and more flexible contracts should be offered to keep young people on their farms, it seemed a different world from that of fifty years before.[100] Even the prospect of eventually occupying a hill farm lost some of its attraction for labouring families; thousands of Rebecca's children preferred, as one of them put it, to move 'out of darkness and into the light'. The main contribution of the Welsh countryside to the wider British economy in the second half of the nineteenth century was the export of young meat and humans.

MYTHS

As the migrants' wagons set out on the last journey, and as the mountain cottages slowly crumbled into the earth, it became harder to explain, or even recall, the exploits of Rebecca's children. This was not entirely accidental, for each age hides some of its history and forges its own myths. In the second and third quarters of the nineteenth century Wales needed its defenders, and it suited them, and Dissenting Liberalism, to write a past that was in keeping with the coming heroic age. Wales appeared in the books of Thomas Phillips and Henry Richard as the most loyal, peaceable, religious, and moral part of Britain. Even her protests were respectable. 'There have been only two considerable exceptions to this rule,' wrote Richard in 1866, 'and these are more apparent than real.' Unlike the Newport rising, the Rebecca riots could not be blamed on English outsiders, but they were hardly a challenge to the lawful authority. Gates only had been destroyed, and the worst excesses had been committed by just a 'few evil-minded persons'.[101] At meetings called to protest against the report on Welsh education, parish after parish declared that it had taken no part in the 'foolish' enterprise. Even Llanelli had been 'innocent'. Thomas Lloyd, Sunday-school teacher of Hope chapel, Cardigan, admitted that there had been Rebeccaites in his district, but they had been concerned only with road taxes, had committed no outrages or destruction of property, and the movement had been put down by the efforts of 'Dissenting ministers and their flocks'.[102] At their worst, the riots were no more than an aberration, the first rural spasm since the Act of Union, and probably the last. In one of the first studies of Rebecca, devoted entirely to the attacks on tollgates, David Davies said that time had brought dignity to the affair, and shown Wales to be a remarkably peaceful place.[103] 'The normal condition of the Principality is one of profound calm,' confided Richard, 'rarely ruffled even by a breath of popular discontent.' Not for the first time a great protest movement filtered through respectable memory like seed through a riddle. One or two of the people who appeared before the Land Commission of the 1890s seemed unaware that their grandfathers had been violent criminals.[104]

It took some time for the events of 1839–43 to be forgotten, but during the next twenty-five years there were repeated assertions that the countryside had returned to its usual peaceful state, and that relations between landlords, tenants, and servants were almost as good as before. Compared to many other regions of Britain this one was less disturbed by tenants' rights and labour unionism. Some attributed this to economic improvement, the slow pace of change, and the constant relief of migration, but others believed that the reactions described above had their influence. It was claimed, for instance, that the initial concessions on roads, rates, and rents and the subsequent efforts by landlords and clergymen in the fields of education, religion, and charity now had their reward. Judges in these years congratulated both the gentry and rural population generally for having learnt the lessons of the turbulent past. It was their contribution to Britain's mid-Victorian stability.

In fact, as the generations passed the region developed a reputation for being the home of true Welshness, a land of beauty and 'contented, industrious and religious people'. There had always been a tendency to contrast the natural relationships of the countryside with the unpleasant crime, violence, and confrontational politics of industrial communities, and now, in the mid-Victorian years, the temptation became irresistible. Indeed, for the thousands who had left this land, the memory of 'yr hen wlad' grew ever more romantic with the passing decades. Of course, the turmoil over the elections of 1868 and 1874, the stories of the Great Depression, and a rare union protest by the forgotten agricultural labourers were reminders that all was not well, but some dreams of rural paradise remained.

Nowhere, it was suggested, was the Victorian conquest of crime and criminal protest more complete than in this region. Judges on the south Wales circuit were greatly impressed by the small number of cases at Assizes and Quarter Sessions in the third quarter of the nineteenth century.[105] At Cardigan county gaol, for example, there were fewer than a hundred committals a year between 1853 an 1861, half the figure of the late 1840s. The peasantry, under the guidance of a more responsible generation of landowners and ministers, were apparently setting their face not only against the evils of drink and pre-marital sex, but also against thieving, poaching, the *ceffyl pren*, the other forms of illegal protest. 'It is really almost painful to see the worthy [police]men . . . parading about the small towns and villages . . .', wrote Henry Richard at this time, 'looking in vain for some small job . . .' 'When compared with the morality of England, Scotland and Ireland, [that of Wales] stands very high', added his colleague Thomas Rees of Swansea. 'No landowner . . . has cause to fear the dagger of the assassin, the fire of the incendiary or the rude assaults of the infuriated mob.'[106] Newspapers, which closely monitored the recorded pattern of crime, were particularly proud of 'the absence of disturbances'; when London had its garrotters, the north of England its Reform riots, and East Anglia its arsonists, the counties of south-west Wales were a haven of tranquillity.

A Slow Death

On closer examination this myth of social peace was only partly true. Although there was a reduction in ideological and physical threats to the State in the mid-Victorian years, there was much illegality and tension beneath the surface. As almost all the studies of Rebecca have begun from the assumption that peace and harmony were the natural state of the Welsh countryside, such a reappraisal casts new light on Rebecca and nineteenth-century rural society generally. The region was less troubled than Ireland, and perhaps other parts of Britain, in this period, but it was not as peaceful as was claimed, nor 'free from agrarian crime'.[107] Even forms of Rebeccaism were present in this society, though more research is needed before we know for certain just how extensive these were.

The amount of criminal activity is impossible to establish; certainly, the criminal statistics were low in the 1850s, but they did begin to rise in the later 1860s, and continued upwards until the early 1880s. They reveal that drunkenness, disorderly conduct, and several other types of crime dealt with at the Petty Sessions were more common than many assumed. Murders were rare, but not attacks on people. Women and children were as likely to be assaulted as in the years before Rebecca. During this later period—1855 is an excellent example—desertion of babies, concealment of births, and the killing of children comprised about a quarter of indictable female crime, notably in Pembrokeshire. Mary Prout, who was given twenty years' penal servitude in 1864 for dropping 6-week-old Rhoda down an Amroth coalpit, was perhaps the most famous offender, but there were scores of others who quietly smothered and choked their young. One has the impression from the available evidence that fewer bailiffs and policemen were attacked in the 1860s than in the early 1840s, but informers, and those put into a farm after an ejectment order had been executed, were still justifiably nervous. In the immediate aftermath of the riots some of the community's anger was also directed against excise officers. In Cardiganshire villages the gangs who fought the officers were known as the 'malt Rebeccaites', and they often acted and dressed the part.[108]

Nor was property as safe as suggested. Sheep-stealing, the taking of vegetables and underwood, and poaching were popular, not least during the worst economic years of the 1850s and 1860s. Malicious damage was always more common than people admitted; there are numerous accounts in the press of incendiarism, animals killed and maimed, fences broken, and cottages torn down. Amongst those who had the unpleasant experience of finding suspended lambs, hamstrung cows, and poisoned pigs on their premises were A. J. Ackland of Boulston, Robert Waters of St Clears, and Daniel Davies, the Llangeler auctioneer.[109] Of course, there were worse things, and a few unfortunate hill farmers and shepherds had to watch as disguised gangs destroyed their homes, and the hay and turf which they had so laboriously collected. In 1850 a crowd on Pencarreg Mountain flattened a dwelling house, a beast house, and fences which had been erected on common land ten years before.[110] Parish and common pounds were another target, especially in the district between Carmarthen and Llanelli; scarcely a month went by without attempts to rescue animals taken because of trespass or for rent and other dues. At Kidwelly in 1854 the pound gate was ceremoniously thrown in the river, and six years later Rebecca appeared at Llangendeirne to carry out similar work.[111]

Three crimes in particular reveal that rural behaviour had not changed as drastically as some ministers claimed. Despite strenuous efforts by chief constable Scott and his men in 1844 to stop the 'dangerous custom', there were more than twenty reports in the *Carmarthen Journal* of carrying the *ceffyl pren* between 1845 and 1862, and there were other incidents recorded elsewhere. Many of these took place, as in the late 1830s, close to Cardigan and up the Teifi valley, but there were also cases at, for example, Llandeilo, Abergwili, and Milford. The usual targets were men and women accused of sexual impropriety, and those guilty of wife-beating. Young people and certainly females were now more prominent in the baiting crowds. What happened at Solva in 1856 was typical; a gang of forty, some with blackened faces, dragged a 72-year-old miller out of his house, and carried him full length on a ladder, the sound of horns and drums beating out his crime of seeing too much of a mariner's wife. Inspired by a misplaced belief that the judge in this case was favourable to the custom, two years later 1,000 people carried effigies of adulterers through the town of Fishguard, and rendered superintendent George Jones speechless with rage. Nor were the constabulary immune from the judgement of the crowd; in one of the many variations of the *ceffyl pren* people arrive late one night in April 1846 at the home of a Pembrokeshire policeman, exposing their persons and making obscene noises. They said that they wanted 'to tear down the house, and you, limb from limb, and to wash our hands in your blood'. They destroyed a hawthorn tree, and held a torch near a mow.[112]

Despite the confidence inspired by the new police, the threat of incendiarism never disappeared from the countryside. In the twenty years after Rebecca there were more than fifty serious incidents reported. This did not match the outbreaks of arson in rural England, but its survival is worth documenting. Some of the fires were acts of individual malice, by angry servants, tenants, and vagrants, but others were more closely associated with the protests of the Rebecca era. William Chambers was subjected to a number of attacks soon after the riots came to an end; in the spring of 1845, for example, the burning of one of his farms and its cottages was seen for miles around.[113] Other fires were started to intimidate landlords and tenants, and these were often accompanied by the sending of threatening letters, even with the Becca signature. At the beginning of 1859 there was a rash of fires and anonymous letters in the Llandeilo neighbourhood, and a committee appointed to help the victims. Sometimes buildings were set ablaze after an insolvent farmer left the premises, or, more commonly, just before a new tenant entered into his property and when there was a long delay in renewing a tenancy. On a rare occasion the attack was carried out by a mob, and there were hints that opposition to tithes, church rates, and other 'oppression' was behind the worst of these crimes. Amongst those who suffered, and offered substantial rewards for information about incendiarism, were John Beynon of Adpar Hill, David Lloyd Harries of Llandovery, William O. Brigstocke of Gellidywell, Edward Lloyd Williams of Gwernant, and the vicars of Llanarth, Llandeilo, and Pembrey.[114]

Some of the acts of incendiarism were related to the poaching war which has yet to find its historian. Ironically, one of the tests of the improvement in a Welshman's character was his apparent dislike of the crime of poaching. In truth, the taking of rabbits and birds was a common pastime, and in Cardiganshire alone there were several murderous affrays with 'foreign' gamekeepers in the 1860s. On the rivers it was a similar story, though the fishermen claimed greater public support. Attempts to control fishing, and stop poaching, on the Teifi, Tywi, Nevern, and other rivers met the firmest resistance. During the quarter of a century after the Rebecca riots, the landowners of the region were accused of turning the rivers, and their banks, into private property, just as they had done with the open moors. Sir James Drummond, Lord Cawdor, Captain Lloyd of Dolhaidd, William Brigstocke, John Beynon, and Charles Bishop formed fishing associations, successfully pressed the government for restrictive legislation, and brought the police into the conflict. The fishermen held meetings and sent petitions. 'However high the Conservators deem themselves to be as landed proprietors we believe ourselves to be no less men than they are,' said a spokesman.[115] In the last resort the fishermen ignored all regulations. Rivers were night poached by men with blackened faces and spears. Weirs and nets were destroyed, and the torch put to gentlemen's property, usually in the early hours of the morning under the guidance of a Rebecca figure. 'You must know the feeling I entertain for the poor people that are wronged', ran one of her Llechryd letters in 1862. 'My strength is still the same.'[116]

Of course, it was not. Although Rebeccaism appeared in a variety of guises during the 1850s and 1860s, it was not on the scale of the early 1840s. The countryside was never as peaceful as Henry Richard suggested, but the Rebecca riots were still unique. If 'the days of bullying and threatening violence' were not, as was claimed in 1854, 'past and gone', the direct action of this later period was no longer part of a great community protest movement.[117] The nearest approach to this was the gathering storm of Liberal politics. The major political issues had been discussed in the meetings of 1843, and the interest in them was kindled by the expanding Welsh press. In 1848 there were petitions in favour of financial and political reform from the largest towns, and amendments in support of greater religious liberty and universal suffrage were moved at public meetings during the next few years. At the same time the ballot received considerable support, local branches of reform societies being established to press for its adoption. One must not, however, exaggerate all this; leading reformers were disappointed by the scale of open commitment to political change in the region, and by the political organization of the chapels. As the gentry feared, much of the agitation on behalf of Liberal politics was led by David Rees of Llanelli, Henry Davies of Narberth, David Davies of Haverfordwest, and other Dissenting ministers, but some of their colleagues remained surprisingly indifferent to reform.

At the parish level, where attitudes were also changing, the mid-Victorian years witnessed some fascinating developments and debates. A few of the people who acted as community leaders during the Rebecca years remained prominent in the conflicts of this later period. One issue, which the riots had highlighted, was the status of the language. As the bureaucracy of the British state broke into the various preserves of local sovereignty, so there were more demands for Welsh-speaking judges, policemen, clerks and agents, water bailiffs, Poor Law officers, workhouse and gaol chaplains, and medical men in the Carmarthen lunatic asylum. Parish leaders argued that the appointment of outsiders, and especially English-speaking officials, whatever their professional merits, only reduced the quality of local services.

Other areas of tension remained much as they had been in the 1830s and 1840s. There was continued opposition to the better paid Poor Law appointments, especially when, contrary to expectations, the rate rose sharply in some places through the later 1850s and 1860s. At Aberystwyth, Cardigan, Llandeilo, and Narberth the courts were overwhelmed at times with defaulters, people unable to pay poor, highway, and county rates. Occasionally the failure to pay local taxes amounted to an organized campaign, and this was true of financial contributions to the Church and titheowners. The church rate was the dominant issue in parish politics in the post-Rebecca period; dozens of vestries and chapels sent petitions to Parliament against compulsion, a few Rebecca letters circulated, and crowds with black flags turned non-payers into public heroes.[118] The Liberation society tried, with limited success before the mid-1860s, to widen the protest, and gained the lasting hostility of Anglican landowners.[119]

The state of landlord–tenant relationships in the generation between 1844 and 1868 has been confused by the subsequent debate over evictions. Things were never quite as black or as white as the two competing sides maintained in the Land Commission report. In view of the number of tenants who voted with their Tory landowners in 1868 and 1874, one has to take seriously the claims that the gentry made a greater effort after the riots to remain on their estates and improve their social standing. In 1856, when George Rice Trevor received a testimonial for his services to the county, it was said that he and his friends had 'regained the esteem' lost thirteen years before. This was not completely true for it seems likely, as William Powell of Nanteos indicated in 1846, that respect for the great landed families had been weakened by the riots, and possibly by the subsequent discussions over tenant rights and the evils of the Game Laws. Those who praised the attempt by Trevor and others to improve class harmony also forgot that not everyone welcomed its direction. The wish to remove tension had a harsh side. Even before 1868 certain land agents and clergymen were pressing their masters to be religiously and politically careful in their choice of tenants. It has become customary to blame Dissent for the growing independence of the peasantry, but there were signs that the landlords themselves were in part responsible for the troubles that came upon them. In the twenty-five years after the riots there was a greater solidarity, determination, and Conservatism amongst the gentry of west Wales. Even men with a radical Whig pedigree felt the need to demonstrate the extent of their legitimate power, worried as they were by the long-term prospects of tenant independence and political change. 'The struggle', wrote Edward Lloyd Hall Fitzwilliams to his son in 1873, 'is beginning even here . . .'[120]

The elections of 1859 and 1865 did not fully prepare the gentry for the electoral politics unleashed by the second Reform Act. The Conservative vote held up fairly well in parts of the region in 1868, but the shock of Liberals heading the polls in two of the three counties made a considerable impression. Amongst the most disappointed landowners were political moderates, people like David Pugh, the independent Carmarthenshire candidate, the Powells of Maesgwynne, and the Tory Lloyds of Glansefin, who had played a careful part in the Rebecca riots and had been at the centre of educational, cultural, and recreational paternalism thereafter. It was a deep sense of hurt, as much as the rough handling of Rees Goring Thomas and his friends, and fears of the ballot and church Disestablishment, which precipitated the evictions of, and the threats against, Liberal tenants, shopkeepers, and ministers in the next few years. Although the scale of the intimidation was deliberately exaggerated, the people of south-west Wales had more reason to feel aggrieved than voters elsewhere in the Principality. On the farms of Alltyrodyn, Llanina, and Blaenpant, and in parishes where Rebecca had so often ridden, tenants who had voted for the first time clung 'shivering' to the land of their fathers, aware, in some cases, that estates were about to be sold off in parcels. For a few who had spoken at Liberal meetings, and chaperoned fellow tenants to the voting booths, retribution was as swift and silent as a signature.[121]

The retired surgeon Dr Enoch Davies of Bryn Teifi, near Llandysul, reflecting on these events, told the Land Commission that he could find in them only one consolation. 'Notwithstanding the fact that 50 families [of Cardiganshire] were rendered homeless [by the evictions] . . . there was not a single act of incendiarism or of violence committed.'[122] In fact, there were a few Rebecca-type letters and incidents during the upheaval of 1868–74, something repeated on a larger scale in the tithe riots twenty years later, but it was the constitutional nature of change in this period which makes it so important. Some of the old farmers who attended the hustings in the 1860s had learned the agenda of rural radicalism at the mountain conventions of 1843, and their sons were to present it, in its extended form, to the Land Commission in the 1890s. One speaker at an exuberant election rally in 1868 wanted to erect a statue of Gladstone near the place where the stone tribute to Rebecca had once stood, but his point was lost in the cheers of victory. Michael Thomas of Gwynfe had looked forward in 1843 to the day when Rebecca's children could give their mother a decent burial. It had finally arrived.

Notes

Chapter 1

1. For a description of the pillars, see *The Times*, 4 Sept. and 14 Oct. 1843.
2. Ibid. 27 June 1843.
3. C. S. Read, 'On the Farming of South Wales', *Journal of the Royal Agricultural Society of England*, 10 (1849), 122–65. Compare J. Gibson, *Agriculture in Wales* (London, 1879), and PP 1882, XV, Royal Commission on Agriculture, A. Doyle's Report on Wales.
4. The yield of crops was well below that in England and Scotland. D. W. Howell, *Land and People in Nineteenth-Century Wales* (London, 1977), pp. 18–19.
5. See the debate in *CJ*, 6 May 1864.
6. Quoted in an interesting discussion of Welsh problems in *W*, 22 Dec. 1843.
7. See the information on wills in the first chapter of the wide-ranging study by R. J. Colyer, *The Welsh Cattle Drovers* (Cardiff, 1976).
8. *The Times*, 9 Oct. 1843. Compare T. Cooke's comment in NLW, 21029 C, letter of 24 Apr. 1843.
9. Number based on lists published regularly in the local newspapers.
10. The phrase is taken from NLS MS 2844, letter from Col. Love, 4 Jan. 1844.
11. Cited in D. Jenkins, *The Agricultural Community in South-West Wales at the Turn of the Twentieth Century* (Cardiff, 1971), p. 280. Howell, op. cit. n. 4 above, is the essential book on price movements of the 19th century.
12. *CJ*, 12 Feb. 1836.
13. Quoted approvingly in *W*, 22 Dec. 1843.
14. The words are those of Hugh Williams, who will appear later in the story. The phrase hints at the anti-slavery agitation of the region. See G. E. Owen, 'Welsh Anti-Slavery Sentiments, 1790–1865', MA thesis, Aberystwyth, 1964.
15. This section on other industries is based on many sources, including newspapers and census returns, and owes a good deal to the work of Geraint Jenkins. See, e.g. his books *The Maritime Heritage of Dyfed* (Cardiff, 1982), and *The Welsh Woollen Industry* (Cardiff, 1969).
16. *W*, 21 July 1843. Context provided by W. J. Lewis, *Lead Mining in Wales* (Cardiff, 1967).
17. There is now a wealth of material on the development of heavy industry in the region. Two recent works which add considerably to our knowledge about two of the main industries are M. V. Symons, *Coal Mining in the Llanelli Area*, i (Llanelli, 1979), and R. T. Toomey, 'Vivian and Sons, 1809–1924', Ph.D. thesis, Swansea, 1979. A good starting-point remains PP 1846, XXIV, Report of the Commission on the State of the Mining Districts.
18. Confirmation of the comparative industrial importance of the Welsh counties can be gained from the statistics provided in the census of 1831,

1841, and 1851. Some of this is tabulated in a most helpful study by L. J. Williams, *Digest of Welsh Historical Statistics*, 2 vols. (Cardiff, 1985). John Jenkins's comment is in *W*, 23 Dec. 1842.
19. PP, 1895, XL, Royal Commission on Land in Wales and Monmouthshire (Land), vol. iii, qu. 47, 768.
20. The best starting-point for the story of road transport is still D. Williams, *The Rebecca Riots* (Cardiff, 1955). Supplement this with A. H. T. Lewis, 'The Development and Administration of Roads in Carmarthenshire, 1763–1860', MA thesis, Swansea, 1968.
21. Morris's comments appear in several places. See the interesting reflections by him, David Rees, and others at the opening of the Llanelli Atheneum. *CJ*, 29 Jan. 1858.
22. To set this information on population in context, see E. A. Wrigley and R. S. Schofield, *The Population History of England, 1541–1871* (London, 1981). It is worth pointing out that visitors to the region in 1839–43 were not as impressed with the pressure of population as historians have been. Williams, op. cit. n. 20 above, pp. 90 and 96.
23. The point is made in Muriel Evan's pioneering articles: 'Society in Nineteenth-Century Tre-lech a'r Betws', 'The Parish of Langyndeyrn 1851: A Population Study', and 'An Aspect of Population History in Carmarthenshire', *Carmarthenshire Antiquary*, 16 (1980), 67–79, 17 (1981), 79–95, and 19 (1983), 53–60.
24. J. Downes, 'Notes on a Tour of the Disturbed Districts in Wales', *Blackwood's Magazine*, 54 (1843), 766. The material on the population of these towns is based on detailed studies of the census of 1851. Some of these have been typed up into bound volumes, available at the Record Offices at Aberystwyth, Carmarthen, and Haverfordwest.
25. A point confirmed by Lady Frankland Lewis when she arrived there with her husband in Oct. 1843. NLW MS 16582 C.
26. *Morning Chronicle*, 5 Oct. 1843.
27. PP 1847, XXVII, Report of the Commission on the State of Education in Wales (Education), CGP, pp. 2–3.
28. *CJ*, 8 May 1846.
29. *The Times*, 7 Aug. 1843.
30. The marriage registers, which have also been typed into bound volumes, are thin on occupational detail. The parish registers of Llangendeirne and St Ishmael's, in the Carmarthen Record Office, are two of the best exceptions to the rule.
31. On the making, and nature, of the élite in Wales, see e.g. G. H. Jenkins, *The Making of a Ruling Class: The Glamorgan Gentry, 1640–1790* (Oxford, 1983) and D. W. Howell, *Patriarchs and Parasites* (Cardiff, 1986) and op. cit. n. 4 above. The *Welshman* comments on the growth of capitalists in its edn. of 13 Sept. 1844.
32. A book on the rise of the middle class, comparable with that of Edward Thompson on the working class, is still awaited. In the mean time there has been an important study by W. D. Rubinstein, *Men of Property* (London, 1981).
33. W. Jones, *A Prize Essay on the Character of the Welsh as a Nation* (Anglesey, 1841), p. 114.
34. Compare the experience of David Jones, wealthy Llandovery banker, who bought Pantglas, Llanfynydd, in 1822. Major Francis Jones, *Historic Carmarthenshire Houses and their Families* (Carmarthen, 1987), p. 141. This is an invaluable volume.

35. These paragraphs are based on an intensive study of the newspapers of the period, and of the census returns for Carmarthen, Llanelli, Llandovery, Aberystwyth, Cardigan, Haverfordwest, and Pembroke.
36. A point made by I. G. Jones, *Explorations and Explanations* (Llandysul, 1981), p. 173.
37. Archdeacon Beynon, in defending and recommending the Carmarthen Cymreigyddion, argued that one of its main purposes was to civilize the lower sections of society. *CJ*, 4 Mar. 1828. For Chadwick, see HO 45/454, letter of 11 July 1843.
38. See e.g. CRO, Acc. 4282 A, TW 8, letters from L. Lewis, 12 Aug. 1844, and from J. Lloyd Price, 12 Sept. 1844.
39. The census of 1841, used for these paragraphs, is not as reliable as one would hope. Parish information taken from copies of the census returns in the Carmarthen Public Library.
40. The considered modern judgement on this and other aspects of marriage and population history is Wrigley and Schofield, op. cit. n. 22 above.
41. See e.g. the summary of regional statistics in PP, 1844, XIX, Sixth Annual Report of the Registrar General. Compare P. Laslett, K. Oostverven, and R. M. Smith (eds.), *Bastardy and its Comparative History* (London, 1980).
42. PP. 1836, XXIX, pt. I, Second report of the Poor Law Commissioners, Appendix D, Table 3. There was a rather unedifying debate throughout the mid-19th century on the issues of illegitimacy, and sex before marriage, which reached perhaps its lowest point in the *Morning Herald*, 6 Dec. 1843. Clergymen concentrated on the second issue, and Henry Richard and others on the first. Their importance for the story of Rebecca will be examined later in this book.
43. W. Brigstocke admitted that 'misconduct' after marriage was rare. PP 1847, XXVII, Education, BCR, p. 94. Compare K. H. Connell, *Irish Peasant Society* (Oxford, 1968), pp. 62 and 82.
44. PP 1847, XXVII, Education, CGP, p. 421.
45. Amongst the parish registers consulted by the author are those of Meidrim, Llanddarog, Llanarthne, St Ishmael's, and Llangendeirne. All are in CRO.
46. Precise details on births, deaths, and marriages, and on illnesses, in the seven districts of the region are given in the Registrar-General's annual reports. PP 1839, XVI, and following years.
47. Information on these paragraphs taken from the census of 1851.
48. Information from the census of 1841 and 1851, and from the marriage registers of Meidrim, Llanddarog, Llechryd, Troedyraur, and Lampeter Velfrey. All now typed up in large bound volumes, available at each of the Record Offices.
49. See e.g. M. I. Williams, 'Seasonal Migration of the Cardiganshire Harvest Gangs to the Vale of Glamorgan in the Nineteenth Century', *Ceredigion*, 3: 2 (1957), 156–9.
50. *CJ*, 15 Dec. 1843.
51. South-west Wales was especially affected by emigration, as one can see from the census of 1841, and from many newspaper reports, as *CJ*, 29 Apr. and 3 June 1842. The *Morning Herald* reporter believed that colonization was the only answer to the problems of the region, 9 Oct. 1843.

Notes 381

Chapter 2

1. PRO, HO 45/454, letter of 23 July 1843.
2. The sources for these tables and paragraphs are PP 1874, LXXII, Return of Owners of Land; J. Bateman, *The Great Landowners of Great Britain and Ireland* (London, 1876); and G. C. Brodrick, *English Land and Landowners* (London, 1881).
3. Examples in these paragraphs taken from either the census returns of 1851 or PP 1844, XVI, Report of the Commission of Inquiry for South Wales (South Wales).
4. Michael Bowen in fact rented some land from his father. The family were, however, considerable freeholders. For this and other information on the proprietors and tenant farmers of Tre-lech a'r Betws, see M. B. Evans, 'The Community and Social Change in the Parish of Tre-lech a'r Betws during the Nineteenth Century', MA thesis, Aberystwyth, 1980.
5. Information on sales, and the reception given to them, from *CJ* and *W* (1825–50). Also useful is Francis Jones, *Historic Carmarthenshire Houses and their Families* (Carmarthen, 1987).
6. For some background on Slebech, see B. Ll. Morris, *The Slebech Story* (Haverfordwest, 1948).
7. For the story of 'Rhyfel y Sais Bach', and further references, see D. J. V. Jones, *Before Rebecca* (London, 1973), pp. 48–50.
8. For some background information on enclosures in Wales during the period, see PP 1844, V, 'Report of the Select committee on Commons' Inclosure; *W*, 20 Oct. 1843 and 13 Sept. 1844; D. Williams, *The Rebecca Riots* (Cardiff, 1955), pp. 79–85; Jones, op. cit. n. 7 above, pp. 45–50; and A. E. Davies, 'Enclosures in Cardiganshire, 1750–1850', *Ceredigion*, 18: 1 (1976), 100–40.
9. NLW, Crosswood, 1174, case and counsel's opinion, 10 June 1797.
10. *W*, 28 July 1843, and *CJ*, 27 July 1838.
11. *W*, 24 Feb. 1843.
12. Ibid. 22 Dec. 1843. See Williams's comment, op. cit. n. 8 above, p. 312 n. 138. Was this the same D. Gower of Kidwelly who was indicted for a riot and attack on the house of J. King? PRO, Assizes 71/7.
13. For the use of the Rural Police against the one-night houses, see the report of the Cardiganshire Quarter Sessions in *W*, 25 Oct. 1844.
14. *CJ*, 11 Aug. 1815, and D. J. V. Jones, op. cit. n. 7 above, p. 48.
15. *CJ*, 6 Oct. 1843. See J. Lloyd Davies's bitter comment on this enclosure, in ibid. 10 Dec. 1847.
16. *The Times*, 9 Sept. 1843.
17. Taken from information given within the census return of 1851 for the parish of Abernant, a copy of which is available at the Carmarthen Public Library. Like all such census details, it is not totally reliable. Compare other lists for different areas, as in PP 1847, XXVII, Education, CGP, p. 234, and D. Jenkins, *The Agricultural Community in South-West Wales at the Turn of the Twentieth Century* (Cardiff, 1971), p. 269.
18. Subletting was not general in the region, according to the *Morning Herald*, 7 Oct. 1843. The *Times* reporter (as in 5 Sept. and 21 Oct. 1843) thought that it was more common than people admitted or realized.
19. The important reference works for this part of the book are PP 1896, XXXIV, Land, and D. W. Howell, *Land and People in Nineteenth Century Wales* (London, 1977). For a little of the long debate over whether leases

were offered, or wanted, see *The Times*, 28 Sept. 1843; *CJ*, 19 Nov. 1847; and *W*, 13 Nov. 1868.
20. For a study of the Cawdors, which touches on some of these aspects, see M. Cragoe, 'The Golden Grove Interest in the Politics of Carmarthenshire, 1804–21', MA thesis, Swansea, 1987.
21. CRO, Cawdor 2/143.
22. PP 1895, XL, Land, vol. III, qu. 44871.
23. CRO, Cawdor 2/206.
24. NLW, Nanteos R. 82, and MS 21029 C, letters from T. Cooke, 23 Jan. and 8 Oct. 1842. Compare PP 1844, XVI, South Wales, App. 5, p. 446.
25. Howell, op. cit. n. 19 above, pp. 9–11, and PP 1896, XXXIV, Land, pp. 371–3.
26. See the comment by Sophia C. Paynter in NLW, Behrens 501(a), letter of 9 Jan. 1844.
27. *The Times*, 21 Oct. 1843. There is a great amount of information on rents. See e.g. ibid. 27 July 1843, 7 and 14 Aug. 1843; *Morning Herald*, 27 Nov. 1843; PP 1844, XVI, South Wales, p. 27; and CUL, Peel Papers (Microfilm), letter from J. Graham, 4 Sept. 1843. For cases of spiteful destruction by tenants under notice to quit, see *CJ*, 8 Jan. and 12 Nov. 1841, and 28 Oct. 1853.
28. For the background, see PP 1870, XIII, Third Report of the Commission on the Employment of Children, Young Persons and Women in Agricuture (Women and Children). On the disadvantages of female labour, see *CJ*, 4 Nov. 1864.
29. Jenkins, op. cit. n. 17 above, pp. 88–91.
30. *The Times*, 30 Sept. 1843.
31. Even at the end of the century, accommodation for these people had improved but little. PP 1896, XXXIV, Land, p. 640. For a good general survey of conditions then, see M. Birtwhistle, 'Pobol y Tai Bach', MA thesis, Aberystwyth, 1981.
32. One claim put the perks at more than half the value of the money wages. *CJ*, 8 May 1868.
33. Jenkins, op. cit. n. 17 above, pp. 51–72; *The Times*, 19 Sept. 1843; PP 1870, XIII, Women and Children, Cardiganshire, p. 129; and PP 1893–4, XXXVI, Royal Commission on Labour: The Agricultural Labourer, Wales, pp. 18–19.
34. *Morning Herald*, 20 Nov. 1843.
35. *W*, 4 Aug. 1843.
36. PP 1844, XVI, South Wales, p. 219.
37. *W*, 29 Nov. 1844.
38. When he addressed a public meeting T. Lloyd apologized for his lack of fluency in Welsh, *The Times*, 26 June 1843.
39. *CJ*, 13 Oct. 1843.
40. Ibid. 25 Aug. 1843.
41. Ibid. 17 Nov. 1843.
42. NLS, Brown MS 2844, letter from Col. Love, 5 Feb. 1844; *CJ*, 21 May 1852; and *The Times*, 22 June 1843.
43. *CJ*, 16 Aug. 1822.
44. Based on information gleaned from PP 1844, XVI, South Wales, p. 209, and CRO, Vestry Minutes of Tre-lech a'r Betws (CPR 12/26), Merthyr (CPR 14/14) and Llangadog (CPR 49/20).
45. CUL, Peel Papers (Microfilm), letter from J. Graham, 4 Sept. 1843.
46. *CJ*, 28 July 1843.

Notes

47. T. Phillips, *Wales* (London, 1849), p. 30. Contemporaries constantly used the term 'peasantry' when describing the people of south-west Wales. For some of the problems, and advantages, in using this term, see the brief survey by M. Reed, 'Nineteenth-Century Rural England: A Case for Peasant Studies?', *Journal of Peasant Studies*, 14: 1 (1986), 78–99.
48. *Morning Chronicle*, 10 Oct. 1843.
49. HO 45/454, letter from E. Laws, 26 June 1843.
50. See ch. 4 for further details on such crimes. Compare the Llangeler case, reported in *CJ*, 14 Mar. 1862.
51. *Cambrian*, 20 Mar. 1846.
52. *Swansea Journal*, 30 Aug. 1843.
53. HO 45/454, letter of 15 June 1843.
54. *W*, 15 Dec. 1843, and PP 1847, XXVII, Education, BCR, p. 64.
55. PP 1847, XXVII, Education, BCR, p. 94.
56. Based on a study of the newspapers and journals of the period. To set it in context, see e.g. T. M. Owen, *Welsh Folk Customs* (Cardiff, 3rd edn. 1974).
57. *CJ*, 6 Dec. 1844 and 1 Jan. 1845.
58. The essential background is the text, tables, and appendices of PP 1847, XXVII, Education.
59. Ibid. CGP, p. 28.
60. Ibid. BCR, p. 160.
61. Ibid. CGP, p. 230.
62. Ibid. BCR, App. D., p. 262.
63. Ibid. CGP, p. 244.
64. Ibid BCR, p. 75.
65. *The Times*, 3 July 1843. For the opposite claims, discussed in the next lines, see *CJ*, 28 June 1839.
66. Information on churches, chapels, and attendances has been chronicled by I. G. Jones and D. Williams, *The Religious Census of 1851: South Wales* (Cardiff, 1976).
67. PP 1844, XVI, South Wales, p. 258.
68. D. M. James, 'Some Social and Economic Problems of the Church of England in the Diocese of St. David's 1818–74', MA thesis, Aberystwyth, 1972.
69. Two useful research studies here are T. Evans, 'Political Thought in Wales, 1789–1846', MA thesis, Aberystwyth, 1924, and R. D. Rees, 'A History of South Wales Newspapers to 1855', MA thesis, Reading, 1955.
70. *W*, 12 Apr. 1844.
71. Ibid. 22 Sept. 1843.
72. *CJ*, 16 Feb. 1849.
73. NLS, Brown MS 2844, letter from Col. Love, 6 Feb. 1844.
74. The identification of these key figures is taken mainly from a study of the very extensive holdings of Quarter Session, turnpike trust, and Poor Law material in CRO.
75. Information from the PP 1844, XVI, South Wales, and from vestry books. See e.g. NLW, Parish Records, Llanfihangel-ar-arth Vestry Minutes 1831–93, and CRO, CPR 35/15, Vestry Book of Llanarthne 1835–72.
76. *Swansea Journal*, 18 Oct. 1843. It is a little difficult to identify J. Lewis, but he appears to be the man of Llwynmendissa, with some 100 acres and a rateable value of £75. CRO, CPR 49/20, and Carmarthen Public Library, Llangadog, census of 1841 and 1851.
77. CRO, CPR 3/22, Vestry Minutes of Llangunnor.

78. CRO, Acc. 4282 A, TW 11, letter from J. Lloyd Price, 22 Sept. 1844.
79. PP 1847, XXVII, BCR, p. 83.
80. HO 45/1611, copy of a letter from W. Day, 9 July 1843.
81. PP 1895, XL, land, vol. iii, qu. 44303.
82. HO 45/347A, second report of person secretly engaged to obtain information.
83. *The Times*, 30 Sept. 1843.
84. See e.g. his comments in PP 1844, XVI, South Wales, p. 5. Compare LEX's interesting letter on the changes in economic and social relationships. *W*, 30 Dec. 1842.
85. Several reports of his speech. *Morning Herald* and *The Times*, 9 Sept. 1843.
86. *The Times*, 18 Sept. 1843.
87. NLW, MS 21209 C, letter of 3 Mar. 1844.
88. See e.g. *The Times*, 30 Sept. and 6 Oct. 1843; R. Colyer, 'The Gentry and the County in Nineteenth-Century Cardiganshire', *Welsh History Review*, 10: 4 (1981), 514; and D. W. Howell, 'An Example of Coercive Landlordism', *Carmarthenshire Historian*, 15 (1978), 63–9.
89. *W*, 17 Nov. 1843, quoting from the *Atlas*.
90. *The Times*, 19 Aug. 1843.
91. *CJ*, 29 Sept. 1843, and CUL, Graham Papers (Microfilm), letter from J. Graham, 14 Sept. 1843.
92. Tense altered. Report in *W*, 15 Sept. 1843. There was a long and acrimonious debate over the quality of magistrates in *The Times*, the *Morning Herald*, and within Parliament. Hansard, vol. 70 (1843), cols. 1440 and 1492.
93. CUL, Graham Papers (Microfilm), letter of 21 Dec. 1843.

Chapter 3

1. *W*, 1 Sept. 1843.
2. *The Times*, 26 June and 10 Aug. 1843.
3. *CJ* and *The Times*, 13 Oct. 1843. Compare the comment in *W*, 21 July 1843.
4. *W*, 10 Nov. 1843. For claims that the poverty about Carmarthen was without equal in the British Isles, see ibid. 15 Dec. 1843.
5. See the changing perceptions in the *Morning Chronicle*, 2 and 10 Oct. 1843, and the *Morning Advertizer*, 28 Sept. 1843. On the popular notion that many years of hardship had brought out the worst characteristics of the Welsh peasants, and created Rebecca, see the *Morning Herald*, 6 Dec. 1843.
6. Sarah Jacob's story is in *W*, 19 Mar. 1869. Anne Davies's is in ibid. 25 Oct. 1844.
7. NLW, Cilgwyn 35, p. 61. Cited by J. H. Davies, 'The Social Structure and Economy of South-West Wales in the Late Nineteenth Century', MA thesis, Aberystwyth, 1967, p. 24. *The Times*, 20 Nov. 1843.
8. PP 1846, XXIV, Report of the Commissioners on the State of the Mining Districts, Pembrokeshire, p. 40.
9. PP 1865, XXVI, Seventh Report of the Medical Officer of the Privy Council, Appendix 9, pp. 498–9. To put it in context, see the excellent article by I. G. Jones, 'The People's Health in Early Victorian Wales', *Transactions of the Honourable Society of Cymmrodorion* (1984), 115–47.

Notes

10. PP 1842, XVII, Commission on the Employment of Children in Mines and Manufactories, II, South Wales, pp. 705–5, and PP 1895, XL, Land, vol. v, qu. 72552. Dr A. Armstrong gives a rather more favourable picture of the labourer's diet and life, but agrees that at the beginning of the nineteenth century, calory intake reached Third World proportions. *Farmworkers: A Social and Economic History 1770–1980* (London, 1988), 42 and 99–101.
11. See one such letter in the *Swansea Journal*, 4 Oct. 1843.
12. Paragraph compiled from the many first-hand descriptions of the Welsh in the London newspapers of the autumn of 1843.
13. *The Times*, 6 Oct. 1843.
14. Were these the worst-housed labourers in Britain? PP 1882, XV, Royal Commission on Agriculture, A. Doyle's report, p. 56.
15. *CJ*, 20 June 1845.
16. PP 1844, XIX, Sixth Annual Report of Registrar General. Not all the cases of suicide were reported in the newspapers and the first police records of this period, but a glance at the *Carmarthen Journal* of 1849, for example, indicates just how inadequate the above Welsh return was. Historians have perhaps underestimated the suicide rate in Wales. O. Anderson, *Suicide in Victorian and Edwardian England* (Oxford, 1987), pp. 99–101.
17. *CJ*, 16 Mar. 1849.
18. Ibid. 1 May 1846. Compare ibid. 11 Apr. 1845, 27 Mar. 1846, and 29 Jan. 1847.
19. Ibid. 30 Apr. and 7 May 1847.
20. Ibid. and *W*, 26 Jan. 1844.
21. For the comment, see *W*, 14 Apr. 1843, and for the cases, see *CJ*, 2 Apr. 1847 and 25 Sept. 1848.
22. *CJ*, 26 Mar. and 2 Apr. 1847. There were many similar cases, as in 22 Mar. 1850, and 28 Mar. 1851. For the comments by Ashley, Graham, and others on the 'shameful neglect' of the insane in Wales, see *W*, 25 Aug. 1843, and Hansard, vol. 76 (1844), cols. 1268 and 1274.
23. Much of the material, admittedly of limited value, for this period can be found in the parliamentary returns of pauper lunatics, and in the newspaper comment on the figures. See e.g. PP 1844, XVIII, Appendix to the Report on Lunacy, and *CJ*, 16 July 1847 and 2 Dec. 1864, and *W*, 26 Mar. 1869.
24. *Morning Chronicle*, 16 Oct. 1843, and *Swansea Journal*, 18 Oct. 1843.
25. CUL, Graham Papers (Microfilm), letter of 19 Sept. 1843.
26. PP 1844, XVI, South Wales, p. 360.
27. *The Times*, 25 Aug. 1843.
28. *CJ*, 6 Nov. 1835.
29. Ibid. 15 Mar. 1839.
30. Ibid. 4 Aug. 1843.
31. Letter cited in D. J. V. Jones, 'Distress and Discontent in Cardiganshire 1814–19', *Ceredigion*, 5: 3 (1966), 282.
32. HO 40/46, letter from A. Brackenbury, 16 June 1821.
33. See e.g. the comments on Llanllawddog, Llanpumpsaint, Llanycrwys, Spittal, and several other places in PP 1847, XXVII, Education.
34. On the problems of work, see the speeches recorded in *The Times*, 24 July and 6 Oct. 1843.
35. *CJ*, 13 Mar. 1840. The winter of 1846–7, a time of potato famine, was bad. See the letter on this in ibid. 19 Feb. 1847.
36. Evidence on wages taken from many sources. There are literally scores of

references to them in PP 1847, XXVII, Education, and in the reports of *CJ* and *W* for 1843.
37. The *Times* reporter claimed that payment in food was a necessity for such a poorly paid and fed workforce. *The Times*, 20 Nov. 1843.
38. See e.g. *CJ*, 31 Aug. 1832 and 31 Dec. 1841.
39. See e.g. ibid. 26 June 1839.
40. See e.g. *W*, 22 and 29 Sept. and 22 Dec. 1843.
41. Ibid. 29 Sept. 1843.
42. *CJ*, 15 Dec. 1843, and NLW, MS 3141 F, Poor Law Correspondence, vol. iii, case of W. Thomas.
43. See e.g. *CJ*, 12 June 1846, 12 Mar. 1847, and 12 Jan. 1849.
44. D. J. V. Jones, 'The Criminal Vagrant in Mid-Nineteenth Century Wales', *Welsh History Review*, 8:3 (1977), 312–44.
45. See e.g. one of her letters in *CJ*, 27 Oct. 1843.
46. NLW, Llwyngwair MS 8, cited in R. J. Colyer, 'The Gentry and the County in Nineteenth-Century Cardiganshire', *Welsh History Review*, 10: 4 (1981), 499.
47. *CJ*, 24 Nov. 1843.
48. CRO, CPR 12/26, Tr-lech a'r Betws Vestry Minutes, 1819–37 (document in wrong section).
49. The best county study on poor relief is A. M. E. Davies, 'Poverty and its Treatment in Cardiganshire, 1750–1850', MA thesis, Aberystwyth, 1968.
50. *CJ*, 10 Feb. 1843, and *W*, 2 June 1843.
51. See e.g. PP 1844, XVI, South Wales, pp. 30 and 168.
52. Much of these chapters is based on an intensive study of the annual reports of the Poor Law Commissioners, PP 1835, XXXV onwards, and of the Poor Law and Parish Records in the record Offices at Aberystwyth, Haverfordwest, and Carmarthen.
53. PP 1836, XXIX, pt. 1, Second Report of the Poor Law Commissioners, Carmarthen Union, pp. 367–8.
54. CRO, CPR 14/14. Merthyr Vestry Minutes, 1802–42.
55. PP 1836, XXIX, pt. 1, Second Report of the Poor Law Commissioners, Carmarthen Union, p. 369.
56. PP 1844, XVI, South Wales, p. 272.
57. Based on the surviving vestry minutes of the period, and on reports in the local press and periodicals. For some government returns on the payment and non-payment of church rates, see PP 1845, XLI; 1851, IX; 1852, XXXVIII; and 1856, XLVIII.
58. CRO, CPR 3/22. Vestry Minutes of Llangunnor, p. 64.
59. Delyth James claims that there was no redress, short of imprisonment, for non-payment of church rates. 'Some Social and Economic Problems of the Church England in the Diocese of St David's 1818–74', MA thesis, Aberystwyth, 1972, p. 197.
60. The important work here is E. J. Evans, *The Contentious Tithe* (London, 1976).
61. *The Times*, 13 Oct. 1843.
62. PP 1844, XXXI, Report of the Tithe Commissioners, p. 3.
63. There is a great amount of information on all the disputed tithe cases. Many of the statements before the commissioners in 1843, and at the public meetings of that year, were about tithes. PP 1844, XVI, South Wales, and *CJ* and *W*, July–Dec. 1843. There are also many references in the estate papers. See e.g. accounts of the Penbryn case in CRO, Evans (Aberglasney), esp. 27/597.

Notes

64. PP 1844, XVI, South Wales, p. 360.
65. *CJ*, 17 Nov. 1843.
66. *The Times*, 30 Sept. 1843.
67. The story behind this paragraph can be followed in the annual reports of the Poor Law Commissioners; in PRO, MH 12/15783 and 15818; Davies, op. cit. n 49 above; NLW, MS 3141 F, Poor Law Correspondence; and the local vestry minutes.
68. *Morning Herald*, 20 Oct. 1843; *W*, 13 Oct. and 3 Nov. 1843; and *Swansea Journal*, 30 Aug. 1843.
69. PP 1844, XVI, South Wales, p. 233.
70. *Swansea Journal*, 30 Aug. 1843.
71. , HO, 45/1611, copy of a letter of 9 July 1843. Day claimed that they really despised the poor.
72. NLW, MS 3141 F, Poor Law Correspondence, vol. iii, case of 18 July 1843. See the interesting petition on such treatment, in *CJ*, 30 Mar. 1837.
73. See the useful tabulation in PP 1844, XVI, South Wales, Appendix 5, pp. 457–8.
74. Pembrokeshire Record Office, SPU/PE/1/2, Minutes of the Pembroke Union, 5 and 19 Apr. 1843.
75. PP 1844, XVI, South Wales, p. 57.
76. See e.g. the many cases in the first years of CRO, QS Minute Book, Oct. 1837–Jan. 1843.
77. PP 1836, XXIX, pt. 1. Second Report of the Poor Law Commissioners, Carmarthen Union, p. 369. Compare e.g. PP 1844, XVI, South Wales, p. 229; *The Times*, 16 Oct. 1843; and *W*, 13 Oct. 1843.
78. *Swansea Journal*, 18 Oct. 1843.
79. Pembrokeshire Record Office, SPU/PE/2/1, Letter Book of the Pembroke Union, 1837–52; letter from the Clerk of the Union, 26 Jan. 1843.
80. PP 1844, XL, Return of Poor Law Expenditure, 1841–3. And PP 1845, XXVII, Eleventh Report of the Poor Law Commissioners, Wales, Table 5(a).
81. PP 1844, XVI, South Wales, p. 52.
82. *W*, 27 Oct. 1843.
83. Information on the county stock taken from various Parliamentary Papers on Local Taxation and County Treasurers' Accounts, from the Quarter Session Records in the Pembrokeshire Record Office and in the CRO, and from *The Times* reports, June–Dec. 1843.
84. There is a huge amount of material on the administration of roads and bridges. The best starting-point is HO 45/454B, Report of G. H. Ellis, 2 Nov. 1843, and in the same file there is a report on the finances of the trusts. PP 1844, XVI, South Wales, has much interesting information, and one can follow the stories of individual trusts in the books kept in the county Record Offices and in the National Library. The author used CRO, TT 2, Main Trust Order Book, 1830–45, and NLW, Cardiganshire County Roads, B.3, Account Book of 1839–44, a good deal. Further information can be obtained from the Quarter Session Records, and from the many newspapers listed in the bibliography. A. H. T. Lewis, 'The Development and Administration of Roads in Carmarthenshire, 1763–1860', MA thesis, Swansea, 1868, is the best secondary account.
85. PP 1844, XVI, South Wales, p. 141.
86. *The Times*, 30 Sept. 1843.
87. PP 1844, XVI, South Wales, p. 111.
88. On this section, see e.g. ibid. pp. 24 and 104, and *The Times*, 19 July 1844.

89. *The Times*, 19 Aug. 1843. Dr Howell insists that the importance of tolls must not be underestimated. 'The Rebecca Riots', in T. Herbert and G. E. Jones (eds.), *People and Protest: Wales 1815–80* (Cardiff, 1988), 115.
90. Ibid. 25 Aug. 1843.
91. Ibid. 18 Aug. 1843.
92. *Morning Herald*, 18 Sept. 1843.
93. Price movements described in this section are taken from the weekly returns in the *Carmarthen Journal*. For the national background, see D. W. Howell, *Land and People in Nineteenth-Century Wales* (London, 1977), Appendices 1 and 2.
94. Several accounts. See e.g. *CJ*, 30 Dec. 1842.
95. *The Times*, 18 Aug. 1843.
96. On labour cuts and the problems of farm workers at Abernant and elsewhere at this time, see e.g. PP 1844, XVI, South Wales, p. 78, and *The Times*, 24 July and 3 Aug. 1843.
97. W, 4 Aug. 1843.
98. CUL, Graham Papers (Microfilm), Bundle 63, half-letter of July 1843.
99. See e.g. D. J. V. Jones, *The Last Rising* (Oxford, 1985), pp. 221–2.
100. *Morning Herald*, 9 Sept. 1843.
101. NLW, Nanteos, R. 82.
102. NLW, Edwinsford, 4666, Rental 1843–6.
103. NLW, Slebech, 7432, and CRO, Coedmore D/LL/2500, and Derwydd CA/19.
104. CRO, Cawdor 2/108, and 143. Compare Castell Gorfod, B. 80(a), letter from J. T. Wedge, 25 July 1843.
105. *The Times*, 18 Aug. 1843.
106. PP 1844, XVI, South Wales, p. 58.
107. For the landlords' viewpoint, at such difficult times, see Lewis Evans's comments in ibid. 75.
108. See e.g. PP 1844, XXXVIII, return of those imprisoned for debt.

Chapter 4

1. To set the chapter in context, see C. Emsley, *Crime and Society in England 1750–1900* (London, 1987); V. A. C. Gatrell, B. Lenman, and G. Parker (eds.), *Crime and the Law: The Social History of Crime in Western Europe since 1500* (London, 1980); D. J. V. Jones, *Before Rebecca* (London, 1973) and *Crime, Protest, Community and Police in Nineteenth-Century Britain* (London, 1982).
2. See e.g. *CJ*, 5 July 1839 and 30 June 1843, and HO 40/51, letter from T. S. Biddulph, 6 July 1839. Pat Molloy accepts the traditional view, *And They Blessed Rebecca* (LLandysul, 1983), p. 15.
3. When the calendars are actually examined the claims of the judges and newspapers are not always substantiated, but the number of cases was often small. See the claims in *CJ*, 10 Sept. 1830, 2 Aug. 1833, and 20 July 1838. Compare the *Quarterly Review*, 74 (1844), 124.
4. For more detail on the Welsh crime rate, and opinions on the Welsh and crime, see D. J. V. Jones, 'The Welsh and Crime, 1801–1891', in C. Emsley and J. Walvin (eds.), *Artisans, Peasants and Proletarians 1760–1860* (London, 1985), pp. 81–103.
5. CUL, Graham Papers (Microfilm), letter of 14 Sept. 1843. Compare NLS, MS 2844, Letter from Col. Love, 22 Feb. 1844.
6. Percentages taken from the judicial statistics in the Parliamentary Papers, which first appeared in 1805 and were improved in 1835–6, and

Notes 389

1856–7. Many of these statistics have been collected and tabulated in D. J. V. Jones and A. Bainbridge, 'Crime in Nineteenth-Century Wales', Social Science Research Council Project, 1975. The Welsh juries were regarded as a problem by government. CUL, Peel Papers (Microfilm), Add. MS 40, 449, letter from J. Graham, 16 Sept. 1843.

7. See the useful return of Petty Session Divisions and magistrates in the three counties, in HO 454, fo. 777 and 837. I have used the contemporary spelling.
8. HO 45/642, letter of 8 Apr. 1844.
9. NLS, MS 2843, letter of 21 Dec. 1843.
10. HO 45/454, letters of 21 June and 26 July 1843.
11. HO 45/453, letter of 30 Sept. 1843. The story of policing and crime in one place, that of the town of Carmarthen, has been well told in P. Molloy, *A Shilling for Carmarthen* (Llandysul, 1980), and *Four Cheers for Carmarthen* (Llandysul, 1981).
12. *CJ*, 21 Nov. 1845.
13. Pembrokeshire Record Office, PQ/AG/1–7, Prison Register and Register of Felons. There were some 409 listed for the year ending Oct. 1834, and 406 for the year ending Oct. 1844. See, too, PP 1854, LIII, Return of Police and Prisoners.
14. See e.g. *CJ*, 7 Apr. and 30 June 1848, and 17 Oct. 1851.
15. Jones, op. cit. n. 4 above.
16. 'R.V.' in *CJ*, 25 Aug. 1843.
17. PP 1847, XXVII, Education, BCR, p. 90.
18. *The Times*, 19 Aug. 1843, and PP 1844, XVI, South Wales, p. 134. See the interesting demands for cheap arbitration and local courts, at Tre-lech and Eglwyswrw. *W*, 29 Sept. and 17 Nov. 1843.
19. Detailed studies of these records are urgently needed. The manorial records of the Cawdors alone are extensive. See e.g. the fascinating information in CRO, Cawdor 2/ 257–9.
20. *CJ*, 22 Nov. 1844.
21. HO 45/454, letter from J. Walters and an anonymous writer, both similar and both dated 4 Oct. 1843. See, too, *CJ*, 12 Sept. 1851.
22. HO 45/642A, letter of 23 Sept. 1844.
23. NLS, MS 2843, letter of 11 Nov. 1844.
24. *Quarterly Review*, 74 (1844), 138. Compare *CJ*, 7 Apr. 1837.
25. *CJ*, 10 Apr. 1840.
26. PP 1844, XVI, South Wales, pp. 207–8, and HO 52/43, letter from C. Norris, 4 July 1839.
27. *CJ*, 14 Nov. 1845.
28. D. Parry-Jones, *Welsh Country Upbringing* (London, 1948), p. 98.
29. The age, literacy, and occupational details of offenders were taken from Assize and Quarter Session papers, Home Office and Transportation lists, as well as from newspaper evidence. They are of doubtful value. The registers for Haverfordwest and Carmarthen gaols are amongst the best, but from cross-checking with other material it is clear that different people sometimes gave a different age and occupation for the same offender. The Felons' Register for Carmarthen is in CRO, Acc. 4916. After 1856 the evidence on these matters is tabulated in the annual parliamentary returns, but these are also poor. Most prisoners in south-west Wales at this later time were either 'labourers', if male, or 'domestic servants', if female.
30. *CJ*, 24 Nov. 1843.

31. Ibid. 7 Apr. 1848.
32. Ibid. 17 Nov, 1843. When the statistics on origins were first published in 1856, 72% of the prisoners in the three counties were Welsh. For a little on this, see D. J. V. Jones, 'The Criminal Vagrant in Mid-Nineteenth Century Wales', *Welsh History Review* 8: 3 (1977), 312–44.
33. *W*, 8 Dec. 1843. See also *CJ*, 31 Oct. 1834.
34. CRO, Acc. 4916.
35. For just a note on three of these characters, see *CJ*, 30 Apr. and 26 Nov. 1841, and 19 May 1843.
36. The evidence on recidivism is notoriously poor for this period. The Felons' Register of Carmarthen (CRO, Acc. 4916) is unusual in that it provides a brief note on whether Rebeccaites and other criminals have been in custody before, but it is not completely reliable. One study which suggests that most offenders of the time had no previous criminal record is D. Philips, *Crime and Authority in Victorian England: The Black Country 1835–60* (London, 1977), p. 287.
37. The evidence for this is the Quarter Session Minute Books which provide details of prosecutors as well as offenders. Pembrokeshire Record Office, PQ 1/10, Order Book, 1839–45, and CRO, Minute Book 1837–43.
38. *Cambrian*, 20 July 1844.
39. CRO, Quarter Session Minute Book 1843–9, Mar. 1844.
40. *CJ*, 7 July 1843.
41. Ibid. 27 Dec. 1833 and 3 Nov. 1843.
42. Ibid. 24 May 1844.
43. Ibid. 20 Feb. 1846.
44. On this paragraph, see e.g. ibid. 13, 20 Mar., and 12 July 1835, 21 Oct. 1840, and 7 July 1843.
45. Ibid. 13 Mar. 1835, 9 Nov. 1838, 21 July 1843, 13 Dec. 1844, and 28 Mar. 1845.
46. Ibid. 13 Mar. 1835.
47. See e.g. ibid. 22 Feb. and 7 Mar. 1828, and 18 Nov. 1836.
48. Ibid. 7 Jan. 1853. The illegal contribution of women and children to the family economy was greater than we realize.
49. Ibid. 11 Feb. 1842 and 26 Jan. 1844.
50. Ibid. 26 Feb. and 26 Mar. 1847, and CRO, Quarter Session Minute Book, 1843–9, Oct. 1844.
51. See e.g. *CJ*, 9 Aug. 1833, and 23 Feb. 1844.
52. For some of this paragraph, see ibid. 8 July 1842, 24 Mar. and 29 Dec. 1843. Compare D. J. V. Jones, 'Life and Death in Eighteenth-Century Wales: A Note', *Welsh History Review*, 10: 4 (1981), 540. See the letter on the difficulty of estimating the amount of this crime, *CJ*, 29 Oct. 1830.
53. *CJ*, 26 July 1839.
54. Ibid. 22 July 1831, 9 Mar. 1832, 16 and 30 June 1843. J. Gibson, *Agriculture in Wales* (London, 1879), p. 35.
55. These paragraphs on offences against the Game and Fishing Laws are based on a study of the reported Petty Sessions cases, and the smaller number at Quarter Sessions and Assizes, for 1820–70. For the *Welshman*'s view of the Game Laws, and for a little on the cases mentioned, see *W*, 3 Nov. 1843, 6 Sept., 22 Nov., and 13 Dec. 1844. For a comparison, see J. G. Rule, 'Social Crime in the Rural South in the Eighteenth and Early Nineteenth Centuries', *Southern History*, 1 (1979), 135–53.
56. *CJ*, 28 Dec. 1838, and 8 and 15 Aug. 1845.

Notes

57. Ibid. 3 Nov. 1837.
58. D. J. V. Jones, 'The Second Rebecca Riots', *Llafur*, 2: 1 (1976), 32–56.
59. *W*, 12 May 1843.
60. *CJ*, 21 Jan., 18 Feb., and 18 Mar. 1842, 21 and 28 Apr. 1843, and 3 Jan. 1845.
61. *CJ*, 19 June 1829.
62. PP 1839, XIX, Report on the Constabulary Force in England and Wales, p. 42.
63. To set this context, see Jones, op. cit. n. 4 above.
64. *CJ*, 12 and 19 Dec. 1851.
65. Reports of case in *Cambrian*, 22 July 1843, and in *CJ*, 21 July 1843.
66. *W*, 25 Nov. 1842, and *CJ*, 29 Dec. 1843.
67. See e.g. *W*, 13 June 1845.
68. *CJ*, 24 Mar. 1843, and *W*, 6 Sept. 1844.
69. *CJ*, 12 Apr. 1822.
70. Ibid. 30 Jan. 1840, 6 Dec. 1844, and 22 Mar. 1845.
71. Ibid. 18 June 1830, 10 Jan. 1834, and 16 Oct. 1835.
72. G. Phillips, *Llofruddiaeth Shadrach Lewis* (Llandysul, 1986), and HO 45/454, letters from E. C. Lloyd Hall, 23 July 1843, and Anon., 4 Oct. 1843.
73. *CJ* and *W*, 6 Sept. 1844.
74. *W*, 3 Nov. 1843, and 21 June 1844, and *The Times*, 19 Apr. and 4 May 1844. On female crime generally, see L. Rose, *Massacre of the Innocents* (London, 1986), and D. Beddoe, *Welsh Convict Women* (Barry, 1979). Mrs A. Philpin, at Swansea, is about to begin a county study of female crime in the 19th century. Compare R. Sauer, 'Infanticide and Abortion in Nineteenth-Century Britain', *Population Studies*, 32: 1 (1978), 81–93.
75. *W*, 13 Jan. and 24 Mar. 1843, and *The Times*, 7 Nov. 1843.
76. *CJ*, 11 Apr. 1845, and *Cambrian*, 16 Mar. 1844.
77. For some cases, see *CJ*, 22 Apr. 1842, 21 June 1844, and 14 Feb. 1845, and *W*, 8 Sept. 1843.
78. *CJ*, 28 Jan. and 11 Mar. 1842, 22 Dec. 1843, 2 Aug. 1844, 28 Mar. and 4 Apr. 1845; *Cambrian*, 23 Dec. 1843; and *W*, 24 May 1844.
79. Even the keenest advocate of the Act of 1834 accepted this. See PP 1844, XVI, South Wales, pp. 167–8.
80. *CJ*, 27 July 1838. There were scores of similar cases. See e.g. ibid. 5 Mar. and 25 June 1852, and 24 June 1853.
81. NLW, MS 14590 E, No. 44, deposition of J. Thomas, 3 Nov. 1843.
82. *CJ*, 7 Jan. 1842, 2 Feb. and 12 July 1844, 27 Oct. 1854, and 16 Mar. 1855.
83. Ibid. 23 May 1845.
84. Ibid. 2 Feb. 1844 and 22 Aug. 1845.
85. *Cambrian*, 16 Mar. 1844; *CJ*, 8 Jan. 1841; and *W*, 26 July 1844. Compare *CJ*, 2 June 1854 and 14 Aug. 1863.
86. *CJ*, 5 Aug. 1831, 16 Mar. 1838, and 10 Jan. 1845; and *W*, 18 Oct. 1844.
87. *CJ*, 4 Nov. 1842, and *Cambrian*, 22 Mar. 1845.
88. *CJ*, 6 Apr. 1832.
89. Ibid. 25 July 1845. For hostility to the police at e.g. St Clears and Carmarthen, see ibid. 13 June 1845 and 14 Aug. 1846.
90. Ibid. 8 Jan. and 8 Apr. 1836, and 12 Apr. 1839.
91. See e.g. Jones, *Before Rebecca*, op. cit. n. 1 above, p. 57, and *CJ*, 21 June 1816, 18 Jan. 1828, 17 and 24 Mar. 1837; and *W*, 2 Aug. 1839.
92. See e.g. *Cambrian*, 26 July and 9 Aug. 1845; *CJ*, 27 Mar. 1835 and 18 Mar. 1842.
93. See e.g. *CJ*, 3 Jan. 1845, and *Cambrian*, 15 and 22 Mar. and 26 July 1845.

94. *Cambrian*, 15 Mar. and 2 Aug. 1845 and 13 Mar. 1846.
95. *CJ*, 21 Feb. 1834 and 27 Mar. 1835. For one unsuccessful rent case brought by the de Rutzens, see ibid. 23 July 1841.
96. There are numerous ejectment cases in the Assize reports. See e.g. ibid. 23 Nov. 1832 and 12 Nov. 1841.
97. *CJ*, 12 July 1839, 13 Aug. 1841, and 2 Feb. 1844.
98. Ibid. 30 Dec. 1864.
99. Ibid. 17 Jan. 1840 and 12 Nov. 1841.
100. To set it in context, see Jones, op. cit. n. 32 above.
101. There was a considerable amount of such animal maiming and killing, sometimes the result of a private grievance and more rarely perhaps a tribute to a wider feeling of discontent. For just a few cases, see *CJ*, 23 July 1830, 7 Jan. 1831, 15 June 1832, 25 Oct. 1833, 10 Sept. 1841, 18 Apr. and 13 June 1845; and *Cambrian*, 22 Mar. and 2 Aug. 1845. Compare J. E. Archer, 'A Fiendish Outrage? A Study of Animal Maiming in East Anglia 1830–70', *Agricultural History Review*, 33 (1985), 147–57.
102. See e.g. *CJ*, 21 Sept. 1832 and 11 June 1846, and *W*, 9 Dec. 1842.
103. See e.g. *CJ*, 10 and 24 July 1835, 8 Jan. 1836, and 30 Nov. and 28 Dec. 1838.
104. Quoted in *W*, 30 June 1843.
105. *The Times*, 12 Aug. 1843. For some background to the custom of *ceffyl pren*, and connections with Rebecca, see Trefor Owen, *Welsh Folk Customs* (Cardiff, 3rd edn. 1974), pp. 168–72; *Quarterly Review*, 74 (1844), 125–7; and *W*, 7 July 1843. See, too, HO 45/454, letter from E. C. Lloyd Hall, 15 June 1843; HO 45/642, letter from G. R. Trevor, 26 Sept. 1844; HO 45/642A, letter from Col. Love, 23 Sept. 1844.
106. HO 52/35, petition of 11 May 1837.
107. *CJ*, 15 June 1832 and 4 Apr. 1851, and HO 52/35, letter from T. Morris, with enclosures, 30 Dec. 1837.
108. For a translation of a *ceffyl pren* speech, see CRO, Acc. 4282 A, TW 4.
109. *CJ*, 8 and 15 May 1835 and 4 Nov. 1836.
110. As can be seen by a substantial collection of letters on the subject in HO 40/40, 43/54, and 52/35.
111. See e.g. HO 52/35, letter from the Mayor of Cardigan, 31 Mar. 1837.
112. PRO, Assize 72/1, and *CJ* 28 July 1837.
113. *CJ*, 6 Apr. 1838, and CRO, Quarter Session Minute Book, 1837–43, Apr. 1838.
114. *CJ*, 22 Mar. and 12 Apr. 1839, and CRO, Quarter Session Minute Book, 1837–43, Apr. 1839. For other *ceffyl pren* cases, see e.g. *CJ*, 13 Sept. 1839 and 13 Mar. 1840, and *W*, 2 Aug. 1839.

Chapter 5

1. The attacks on the tollgates were described in prodigious detail by the magistrates, commissioners, military officers, policemen, newspaper reporters, and others. It is impossible to list all the references to these incidents in the footnotes of one chapter. Frequently there were ten or twenty reports on one incident, with only small but important differences between them. I have been necessarily selective in the choice of footnotes, usually giving references that are especially revealing or new to historians of Rebecca.
2. For some of the complaints and trouble described in this paragraph, see *W*, 30 Nov. 1832, and *CJ*, 17 Dec. 1830, 4 Feb. 1831, 8 June 1832, 6 Sept.

Notes

1833, 28 Nov. and 26 Dec. 1834, 20 Feb. 1835, and 1 Mar. and 28 June 1839.
3. CRO, TT2, Main Trust Order Book, 19 Feb. 1833, and *CJ*, 8 Nov. 1833. For further trouble over a side-bar at St Clears, see *CJ*, 7 Aug. 1835.
4. *CJ*, 13 Sept. 1833.
5. PP 1844, XVI, South Wales, p. 254.
6. See editorial in *CJ*, 8 Sept. 1843.
7. Compare the list in H. Tobit Evans, *Rebecca and Her Daughters* (Cardiff, 1910), pp. 244–7. There were many contemporary lists, often inaccurate. See e.g *Y Diwygiwr* (1843), 257.
8. On this paragraph, see e.g. *W*, 15 Dec. 1843 and 19 Apr 1844; *CJ*, 11 Aug. and 22 Dec. 1843; HO 45/453, letter from Col. Love, 12 Dec. 1843; HO 45/454B, Report of G. H. Ellis, 2 Nov. 1843; PP 1844, XVI, South Wales, pp. 5, 12, 165, 170, 191, and 266; PP 1847, XXVII, Education, CGP, pp. 258 and 420; NLW, Nanteos 1334.
9. Based on typical building costs and compensation of £25–40 as a result of an attack. See e.g. Pembrokeshire Record Office, PQ 1/10, Quarter Session Order Book, 2 Jan. 1844.
10. *CJ*, 26 July 1839. Note the concern, from the beginning of the movement, that no one should recognize another, or inform. *Tarian y Gweithiwr*, 19 Aug. 1886.
11. HO 40/51, letters from J. M. Child and W. B. Swann, 14 June and 22 and 28 July 1839, and from B. Thomas, 11 July 1839. See, too HO 40/52, letter from J. Williams, 19 July 1839.
12. Tobit Evans, op. cit. n. 7 above, pp. 29–31. Both Tobit Evans and Pat Molloy have good accounts of the beginnings of Rebecca. For the latter, see *And They Blessed Rebecca* (Llandysul, 1983), ch. 1.
13. HO 40/51, letter from T. S. Biddulph, 18 July 1839.
14. T. Powell was wiser later than at the time. Powell offered thanks to Beynon at Whitland trust meeting which confirmed the recommendation of 23 July 1839. *CJ*, 4 Oct. 1839. On the recommendations, see *W*, 26 July 1839. For views on the compromise reached, see PP 1844, XVI, South Wales, pp. 4 and 60; HO 45/265, letters from Lord Dynevor, 30 Nov. 1842, and T. Powell, 31 Dec. 1842; HO 45/454B, Report of G. H. Ellis, 2 Nov. 1843; and HO 40/51, letter from J. H. Allen, 2 Aug. 1839.
15. The term was used in *The Times*, 2 Oct. 1843.
16. T. Gwyn Jones, 'Rebeccaism: Bibliography', *Carmarthen Antiquary*, 1 (1943–4), 71; *CJ*, 27 May 1842; and *W*, 30 Dec. 1842.
17. Among the many reports of these events are HO 45/265, letter from T. Powell, 13 Dec. 1842, and HO 45/454, Report of G. Martin, 2 Jan. 1843.
18. *CJ* and *W*, 10 Feb. 1843.
19. For J. Lloyd Davies's anger at the events, and his determination to proceed against the Rebeccaites in the spring of 1843, see *CJ*, 17 Mar. 1843.
20. HO 45/454, letter from J. K. Wilson, 17 Jan. 1843.
21. See the Rebecca letters in HO 45/265, fos. 100 and 103.
22. Accounts in *W* and *CJ*, 17 Feb.–17 Mar. 1843, and F. Green, 'Rebecca in West Wales', *West Wales Historical Records*, 7 (1918), 27–33, and HO 45/454, letters from T. Powell, 15 Feb. 1843, and H. Leach, 8 Apr. 1843.
23. See e.g. HO 45/454, letter from J. Lloyd Davies, 21 Apr. 1843. Note the government comment on the letter, which gives Graham's brusque response to such fears.
24. See e.g. *CJ* and *W*, 10 and 17 Mar. 1843.

Notes

25. This occurred in July. There is a wealth of detail on the Colby affair. HO 45/454, letters from J. Colby and H. Owen, 1 and 5 Aug. 1843; *CJ*, 20 Oct. 1843; and *The Times*, 21 Oct. 1843.
26. The best accounts of events described in this paragraph are *CJ*, 21 Apr. 1843, and HO 45/454, letter from J. Lloyd Davies, 21 Apr. 1843.
27. See e.g. HO 45/453, a report of 30 June 1843; HO 45/454, letters from J. Lloyd Davies, 14 and 17 June 1843; and *W*, 23 June 1843.
28. CRO, Acc. 4916, PRO Assize 71/8, and *Cambrian*, 20 July 1844.
29. As the idea of a Hundred police was abandoned, George Rice Trevor stated his plans for a county police. CRO, Quarter Session Minute Book, 28 June 1843. See also *CJ*, 30 June 1843. J. Lloyd Davies's views on the value of a county police seem to have varied. NLS, MS 2844, letter from Col. Love, 4 Jan. 1844.
30. HO 45/454, letter of 14 June 1843.
31. *W*, 16 and 23 June 1843, and *CJ*, 30 June 1843. The farm of Pantyfen was said to have had only 50 acres by 1851. Census of Llanegwad, 1851. Copy in the CPL.
32. One of the best accounts is HO 45/454, examinations of H. Thomas, J. Pugh, and W. Lewis, 27 May 1843. Compare *CJ*, 2 June 1843. Alcwyn Evans identified Michael Bowen, NLW, MS 12368 E.
33. HO 45/454, letter from D. Davies, 12 June 1843.
34. It is impossible to list all the sources for the events described in these paragraphs. On the Talog affair, see *CJ*, 9 and 16 June 1843; PP 1844, XVI, South Wales, pp. 60–3; and HO 45/454, examinations of N. Martin and others, 12 June 1843.
35. For a little on the events preceding the march to Carmarthen on 19 June, see HO 45/454, letters of J. Lloyd Davies, 14, 17, and 19 June 1843, E. H. Stacey, 17 June 1843, and *CJ*, 16 and 23 June 1843.
36. HO 45/454, copy in a letter from J. Wood, 21 June 1843.
37. Amongst the best of a great volume of evidence are accounts in *CJ* and *W*, 23 and 30 June and 6 Oct. 1843, and 22 Mar. 1844; *Cambrian*, 23 Mar. 1844; *The Times*, 22 and 28 June 1843; NLW, MS 12368 E; and HO 45/454, letters from J. Lloyd Davies and E. H. Stacey, 19 and 20 June 1843. Lewis Evans's comments on the events leading to and including the invasion are worth a glance. PP 1844, XVI, South Wales, pp. 73–8. Great efforts were subsequently made to detach the workhouse incident from the main Rebecca story, but it is worth recalling the opposition to the Poor Law which we shall encounter in the next chapter, and the fact that it was not only 'town ruffians' who were arrested in the workhouse yard. The accounts in Welsh of the whole affair are especially interesting, and often balanced. See e.g. *Y Diwygiwr* (1843), 223–5; *Seren Gomer* (1843), 213–17, and *Tarian y Gweithiwr*, 9 Sept. 1886.
38. See e.g. HO 45/454, letters from E. C. Lloyd Hall, 15, 20, and 21 June 1843, H. Vaughan, 16, 21, and 22 June 1843, and R. Jenkins, 24 June 1843. For the trouble at St Clears, see *Swansea Journal*, 28 June 1843.
39. HO 45/454, letter from E. C. Lloyd Hall, 22 June 1843.
40. *Swansea Journal*, 28 June 1843; *W* and *CJ*, 30 June 1843; and HO 45/454, letters from R. Jenkins and H. Vaughan, 24 June 1843.
41. For a good account of the meeting, and the committee's recommendations, see *The Times*, 26 June, 12 and 29 Aug. 1843.
42. For just one account of the fear, and response, at Narberth only, see ibid. 30 June and 3 July 1843.
43. *W* and *CJ*, 28 July 1843, and *The Times*, 27 July 1843.

Notes

44. HO 45/453, letter from Col. Love, 22 July 1843, and *The Times*, 22 July 1843.
45. *The Times*, 24 June 1843. Foster had a colleague in the area, who sometimes wrote reports himself.
46. This was not entirely accidental. At their large public meetings, the organizers invited reporters, supplied them with interpreters, and thanked them at the end of the proceedings. In addition, towards the end of the riots, a fund was launched to purchase a fitting testimonial to Foster.
47. NLW, Nanteos, 1288.
48. HO 45 347A, first report of informer, Aug.–Sept. 1843. G. Rice Trevor said that 99 out of 100 jurymen were Rebeccaites. NLS, MS 2843, letter from 5 Oct. 1843.
49. *W* and *CJ*, 21 July 1843.
50. Information on this story does not always tally. *The Times*, 14 July 1843; *Swansea Journal*, 12 July 1843; *CJ* and *W*, 14 July 1843; and HO 45/454, letter from G. R. Trevor, 11 July 1843.
51. *CJ* and *W*, 21 July 1843. See two, the *Swansea Journal*, 19 July 1843, and *The Times*, 25 July 1843, for splendid information on this and related attacks on gates in industrial areas.
52. Amongst the better information on these attacks, see *Cambrian*, 29 July and 5 Aug. 1843; *The Times*, 14 and 24 July 1843; and *Swansea Journal*, 19 July, 2 and 9 Aug. 1843.
53. There are many, sometimes contradictory, accounts of the arrests. See e.g. the legal evidence in PRO, Assize 71/7, 72/1, and 73/2. On the fascinating reaction to the arrests and the popular threats, see *The Times*, 27 and 31 July and 3 Aug. 1843, and *Swansea Journal*, 2 Aug. and 6 Sept. 1843.
54. *The Times*, 8 Aug., 9 Sept., and 6 Oct. 1843.
55. There are many letters on this case in HO 41/18, 45/454, and 45/642, and for further information on the arrests and the arrested, see e.g. NLS, MS 2843, letters from Col. Hankey, 11 and 14 Nov. 1843; PRO, Assize 76/1; and Pembrokeshire Record Office, PQ/21, Calendars of Assize, 1840–51.
56. For one account, see *Swansea Journal*, 30 Aug. 1843. On the handbill, see Tobit Evans, op. cit. n. 7 above, p. 155. Joshua later assaulted someone, and was given a gaol sentence for forgery. PRO, Assize 76/1.
57. *Swansea Journal*, 16 Aug. 1843, and *The Times*, 27 July 1843.
58. For some of this, see *CJ*, 4 Aug., 15 Sept. 1843, and 30 Aug. 1844; *W*, 4 and 18 Aug. 1843; and *Cambrian*, 23 Mar. 1844.
59. J. Downes, 'Notes on a Tour of the Disturbed Districts in Wales', *Blackwood's Magazine*, 54 (1843), 766–76.
60. There is an astonishing amount of material on the attacks at Porthyrhyd. For just the one incident, on 4 Aug., the best newspaper accounts are the *Swansea Journal*, 9 Aug. 1843, *The Times*, 8 and 10 Aug. 1843, and *CJ* and *W*, 11 Aug. 1843.
61. NLW, MS 21209 C, letter from T. Cooke, 6 Aug. 1843.
62. As in so many accounts, the dates differ slightly. One of the best reports, in *CJ*, 25 Aug. 1843, claims that the attack occurred in the early hours of 19 Aug. 1843.
63. There are many versions of the Llanelli story. See *CJ* and *W*, 4 Aug. 1843; *Swansea Journal*, 16 Aug. 1843; and *The Times*, 17 Aug. 1843. For some legal evidence, see NLW, MS 14590 E, Document 3.

64. Amongst the best accounts are those in *CJ*, 4 and 11 Aug. 1843, and *Cambrian*, 5 and 12 Aug. 1843.
65. Downes, op. cit. n. 59 above, pp. 773–4. There were many versions of how, and why, she died. *The Times* is a good starting-point (edns. of 13, 14, and 25 Sept. 1843). On the impact of the Hendy murder, see *Swansea Journal*, 4 Oct. 1843.
66. There are innumerable references to this riot. David Williams has 100 of them, *The Rebecca Riots* (Cardiff, 1955), p. 349. Add to the list the *Swansea Journal*, 6 and 13 Sept. 1843, *CJ*, 8 Sept. 1843; *The Times*, 9, 14, 16, and 25 Sept. and 30 and 31 Oct. 1843; *Morning Herald*, 14 Sept. 1843; and many other accounts. On the claim that Rebeccaites were killed, or died at home, see *The Times*, 9 Sept. 1843.
67. For just some of this, see *The Times*, 25 Sept. and 5 Oct. 1843; *CJ*, 8 and 15 Sept., 20 Oct., and 10 Nov. 1843; HO 45/454, letter from G. R. Trevor, 13 Nov. 1843; NLW, Nanteos 1273 and 1384; and G. E. Evans, 'The Rebecca Riots', *Transactions of the Carmarthenshire Antiquarian Society*, 23 (1932), 66–7, 71.
68. CUL, Graham Papers (Microfilm). Compare this with the Peel Papers (Microfilm), Add. 40, 449, letter from J. Graham, 17 Sept. 1843.
69. *Planet*, 21 Jan. 1844.
70. On this, see the many lists and letters in CRO, 4282 A; NLW MS 14536 E; HO 45/642C and 642D; and *Morning Herald*, 14 Sept. 1843.
71. On the reasons why a special commission was held at Cardiff, the popular response to that, and the government's reaction to the proceedings there, see CUL, Peel Papers (Microfilm), Add. 40, 449, letters from J. Graham, 16 Sept. and 31 Oct. 1843.
72. *Nation*, 11 Nov. 1843. John Lloyd Price protested at the ruthless treatment of the families of the convicts. CRO, 4284 A, letter of 12 Aug. 1844.
73. HO 45/642A, letter of 31 Mar. 1844. See his gloomy letters in NLS, MS 2844, especially 3 Mar. 1844.
74. *The Times*, 14 Oct. 1843.
75. See next chapter.
76. There is a very detailed first-hand acount of the Cardigan attack in HO 45/642, letters from D. Jenkins, 27 Mar. and 3 Apr. 1844, with enclosures, and HO 45/642A, letters from Col. Love, 28, 30, and 31 Mar. and 6 Apr. 1844, and from W. Davies, 29 Mar. 1844, with enclosures.
77. On George Jones's unpopularity, see *Swansea Journal*, 9 Aug. 1843. For two excellent accounts of the trouble at Dolauhirion and Cwmdwr, see *CJ*, 6 Oct. and 29 Dec. 1843.
78. NLS, MS 2843, letter of 5 Nov. 1843.
79. There is a vast amount of source material on the Rebecca riots in mid-Wales. In addition to the sources listed in Williams, op. cit. n. 66 above, ch. X, see the *Hereford Times* for these months, especially the edn. of 11 Nov. 1843, and the many letters from Sir J. Walsh in Oct. and Nov. 1843 in NLS, MS 2843.
80. I have obtained personal details on 236 Rebeccaites apprehended for their crimes, and 127 of these were accused of tollgate offences. Occupational information has been found on 86 of the latter, mainly from legal sources. The details have been checked, where possible against the census returns. Compare, too, the details given in the next chapter. For the statement of David Williams, op. cit. n. 66 above, p. 75. See the

Notes

interesting four-class model of Rebeccaites in E. C. Lloyd Hall, HO 45/454, letter of 30 Sept. 1843.
81. See e.g. HO 40/51, letter from J. H. Allen, 2 Aug. 1839; HO 45/1611, Report of W. Day, 9 July 1843; PP 1844, XVI, South Wales, Evidence, p. 3, and 1847, XXVII, Education, CGP, p. 244; NLW, MS 21209C, letters from T. Cooke, 9 and 24 Oct. 1843. The information on Rebeccaites in the Felons' Register at the Carmarthen Record Office confirms the impression that few of these people had criminal records.
82. On young people and Rebecca, see HO 45/454, letters from E. C. Lloyd Hall, 22 June 1843, and from J. Walters, 4 Oct. 1843. I have obtained age details on 88 of these Rebeccaites. Rather surprisingly perhaps, if the sample can be regarded as representative, they were on average a few years younger than the people described in the next chapter.
83. See e.g. *The Times*, 29 July 1843; HO 45/265, letter from G. R. Trevor, 27 Dec. 1842; and HO 45/454, letter from J. Walsh, 29 Sept. 1843.
84. PP 1844, XVI, South Wales, pp. 225, and 244; *The Times*, 22 Aug. 1843; HO 45/453, letter from Col. Love, 26 July 1843; and HO 45/454, letter from E. C. Lloyd Hall, 30 Sept. 1843.
85. HO 45/454, letter from W. O. Pell, 13 July 1843; NLS, MS 2843, letter from Col. Love, 21 Dec. 1843; MS 2844, letter from Col. Love, 28 Jan. 1844: *CJ*, 13 Oct. 1843; *Tarian y Gweithiwr*, 19 Aug. 1886; *Seren Gomer* (1843), 345; and G. Borrow, *Wild Wales* (London, edn. of 1901), p. 94.
86. See e.g. *Cambrian*, 30 Mar. 1844, and a note in the *Transactions of the Carmarthenshire Antiquarian Society*, 15 (1921), 39.
87. For one who did, see *The Times*, 29 July 1843.
88. Tobit Evans, op. cit. n. 7 above, pp. 77–8.
89. On this paragraph, see HO 45/454, letter from E. C. Lloyd Hall, 24 June 1843; *The Times*, 22 and 27 June 1843; W. J. Linton, *Memories* (London, 1895), p. 89; and HO 45/453, secret report of a Chartist meeting, 5–6 Nov. 1843. Thomas Cooke said that 'about here' it was generally thought that Hugh Williams was the leader. NLW, MS 21209C, letter of 3 Sept. 1843.
90. There were many reports of a 'very tall man' who acted as Rebecca about St Clears and elsewhere, and of a 'smith' who impersonated the Lady on several occasions, but whether these referred to the same one or two people is not clear. On the early Beccas, see e.g. Tobit Evans, op. cit. n. 7 above, pp. 9–134; Molloy, op. cit. n. 12 above, pp. 27–30; *Tarian y Gweithiwr*, 19 Aug. 1886; HO 45/454, examination of Henry Thomas, 27 May 1843; and *W*, 16 June and 27 July 1843.
91. Spoken 'like a Gentleman'. PRO, Assize, 72/1, evidence of E. Llewelyn. See the letter from J. Lloyd Price, 12 Sept. 1844, in CRO, 4282 A.
92. 'Charlotte', 'Lydia', 'Nelly', and others were present more often than has been suggested. Molloy, op. cit. n. 12 above, p. 94 n. 4.
93. On 'Miss Cromwell', see e.g. *CJ*, 27 Oct. 1843; *W*, 18 Aug., 1 Sept., and 6 Oct. 1843; and *Cambrian*, 23 Dec. 1843.
94. There is much on this. For a selection of sources, see *The Times*, 30 June and 28 July 1843; HO 45/347A, report of spy, No. 5; and HO 45/454, letters from E. C. Lloyd Hall, 26 July, 12 Aug., 2 and 8 Sept. 1843; from J. Lloyd Davies, 14 June and 22 July 1843; from W. D. Jones, 19 Sept. 1843; from J. Cooper, 14 Sept. 1843; and H. Vaughan, 3 Oct. 1843; and from J. Walters, 20 Sept. 1843.
95. NLW, MS 14590E, Document 61, statement by J. Jones, undated.

96. NLS, MS 2843, letter from Col. Hankey, 14 Nov. 1843.
97. Evans, op. cit. n. 67 above, p. 62.
98. HO 45/454, enclosed in a letter from E. C. Lloyd Hall, 21 June 1843.
99. *CJ*, 28 July 1843.
100. See e.g. HO 45/454, letters from H. Vaughan, 24 June 1843, from E. Lloyd Williams, 26 June 1843, and J. Walters, 4 Oct. 1843.
101. See e.g. ibid. letter from E. C. Lloyd Hall, 30 June 1843, and *CJ*, 22 Sept. 1843.
102. There was some concern at the time over the economic plight of the pensioners, and the dangers of this. NLS, 2843, letter from W. H. Kenny, 29 Oct. 1843.
103. HO 45/453, letter from Col. Love, 22 July 1843. On the speed at which messages were passed across the countryside, see *The Times*, 5 Sept. 1843.
104. Captain Napier saw the unusual sight of one Becca in a druid's dress, of white with blue sleeves. *Swansea Journal*, 1 Nov. 1843.
105. See e.g. *W*, 6 Jan. 1843. Compare B. A. Babcock (ed.), *The Reversible World: Symbolic Inversion in Art and Society* (New York, 1978).
106. *Morning Chronicle*, 29 Sept. 1843.
107. There is a wealth of material on intimidation. For a little of it, see *Cambrian*, 20 July 1844; *CJ*, 23 June, 14 July, 6 Oct., and 1 Dec. 1843; *W*, 19 Nov. 1843; *The Times*, 4, 8, and 17 Aug. 1843; NLW, MS 14536 E; and CRO, 4282 A, Letter from G. W. Green, 30 Nov. 1844. Note the fear recorded in *Tarian y Gweithiwr*, 2 Sept. and 21 Oct. 1886. I am grateful to the Librarian of the National Library of Wales for providing me with copies of this paper. The quotation (translated) is from this journal.
108. On this section, see e.g. HO 45/347A, fourth report of informer, Aug.–Sept. 1843; *The Times*, 24 June and 4–6 July 1843; *Swansea Journal*, 13 Sept. 1843; NLS, MS 2843, letter from J. H. Vivian, 15 Oct. 1843; HO 45/265, letter from G. Martin, 21 Dec. 1842; *CJ*, 11 Aug. and 15 Sept. 1843.
109. Quoted in CRO, Bryn Myrddin, 90, p. 23.
110. For opposite views of the value of open meetings in this respect, see *W*, 20 Oct. 1843, and *CJ*, 13 Oct. 1843.

Chapter 6

1. *CJ*, 2 June 1843.
2. *The Times*, 28 and 31 July 1843, and *CJ*, 28 July 1843.
3. The term is from *CJ*, 22 Sept. 1843.
4. For a contemporary note on this, see *Annual Register* (1843), 260, and PP 1844, XVI, South Wales, p. 1.
5. P. Molloy, *And They Blessed Rebecca* (Llandysul, 1983), p. 163, and D. Williams, *The Rebecca Riots* (Cardiff, 1985), pp. 233 and 255.
6. PP 1844, XVI, South Wales, p. 254.
7. *CJ*, 1 Sept. 1843.
8. CUL, Graham Papers (Microfilm), letters of 14, 15, and 22 Sept. 1843.
9. As an example of cases ignored in the table and map, see those mentioned in NLS, MS 2843, letter from W. E. Powell, 11 Nov. 1843. Some people were too ready to attach the Rebecca label to activities which did not interest her. See the account of the Llandovery troubles in *W*, 30 June 1843.
10. They were sent in 1839 and 1843. See e.g. *CJ*, 10 May, 22 Nov., and 13 Dec. 1839, and HO 45/454, enclosures in letters from H. Vaughan, 16

Notes

June and 15 Aug. 1843, and H. Tobit Evans, *Rebecca and Her Daughters* (Cardiff, 1910), pp. 153–4. On the popularity of sending threatening letters, see *CJ*, 12 May 1843.
11. HO 45/454, letter from E. C. Lloyd Hall, 26 Aug. 1843.
12. *CJ*, 13 Oct. 1843, and *Swansea Journal*, 13 Sept. 1843.
13. NLW, MS 21209C, and MS 3294E.
14. HO 45/454, letter from E. C. Lloyd Hall, 7 Sept. 1843, and J. Downes, 'Notes on a Tour of the Disturbed Districts in Wales', *Blackwood's Magazine*, 54 (1843), 767.
15. See e.g. *CJ*, 20 and 27 Oct. 1843; *Cambrian*, 30 Mar. 1844; and *W*, 1 Sept. and 8 Dec. 1843.
16. *The Times*, 28 July 1843.
17. See e.g. CRO, Aberglasney 24/570A; *W*, 15 Sept. 1843; *CJ*, 22 Sept. and 6 Oct. 1843; *Morning Herald*, 12 Sept. 1843; and letters from J. Walters in HO 45/454 and 642.
18. HO 45/454, letter from E. C. Lloyd Hall, 24 Aug. 1843, and *CJ*, 6 Oct. 1843.
19. *Cambrian*, 30 Mar. 1844, and *CJ* and *W*, 10 Feb. 1843, and *W*, 15 Dec. 1843.
20. HO 454/454, letter from E. C. Lloyd Hall, 2 Sept. 1843, and NLW, Nanteos 1332. Compare the account of Rebeccaites visiting farms in Troedyraur for similar purposes, in the *Morning Herald*, 12 Sept. 1843.
21. HO 45/454, letters from W. D. Jones, 19 Sept. 1843, and from Anon., 27 Sept. 1843.
22. See e.g. *W*, 5 May and 16 June 1843, and HO 45/454, letter from H. Owen, 28 June 1843.
23. *W*, 2 Aug. 1839, and *CJ*, 30 Aug. and 13 Sept. 1844.
24. *W*, 5 May 1843, and *CJ*, 21 Dec. 1841.
25. HO 45/454, letters from E. C. Lloyd Hall, 12 and 14 Sept. 1843; *CJ*, 1 Dec. 1843; *Tarian y Gweithiwr*, 4 Nov. 1886.
26. HO 45/454, letter from E. C. Lloyd Hall, 10 and 12 Sept. 1843; *Morning Herald*, 12 Sept. 1843; *Tarian y Gweithiwr*, 4 Nov. 1886.
27. HO 45/642, letter from E. C. Lloyd Hall, 7 Oct. 1843.
28. HO 45/454, letter from E. C. Lloyd Hall, 15 Sept. 1843; *Swansea Journal*, 23 Aug. 1843; *Tarian y Gweithiwr*, 4 Nov. 1886.
29. *The Times*, 27 Nov. 1843; *W*, 1 Dec. 1843; and *CJ*, 1 Sept. 1843 and 18 Apr. 1845. D. Parry-Jones has insights on this. See *My Own Folk* (Llandysul, 1972), pp. 79 and 125, and *Welsh Country Upbringing* (London, 1948, edn. of 1976), pp. 98–9. Compare the numbers of *Tarian y Gweithiwr* 23 Sept.–4 Nov. 1886.
30. *CJ*, 15 and 22 Sept. 1843.
31. HO 45/454, letter of E. C. Lloyd Hall, 24 Aug. 1843; *CJ*, 18, 25 Aug. and 8 Sept. 1843; *Swansea Journal*, 16 Aug. 1843; and *The Times*, 30 Aug. 1843.
32. PRO, Assize 71/7; *Cambrian*, 23 Dec. 1843; and PP 1844, XVI, South Wales, p. 103.
33. *W*, 25 Aug. 1843, and HO 45/454, letter from E. C. Lloyd Hall, 30 Sept. 1843.
34. See e.g. PP 1847, XXVII, Education, BCR, p. 84; *CJ*, 1 Sept. 1843; and HO 45/454, letter from E. C. Lloyd Hall, 2 Sept. 1843. Note the warning on labour payments at the end of the Rebecca's instructions about tenancy and rent.*CJ*, 29 Sept. 1843.
35. *CJ*, 1 and 8 Sept. 1843; *Swansea Journal*, 13 Sept. 1843; and HO 45/453, letter from J. H. Harris, 18 Sept. 1843.

36. NLS, MS 2844, letter from Col. Love, 5 Mar. 1844, and *CJ*, 29 Dec. 1843.
37. *The Times*, 19 Oct. 1843; *CJ*, 20 Oct. 1843; and HO 45/454, letter from G. R. Trevor, 24 Oct. 1843.
38. HO 45/453.
39. *CJ*, 27 July 1838, 25 Jan. and 1 Feb. 1839, 1 Mar. 1844, and Pembrokeshire Record Office, SPU/NA/2/1, Narberth Union Minute Book, 1837–40.
40. See e.g. *W*, 3 Feb. 1843; HO 45/453, letters from Col. Love, 20 July, 16 Aug., and 30 Sept. 1843; HO 45/454, letters from H. Vaughan, 16 June, from H. Leach, 25 June, from G. R. Trevor, 25 June, from W. E. Powell, 29 June, from A. L. Gwynne, 12 Aug., from C. J. Wrigley, 24 Aug., from W. Chambers, 29 June, and from Col. Love, with enclosures, 20 July 1843.
41. See e.g. HO 45/454, letter from G. R. Trevor, 22 Aug. 1843; *Morning Herald*, 26 Sept. 1843; and Ceredigion Record Office, Cardigan Board of Guardians Minute Book, 2 Sept. 1843.
42. *W*, 15 Sept. 1843.
43. Ceredigion Record Office, Aberaeron Board of Guardians Minute Book, 1841–52, Aug. 1843. Other material in this and the next paragraph has been gleaned from this source, and from the Cardigan Minute Books.
44. For some of this, see HO 45/453, letter from Col. Love, 21 Sept. 1843; HO 45/454, letter from E. C. Lloyd Hall, 15 July 1843, and from several others in Aberaeron and Cardigan, Aug.–Sept. 1843; and PP 1844, XVI, South Wales, p. 233.
45. CUL, Graham Papers (Microfilm), letter from J. Graham, 15 Sept. 1843 and Peel Papers (Microfilm), 40, 449, letters from J. Graham, 17 and 21 Sept. 1843. The *Times* reporter had some sympathy with the protesters. Some of his detailed accounts on the subject can be found in the edns. of 24, 29, and 30 June, 22 Aug., 15 and 21 Sept., and 10 and 19 Oct. 1843. See, too, HO 45/1611, Report of W. Day, 9 July 1843.
46. See e.g. the letters from the Queen and Prince Albert demanding action on Rebecca and less tolerance by the government. CUL, Graham Papers (Microfilm), Bundle 62B.
47. *The Times*, 17 Nov. 1843, and *CJ*, 29 Sept. 1843. See the rather smug comments of J. Lloyd Davies, who had failed to get many of his fellow magistrates to cut rents two years before. PP 1844, XVI, South Wales, Evidence, p. 5.
48. HO 45/454, letters from J. Lloyd Davies, 17 June 1843, and W. D. Jones, 19 Sept. 1843; *The Times*, 28 Sept. 1843. On the threats behind the notices on rents now being 'served on the landlords', see *The Times*, 22 July 1843; NLW, Nanteos 1391; and HO 45/454, letter from A. L. Gwynne, 12 Aug. 1843.
49. HO 45/454, letter from W. Peel, with enclosure, in that of Lord Dynevor, 27 June 1843. See note in the *Transactions of the Carmarthenshire Antiquary Society*, 23 (1932), 50–1. A. L. Gwynne received a similar letter, demanding a 30% cut in rents. HO 45/454, letter of 12 Aug. 1843.
50. CRO, Cwmgwili, 705.
51. They were also conscious of the difficulties of taking legal action, and removing tenants, at such a time. *The Times*, 17 Nov. 1843.
52. See e.g. HO 45/454, letter from E. Lloyd Williams, with his address, 14 July 1843, and *CJ*, 15 Sept. 1843.
53. There is much on this. See HO 45/453, letter from Col. Love, 25 Sept. 1843; HO 45/454, letters from E. C. Lloyd Hall, 24 June and 19 Aug. 1843, and from W. D. Jones, 19 Sept. 1843; PP 1844, XVI, South Wales, pp. 225–6.

Notes   401

54. Tobit Evans, op. cit. n. 10 above, pp. 52–3, and PP 1844, XVI, South Wales, pp. 248–57. See also W. Day's comment in his report of 9 July 1843. HO 45/1611.
55. *The Times*, 26 Aug. 1843; *Morning Herald*, 26 Sept. 1843; and *CJ*, 1 Sept. 1843.
56. HO 45/453, letter from Col. Love, 31 Aug. 1843; HO 45/454, letters from R. Jenkins, 24 June, E. C. Lloyd Hall, 19 Aug., and S. P. R. Wagner, with enclosure, 26 Aug. 1843.
57. *The Times*, 19 Oct. 1843, and *CJ*, 20 Oct. 1843.
58. John Jones was later convicted of writing a threatening letter. PRO, Assize 71/7, *CJ*, 29 Dec. 1843; and *The Times*, 20 Oct. 1843.
59. *CJ*, 3 Nov. and 29 Dec. 1843, and *W*, 29 Dec. 1843.
60. *CJ*, 3 Mar. 1843, and HO 45/454, letters from E. C. Lloyd Hall, 21 Sept. and 15 Oct. 1843. Compare the stories in the *Morning Herald*, 13 Nov. 1843, and *CJ*, 10 Nov. 1843.
61. See e.g. *W*, 5 Apr. and 16 June 1843; HO 45/454, letters from Anon., 27 Sept. 1843, and from E. C. Lloyd Hall, 2 Sept. and 15 Oct. 1843; *Morning Herald*, 18 Nov. 1843; *W*, 8 Dec. 1843; *Morning Chronicle*, 21 Oct. 1843; and *CJ*, 1 Sept. and 15 Dec. 1843.
62. *Swansea Journal*, 11 Oct. 1843; *W*, 20 Oct. 1843; *CJ*, 13 Oct. 1843; and *Cambrian*, 6 Jan. 1844.
63. HO 45/454, letters from E. C. Lloyd Hall, 8 and 13 Sept. 1843.
64. *CJ*, 13 Oct. 1843, and *Morning Herald*, 10 Oct. 1843.
65. *W*, and *CJ*, 15 and 29 Sept. and 27 Oct. 1843; CRO, Aberglasney, 24/570A; *Seren Gomer* (1843), 347.
66. For this paragraph, see e.g. *CJ*, 24 Mar., 28 July, and 22 Dec. 1843; *W*, 22 Dec. 1843; and *Cambrian*, 23 Dec. 1843. F. Green, 'Rebecca in West Wales', *West Wales Historical Records*, 7 (1918), 33–7.
67. HO 45/454, letter from W. D. Jones, 19 Sept. 1843; *W*, 20 Oct. 1843 and 20 Sept. 1844; and *CJ*, 20 Sept. 1844. It was said that death was recorded against the names of any buyers at auctions. HO 45/453, report of a Chartist meeting, 5 Nov. 1843.
68. *Swansea Journal*, 4 Oct. 1843.
69. HO 45/454, letter from G. R. Trevor, 13 Sept. 1843; *W*, 12 Apr. 1844, and *CJ*, 19 Apr. 1844.
70. *CJ*, 29 Dec. 1843; *W*, 15 Mar. 1844; *Cambrian*, 16 Mar. 1844; PRO, Assize 76/1; and NLS, MS 2843, letter from Col. Love, 30 Dec. 1843.
71. HO 40/40, letter from H. Vaughan, 23 Oct. 1838, and *CJ*, 30 Nov. 1838.
72. NLW, Slebech 2409. Cited by D. W. Howell, *Land and People in Nineteenth Century Wales* (London, 1977), p. 12. See also PP 1844, XVI, South Wales, p. 55, and NLW, MS 21209C. Compare Cooke's letter of 24 Aug. 1843, where he describes the pressure to reduce rents, and the determination of Rebeccaites that no one should take a vacated farm.
73. *W*, 27 Oct. 1843; *Swansea Journal*, 11 Oct. 1843; and *CJ*, 13 and 20 Oct. 1843.
74. *CJ*, 1 Sept. 1843.
75. CRO, Cwmgwili 705.
76. *Cambrian*, 6 Jan. 1844; *W*, 5 Jan. 1844; and *Morning Herald*, 10 Oct. 1843. Compare NLS, MS 2843, fo. 371, and HO 45/454, letter from G. R. Trevor, 13 Nov. 1843.
77. *The Times*, 4 and 25 Sept. 1843.
78. *CJ*, 17 Nov. 1843, and *The Times*, 17 Nov. 1843.

79. *CJ*, 27 Oct. 1843; HO 45/454, letter from G. R. Trevor, 24 Oct. 1843, and the letter enclosed in that from Lt.-Col. Maberley, 19 Sept. 1843.
80. *CJ*, 6 Oct. 1843. Compare *Morning Herald*, 26 Oct. 1843, and PP 1844, XVI, South Wales, p. 110.
81. HO 45/642A, letter from Col. Love, 22 Oct. 1844. See, too, W. Lloyd Davies, 'Notes on Hugh Williams and the Rebecca Riots', *Bulletin of the Board of Celtic Studies*, 11: 4 (1944), 160–7.
82. *CJ*, 24 Nov. 1843, and *Swansea Journal*, 4 Oct. 1843.
83. HO 45/454, letter from Anon., 25 Aug. 1843.
84. NLW, Bronwydd 6814, and MS 14590E, 57, Voluntary Examination of D. Davies.
85. Tobit Evans, op. cit. n. 10 above, p. 148.
86. See e.g. *W*, 4 Aug. 1843; *The Times*, 5 Aug. 1843; *CJ*, 4 and 25 Aug. and 29 Dec. 1843.
87. *Morning Herald*, 18 Sept. 1843, and *The Times*, 26 Aug. 1843. The question of compensation for farmers blighted by game preservation was raised at one meeting. *CJ*, 6 Oct. 1843.
88. There are many sources for this paragraph. See *W*, 4 Aug. 1843; *CJ*, 11 Aug., 20 Oct., and 1 Dec. 1843; PP 1861, XXIII, Report on Salmon Fisheries, qu. 4664; PP 1844, XVI, South Wales, p. 38; *The Times*, 21 Oct. 1843; NLS, MS 2843, letters from G. Brown, 18 Oct., and from H. Owen, 19 and 24 Oct. 1843; HO 45/454, letters from D. Prothero, 27 July, from Anon., 25 Aug., and from E. C. Lloyd Hall, 30 Sept. 1843.
89. On the Llechryd weir, see e.g. PP 1844, XVI, South Wales, pp. 237–8, *CJ*, 15 and 22 Sept., 6 and 13 Oct., and 15 Dec. 1843, 12 Jan. 1844, and 24 Jan. 1845. The *Welshman* has reports on similar dates. HO 45/454, letters from E. C. Lloyd Hall, W. D. Jones, and T. Jenkins, July, Sept., and Oct. 1843.
90. *W*, 13 Oct. 1843.
91. See e.g. HO 45/454, letter of 26 Aug. 1843.
92. *W*, 12 Apr. 1844.
93. NLW, MS 21209C. On 21 Dec. 1843 he said that Wales was daily becoming more 'like Ireland'.
94. *W* and *CJ*, 23 June 1843. Compare HO 45/454, letter from E. C. Lloyd Hall, 6 Sept. 1843.
95. HO 45/453, letter from Col. Love, 31 Aug. 1843; *Morning Herald*, 18 Sept. 1843; *CJ*, 6 Oct. 1843 and 26 Jan. 1844.
96. See e.g. *W*, 18 Aug. and 15 Sept. 1843; *Swansea Journal*, 13 Sept. 1843; and NLS, MS 2843, letter from G. B. J. Jordan, 22 Oct. 1843. There was some confusion over the reference to a Farmers' Arms. See CRO, Aberglasney, 24/570A.
97. For some of this, see HO 45/454, letter from E. C. Lloyd Hall, 30 Sept. 1843; *W*, 7 July and 27 Oct. 1843; HO 45/642A, letters from G. Martin, 22 Jan., Col. Love, 5 Mar., and report of J. Daniel, 29 Mar. 1844.
98. *CJ*, 4 Aug. 1843. See, too, HO 45/454, letters from E. C. Lloyd Hall, 22 July, and G. R. Trevor, 7 Aug. 1843; *The Times*, 28 July 1843.
99. NLS, MS 2843, letters from J. Walsh, 20 Oct. and 26 Nov. 1843.
100. *W*, 13 Oct. 1843.
101. HO 45/642A, letter from G. Martin, 22 Jan. 1844, with enclosures; NLS, MS 2844, letter from Col. Love, 28 Jan. 1844; HO 45/454, letters from H. Vaughan, 8 Aug. and from C. J. Wrigley, with enclosure, 24 Aug. 1843; *CJ*, 6 Oct. 1843; PP 1844, XVI, South Wales, p. 254.
102. *W*, 15 Sept. 1843; HO 45/347A, letter from G. R. Trevor, 16 Sept.

Notes

(wrongly dated) 1843; HO 45/642A, letter from Col. Love, 23 Sept. 1844.
103. There is much material on the death of T. Thomas. See e.g. *CJ*, 13 Oct. and 22 Dec. 1843, 6 Dec. 1844, and 2 Mar. 1845; *W*, 13 and 20 Oct., 22 and 29 Dec. 1843, and 5 Jan. 1844, and *Cambrian*, 6 Jan. 1844 and 22 Mar. 1845; HO 45/453, letter from Col. Love, 20 Dec. 1843; HO 45/642, letter from Col. Love, 2 Jan. 1844; HO 45/642A, letter from Col. Love, 11 Dec. 1844.
104. HO 45/454, letter of 25 Aug. 1843.
105. *Swansea Journal*, 11 Oct. 1843, and NLS, MS 2843, letter from G. Brown, 15 Oct. 1843. The homes of magistrates in the district of Haverfordwest were also threatened. HO 45/453, letter from J. H. Peel, 26 Aug. 1843.
106. H. M. Vaughan, *The South Wales Squires* (London, 1926), p. 16. Compare HO 45/454, letters from E. C. Lloyd Hall, 23 July and 12 Aug. 1843.
107. NLS, MS 2843, letter from J. Graham, 10 Oct. 1843.
108. Ibid., letter from Col. Love, 21 Dec. 1843.
109. *The Times*, 5 Sept. 1843.
110. *CJ*, 10 Mar. and 5 May 1843. There were also other attacks on his property. See e.g. HO 45/265, letter from T. Powell, 31 Dec. 1842.
111. *W*, 6 Oct. 1843.
112. Ibid. 16 June 1843.
113. PP 1844, XVI, South Wales, p. 224; *CJ*, 11 Aug. and 8 Sept. 1843, 26 Mar. and 3 May 1844; HO 45/453, letter from Col. Love, 8 Aug. 1843; HO 45/454, letter from W. D. Jones, 19 Sept. 1843; *W*, 24 Nov. 1843.
114. *Swansea Journal*, 13 Sept. 1843.
115. *W*, 8 Dec. 1843. There is much material on animosity towards Chambers. There are addresses, letters, and reports on all this in HO 45/453 and 454, NLW, MS 14590E, *CJ*, *W*, and the *Swansea Journal* for Sept.–Dec. 1844, and in PP 1844, XVI, South Wales, p. 130.
116. *CJ*, 22 and 29 Dec. 1843, and 12 Jan. 1844. On his financial and legal affairs, see NLW, Eaton Evans and Williams, 5875–85, and 12211–34.
117. HO 45/454, letter from E. C. Lloyd Hall, 22 Sept. 1843.
118. *CJ*, 10 Jan. 1845. Compare the treatment of the vicar of Oxwich in Gower. A. Lloyd Hughes kindly directed my attention to this. Welsh Folk Museum, St Fagans, MS 2249.
119. NLW, MS 14590E, Document 78; PP 1844, XVI, South Wales, pp. 225–6; *CJ*, 6 Oct. 1843; and *The Times*, 7 Oct. 1843. Morris had financial problems, and Gwynn was accused of rape. *W*, 26 Apr. 1844.
120. HO 45/454, letters from E. C. Lloyd Hall, 26 Aug. 1843, and from Anon., 25 Aug. 1843.
121. See e.g. *CJ*, 25 Aug. 1843; HO 45/454, letters from J. Cooper, 14 Sept. 1843, and from E. C. Lloyd Hall, 19 Aug. and 17 Sept. 1843; and NLS, MS 2843, letter from H. Owen, 17 Oct. 1843.
122. HO 45/454, letters from E. Evans, 4 Sept. 1843, and from E. C. Lloyd Hall, 8 Sept. 1843; PP 1844, XVI, South Wales, pp. 254–7; Tobit Evans, op. cit. n. 10 above, pp. 67–9.
123. *The Times*, 26 Aug. 1843. R. G. Thomas, J. Lloyd Davies, J. W. Philipps, and Captain Pritchard were just four gentlemen whose homes or property were attacked.
124. *W*, 8 and 15 Sept. 1843; HO 45/454, letter of G. R. Trevor, 31 Aug. 1843.
125. *CJ*, 22 and 29 Sept. and 6 Oct. 1843; HO 45/453, letter from Captain Wilson, 22 Sept. 1843; HO 45/454, letters from J. R. Lewes Lloyd, Sept. 1843, from J. Walters, 20 Sept. 1843, and from E. C. Lloyd Hall, 22 and 30

Sept. 1843. Commenting on one atack on his property, one writer said that people tended to blame Rebecca for everything. *Tarian y Gweithiwr*, 7 Oct. 1886.
126. CUL, Graham Papers (Microfilm), letters from J. Graham, 14 and 24 Sept. 1843. W. Peel considered leaving the country, as did others. *Swansea Journal*, 4 Oct. 1843. On the experience of Baron de Rutzen, his receipt of Rebecca letters, and his anger with the authorities, see NLS, MS 2843, letter from H. Owen, 19 Oct. 1843.
127. *CJ*, 12 May, 9 June, 28 July, 1 Sept., and 6 Oct. 1843; HO 45/454, letter from E. C. Lloyd Hall, 30 Sept. 1843.
128. He also sacked all his workmen. *The Times*, 28 Sept. 1843. The story of Rebecca's attack on the Middleton Hall estate can be followed in many of the letters in NLW, MS 21209C, and the columns of the *Swansea Journal*, *The Times*, *CJ*, and *W* for Sept. 1843. Adams promised to consider rent reductions, but he was an uncompromising character in this matter. See NLW, MS 21209C, letters from T. Cooke, 30 June and 12 Oct. 1845.
129. NLW, MS 21209C; HO 45/454, letters from W. D. Jones, 19 Sept., and from E. C. Lloyd Hall, 13 Sept. 1843.
130. *W*, 1 Sept. 1843.
131. HO 45/454, letters from E. C. Lloyd Hall, 13 and 21 Sept. 1843; *The Times*, 30 Sept. 1843; *Morning Chronicle*, 26 Sept. 1843; and *CJ*, 25 Aug. 1843.
132. For just a little on this, see D. Williams, *The Rebecca Riots* (Cardiff, 1955), pp. 246–9, and many of the documents in NLW, 14590E. See also *Swansea Journal*, 23 Sept. 1843; *Morning Herald*, 11 Oct. 1843; and NLS, MS 2844, letter from Col. Love, 6 Jan. 1844.
133. PP 1844, XVI, South Wales, p. 130.
134. HO 45/454, letters from E. C. Lloyd Hall, 20 and 24 Aug. 1843; *Cambrian*, 30 Mar. 1844, and 9 and 23 Mar. 1849. Compare the case of farmer D. Williams, in *CJ*, 1 Dec. 1843.
135. *W*, 8 and 15 Sept. and 8 Dec. 1843.
136. *CJ*, 17 and 24 Nov. 1843, *Cambrian*, 6 Jan. 1844 and 28 Mar. 1845.
137. See e.g. *CJ*, 22 and 29 Dec. 1843, and *Cambrian*, 23 Dec. 1843 and 6 Jan. 1844.
138. *CJ*, 12 Jan. 1844.
139. For some of this, see HO 45/453, letter from Col. Love, 10 Sept. 1843; *W*, 5 Jan. 1844; *CJ*, 1 and 8 Sept. 1843; *The Times*, 24 Aug. 1843; NLW, MS 3294E; HO 45/454, letters from J. Walters, 4 Oct. 1843, and E. C. Lloyd Hall, 2 and 21 Sept. 1843.
140. Benjamin Jones, blacksmith's son, shared with John Jones (Shoni Sgubor Fawr) the distinction of being the most violent man in the movement. He was strongly suspected of murdering Thomas Thomas, and he assaulted several people. His reputation is conveyed in CRO, 4282A, letter from J. Lloyd Price, 4 Dec. 1844.
141. NLW, MS 14590E, Document 45.
142. See e.g. *W*, 20 Sept. 1843; NLW, MS 14590E, Document 50; and PP 1844, XVI, South Wales, p. 255.
143. HO 45/453, letter from Col. Love, 3 Oct. 1843; HO 45/454, letters from E. C. Lloyd Hall, 12 and 24 Aug. 1843; *The Times*, 5 Sept. 1843; and *W*, 1 Sept. 1843.
144. See e.g. *CJ*, 6, 20, and 27 Oct. 1843.
145. See e.g. *CJ*, 21 July 1843; *The Times*, 29 Sept. and 3 Nov. 1843, and NLS, MS 2843, letter from J. Graham, 2 Nov. 1843.

Notes

146. HO 45/454, letter from 26 Aug. 1843.
147. PP 1847, XXVII, Education, BCR, p. 84.
148. W. J. Linton, *Memories* (London, 1895), p. 90.
149. CUL, Graham Papers (Microfilm), letter of 15 Sept. 1843.
150. Ibid., letters of 24 Sept. 1843.
151. HO 45/642A, letter from Col. Love, 27 Aug. 1843.
152. NLS, MS 2843, letter from J. Graham, 27 Oct. 1843. Compare his letter to the Queen, 22 Sept. 1843, in CUL, Graham Papers (Microfilm).
153. *CJ*, 1 Sept. 1843.

Chapter 7

1. PP 1847, XXVII, Education, BCR, pp. 79 and 90. Compare *CJ*, 30 June and 22 Sept. 1843, and *W*, 30 June 1843.
2. J. H. Davies, 'The Social Structure and Economy of South-West Wales in the Late Nineteenth Century' MA thesis, Aberystwyth, 1967, p. 10.
3. *The Times*, 13 July 1843, and *W*, 18 Aug. 1843. On the notion that Chartists and politicians stirred up discontent, see HO 45/642, letter from G. R. Trevor, 24 July 1844. On the idea of a simple non-political and political contrast between the rural and industrial people of South Wales, see *The Times*, 13 July 1843.
4. Hansard, vol. 70 (1843), col. 1420, and *The Times*, 19 Aug. 1843.
5. *The Times*, 18 Sept. 1843. Compare the verbal assault by Michael Thomas on Charles Bishop. Ibid. 13 Oct. 1843.
6. *Nonconformist*, 28 June 1843, and D. T. Evans, *The Life and Ministry of Caleb Morris* (London, 1902), p 245.
7. *Swansea Journal*, 27 Sept. 1843.
8. For background on this, see D. Williams, *The Rebecca Riots* (Cardiff, 1955), pp. 25–30, and D. J. V. Jones, *Before Rebecca* (London, 1973), ch. 5. See, too, *CJ*, 26 Aug. 1831 and 19 Oct. 1832.
9. There is much on this. See e.g. *Charter*, 22 Feb. 1839; *Northern Star*, 23 Feb. 1839; HO 40/46, letter from J. R. L. Lloyd, 8 May 1839; and HO 40/51, letters from J. H. Allen, 18 July 1839, and from W. B. Swann, 28 July 1839.
10. PP 1844, XVI, South Wales, Evidence, p. 44.
11. *W*, 3 Nov. 1843.
12. *The Times*, 24 July 1843, and *CJ*, 28 July 1843.
13. Lewis, a man 'in a conspicuous station' in life, was accused, along with his friends, of not doing more to restore peace. *Morning Herald*, 13 Oct. 1843.
14. *CJ* and *W*, 8 Sept. 1843.
15. *The Times*, 12 Oct. 1843.
16. Ibid. 6 Oct. 1843, and PP 1844, XVI, South Wales, pp. 74–5.
17. See e.g. *W*, 3 Nov. 1843, and *The Times*, 18 Aug. and 1 Sept. 1843.
18. Cited by Williams, op. cit. n. 8 above, p. 156. See *Morning Herald*, 25 Mar. 1844, for its observations on the report of the turnpike commission.
19. There was some annoyance with speeches delivered in English, *Swansea Journal*, 30 Aug. 1843.
20. See e.g. *CJ*, 23 June 1843.
21. The identity of LEX is uncertain. D. Williams suggested it was E. C. Lloyd Hall, though some of the first letters do not have his style. Op. cit. n. 8 above, p. 106. LEX was writing to the *Welshman* as early as 30 Dec. 1842 and 3 Feb. 1843, and as late as 1 Jan. 1869. During the winter of 1842–3,

and in the summer of 1843, he provided the newspaper with a series of letters on farmers' problems and the need for union. On the institution, see W, 29 Mar. 1844, and compare ibid. 11 Aug. and 1 Sept. 1843.
22. For the union, see The Times, 24 and 27 July, 7 Aug., 4, 19, and 25 Sept. 1843; CJ, 28 July 1843; Swansea Journal, 9 Aug. 1843; W, 22 Sept, 1843; and HO 45/347A, second report by informer. Aug.–Sept. 1843.
23. See e.g. reports of meetings in The Times, 28 and 30 Sept. and 13 Oct. 1843. Evan Thomas is in CJ, 6 Oct. 1843.
24. Nation, 6 Apr. 1844.
25. CJ and W, 3 Nov. 1843, and W, 24 Nov. 1843.
26. W, 21 Apr. 1844.
27. The Times, 31 July, 10 and 15 Aug., and 5 Sept. 1843. Foster began a rather unpleasant debate on the role of ministers in the troubles. See the stout reply of J. Jones, in W, 11 Aug. 1843, and of D. Rees, reported in The Times, 29 Aug. 1843.
28. W, 8 Sept. 1843. Compare John Pugh's words in The Times, 16 Oct. 1843.
29. For one list of the Llandyfaelog grievances, see The Times, 9 Sept. 1843. Gravelle's comparison with Germany was made at a later meeting. W, 29 Sept. 1843.
30. CJ and The Times, 13 Oct. 1843.
31. On this, see e.g. W, 22 Dec. 1843, 5 and 26 Jan. 1844, and PP 1844, XVI, South Wales, pp. 210–13.
32. Reports, slightly different, of the speech in The Times, 25 Sept. 1843; Swansea Journal, 27 Sept. 1843, and W, 28 Sept. 1843.
33. CJ, 6 Oct. 1843. Free-trade ideas were affecting the Narberth area, too, The Times, 17 Nov. 1843. Note how the Anti-Bread Tax Circular, 11 July 1843, and League, 30 Dec. 1843, placed the blame for the economic and Rebecca crises on the landlords. Compare Y Diwygiwr (1844), 30–1.
34. The Times, 18 Sept. 1843.
35. HO 45/454, letter from E. C. Lloyd Hall, 26 Aug. 1843.
36. W, 24 Nov. 1843.
37. PP 1844, XVI, South Wales, p. 105. See the interesting speech by Morgan Williams on the state of Chartism in Wales in the spring of 1842. Northern Star, 23 Apr. 1842.
38. For comments on the reporter of the Northern Star, see The Times, 8 Aug. 1843; Swansea Journal, 23 Aug. 1843; and HO 45/454, letter of G. Bird, 19 Aug. 1843. It is instructive to follow the reports in this Chartist newspaper at the height of the Rebecca troubles. Northern Star, 22 July–16 Sept. 1843.
39. W, 3 Nov. 1843, and The Times, 30 Sept. 1843. Col. Love said, surprisingly, that O'Connor had already been to Swansea. HO 45/453, letter of 3 Aug. 1843.
40. Northern Star, 12 Aug. 1843, and Poor Man's Guardian, and Repealer's Friend, Nos. 10 and 11 (1843). The Times reporter heard the rumours of Chartists influencing Rebeccaites, but was dubious about them. The Times, 10 and 18 Aug. 1843.
41. There is much on this, especially by Col. Love, in his many letters, and in the secret reports on Chartist meetings at Merthyr, in HO 45/453. One has the impression, not least from the Northern Star, 19 Aug. 1843, that the national Chartist leadership was less keen than some local radicals to bring together striking copper workers, Rebeccaites and Chartists into one combination. For a little background, see D. J. V. Jones, 'Chartism at Merthyr', Bulletin of the Board of Celtic Studies, 24 : 2 (1971), 230–45.

42. HO 45/347A, reports of informers, 15 Aug.–15 Sept. 1843. There was much fuss, and exaggeration, over the threatening letters sent to the jurymen who convicted the Rebeccaites at Cardiff. NLS, MS 2843, letter from Col. Love, 30 Nov. 1843.
43. See e.g. *CJ*, 1 Sept. 1843, and HO 45/454, letter from E. C. Lloyd Hall, 2 Sept. 1843.
44. *The Times*, 17 Aug. 1843, and *CJ*, 22 Sept. and 6 Oct. 1843.
45. See the many letters from Col. Love, 5 Mar.–11 Dec. 1844, in HO 45/642A.
46. Stephen Evans, in public at least, acted as a rational man, although on one occasion he did state that, while he was opposed to Rebeccaism, he understood why people turned to it. *The Times*, 29 Aug. 1843. Like David Rees and Hugh Williams he must have known more, from family and friends, about the troubles than he ever admitted in public. Note the threat expressed, and carried out, at a Llannon meeting, that unless R. G. Thomas listened to their words they would resort to Rebecca. Ibid. 25 Aug. 1843.
47. At Mynydd Mawr. *Morning Herald*, 18 Sept. 1843.
48. *W*, 15 and 29 Sept. 1843, and *The Times*, 30 Sept. 1843.
49. *W*, 30 June 1843.
50. See the letter in *Swansea Journal*, 13 Sept. 1843. Compare *Morning Herald*, 6 Dec. 1843. See J. Jones's speech at the Llechryd meeting, *W*, 13 Oct. 1843. See, too, the Welsh complaints and bitterness behind the letters and comments in ibid. 15 Sept., 20 Oct., and 22 Dec. 1843, and 25 Oct. 1844.
51. See, e.g. CUL, Graham Papers (Microfilm), letter of 22 Sept. 1843, and HO 45/453, letters from Col. Love, with enclosures, 11 July and 6 Aug. 1843.
52. *W*, 10 Nov. 1843, and *Nation*, 6 Apr. 1844. See, too, *W*, 12 and 19 April 1844. Sir Reginald Coupland was impressed with the slow growth of Celtic nationalism at this time, and by the lack of Welsh interest in the fate of Ireland. *Welsh and Scottish Nationalism* (London, 1954), p. 168.
53. For some of this, see *The Times*, 27 June and 30 Sept. 1843. On Welsh sympathy for the starving Irish, see *CJ*, 15 and 22 Jan. 1847, and on opposition to 'papal aggression', see ibid. Nov.–Dec. 1850.
54. For a comparative study of East Anglian arson, see D. J. V. Jones, *Crime, Protest, Community and Police in Nineteenth-Century Britain* (London, 1982), ch. 2. See *W*, 9 Dec. 1842, 14 July and 15 Dec. 1843, 17 May and 19 July 1844. For tollgate troubles outside Wales, see reports in ibid. 2 Dec. 1842, 25 Aug., 15 and 22 Sept., 10 and 17 Nov. 1843, and 12 Jan. 1844. See the Rebecca letter of 1847 in NLW, MS 14005E.
55. Phrase used in NLS, MS 2843, letter from O. Morgan, 18 Oct. 1843. There were attacks on tollgates, for example, near Cardiff, Llantrisant, and Pontypridd.
56. *CJ*, 10 and 24 Nov. 1843.
57. NLS, MS 2843, letter from J. H. Vivian, 15 Oct. 1843.
58. The comment by the *Times* correspondent is cited by *W*, 29 Sept. 1843, and those by Henry Richard in *Letters and Essays on Wales* (London, 1866), pp. 72–3.
59. *W*, 10 Nov. 1843.
60. See report in 45/454B.
61. On the few attacks, see e.g. the letters from Col. Love and Lt.-Col. Worthen, in HO 45/1431, and HO 45/1809; *W*, 28 Apr. 1848; *CJ*, 26 Dec. 1851; and A. H. T. Lewis, 'The Development and Administration of

Roads in Carmarthenshire, 1763–1860', MA thesis, Swansea, 1968, p. 432.
62. See e.g. *CJ*, 10 Apr. 1846.
63. PP 1844, XIX, pp. 21–5, for complaints in south Wales over bastardy and salaries.
64. There are dozens of letters on the comparative effectiveness of Love and Brown, in CUL, Peel and Graham Papers (Microfilm), and in NLS, MS 2843 and 2844.
65. See e.g. *CJ*, 10 Apr. 1846 and 4 June 1847.
66. Ibid. 5 Sept. 1851.
67. Pembrokeshire Record Office, PQ/AP/2/1.
68. These paragraphs are based on debates in the Quarter Sessions, and the annual reports of chief constables, reported in the newspapers of the period. For some of the reasoning behind, and against, the establishment of a Rural Police, see the reports of a Llanedi meeting and Quarter Sessions in *The Times*, 18 Aug. and 23 Oct. 1843.
69. The campaign to get, and then to remove, the Rural Poice in Cardiganshire produced a large collection of documents. For a selection, see NLS, MS 2843, letters from W. E. Powell, 18 Oct. and 11 Nov., from Col. Love and G. Brown, 18 Oct. 1843. In 1849 it was agreed to press for the removal of the old parish constables from the northern half of the county. *CJ*, 5 Jan. 1849.
70. Letter in *CJ*, 10 Oct. 1851.
71. *CJ*, 21 Oct. 1853 and 9 Jan. 1857.
72. *W*, 8 Sept. 1843. Note Harford's comment in *CJ*, 30 Nov. 1849.
73. *Nation*, 6 Apr. 1844.
74. For background here, see I. G. Jones, *Communities: Essays in the Social History of Victorian Wales* (Llandysul, 1987), chs. 1–5.
75. *CJ*, 18 and 25 July 1845.
76. On revivalism, see the recent thesis of C. B. Turner, 'Revivals and Popular Religion in Victorian and Edwardian Wales', Ph.D. thesis, Aberystwyth, 1979.
77. *CJ*, 22 July 1859.
78. *CJ*, 17 Apr. 1846.
79. CRO, MS 2842A, letter from J. Lloyd Price, 22 Sept. 1844, and J. Downes, 'Notes on a Tour of the Disturbed Districts in Wales', *Blackwood's Magazine*, 54 (1843), 776.
80. *CJ*, 29 Jan. 1847, and the reference to the *Principality* is cited in W. G. Evans, 'The Voluntary Movement in South wales 1843–67', M.Ed. thesis, Cardiff, 1978, p. 208. On the notion of visits to England by young Welshmen, see PP 1847, XXVII, Education, BCR, p. 109.
81. *CJ*, 22 Aug. 1856, 22 July 1859, and 17 June 1864.
82. R. Colyer provides a general background on this. 'The Early Agricultural Societies in Wales', *Welsh History Review*, 12: 4 (1985), 567–81. Trevor's comments are in *W*, 20 Sept. 1844.
83. For a little on this, see *CJ*, 31 Dec. 1847, 21 Sept. 1849, 4 Oct. 1850, 7 Oct. 1853, 4 Apr. 1856, 6 Nov. 1857, and 6 May 1864.
84. Ibid. 10 May 1850 and 23 Oct. 1868.
85. Ibid. 24 Sept. 1849.
86. Term used in ibid. 7 Jan. 1848.
87. According to R. Davies infant mortality was still high in this countryside at the turn of the century. 'In a Broken Dream', *Llafur*, 3: 4 (1983), 24–33.

Notes 409

See the interesting contemporary question as to whether public health or education was the first priority. *CJ*, 30 Oct. 1846.
88. *CJ*, 30 Oct., 27 Nov., and 25 Dec. 1846.
89. For some of this, see ibid. 18 Mar. 1853, 18 Sept. 1857, and 22 Oct. 1858.
90. See e.g. ibid. 2 Mar. 1849, 8 Nov. 1850, 26 Dec. 1852, and 7 Mar. 1865.
91. See the mocking letter in ibid. 4 Mar. 1864.
92. PP 1847, XXVII, Education, CGP, p. 437, and *CJ*, 25 Oct. 1850.
93. See the critical appraisal of their exclusiveness in *W*, 12 Feb. 1869. This paragraph, like others in this section, is based on an intensive study of the *Welshman*, the *Carmarthen Journal*, and the Parliamentary Papers of the years 1844–69.
94. *CJ*, 13 and 20 Aug. 1858.
95. The winter of 1846–7 was bad for potato-eating labourers, and there were signs in the summer and autumn of 1849 of hard times to come for the farmers. Ibid. 5 Feb. 1847, and 4 July–17 Aug. 1849.
96. 'Signs of decay' in Carmarthen had been spotted in the *Carmarthen Journal* as early as 24 Apr. 1846.
97. In view of the level of migration during the 1840s, the importance of the railways in the Rebecca story has been perhaps exaggerated. Williams, op. cit. n. 8 above, p. 292, and P. Molloy, *And They Blessed Rebecca* (Llandysul, 1983) p. 21.
98. See e.g. *CJ*, 18 July and 8 Aug. 1845, 21 May 1852, 24 Feb. 1854, 15 Apr. 1864, and 12 Mar. 1869.
99. Ibid. 2 Aug. 1850.
100. Ibid. 27 Sept. 1867 and 8 May 1868.
101. Richard, op. cit. n. 58 above, pp. 72–3.
102. *CJ*, 14 Apr. 1848. Compare ibid. 3 Mar. 1848, and PP 1844, XVI, South Wales, pp. 206, 213, and 352.
103. D. Davies, 'Rebecca and her Daughters', *Red Dragon*, 11 (1887), 240.
104. See the interesting debate over the violent history of a community, in PP 1895, XL, Report on the Land, vol. iii. qus. 41, 413–16.
105. Praise for the natives was lavish when vagrants were responsible for the few cases at the higher courts. See e.g. *CJ*, 17 Mar. 1848 and 13 July 1860. The Welsh pattern of indictable and non-indictable crime can be seen in D. J. V. Jones, 'The Welsh and Crime, 1801–1891', in C. Emsley and J. Walvin (eds.), *Artisans, Peasants and Proletarians 1760–1860* (London, 1985), p. 90, and the county breakdown for 1858–92 in D. J. V. Jones and A. Bainbridge, 'Crime in Nineteenth-Century Wales', Social Science Research Council Project, 1975, pp. 93–109.
106. Richard, op. cit. n. 58 above, p. 73, and T. Rees, *Miscellaneous Papers on Subjects Relating to Wales* (London, 1867), pp. 15–17.
107. Term from PP 1895, XL, Report on the Land, vol. iii, qu. 44, 317.
108. Term borrowed from a case reported in *CJ*, 24 Apr. 1846.
109. Ibid. 22 Mar. and 8 Nov. 1850, and 2 May 1851.
110. Ibid. 22 Mar. 1850. Compare ibid. 27 July and 18 Sept. 1849.
111. Ibid. 14 Apr. 1854 and 11 May 1860.
112. Ibid. 7 Mar. 1856, 3 July 1846, and copy of a letter from G. Jones, 27 Mar. 1858, to his chief constable, kindly provided for me by Mr Winston Jones of Haverfordwest.
113. *W*, 21 Mar. 1845.
114. For a selection of these, see *CJ*, 19 Mar. 1847, 26 May 1848, 20 Sept. 1850, 28 Oct. 1853, 27 Apr. 1855, 27 Mar. 1857, and 13 Nov. 1863.

115. Ibid. 26 Apr. 1867.
116. Ibid. 27 June 1862. See, too, the Rebecca reference in ibid. 7 Sept. 1855. The full story of these troubles can be followed in the local press, and in the annual reports of the inspectors of the salmon fisheries in the Parliamentary Papers.
117. Terms from *CJ*, 14 Apr. 1854.
118. See the Rebecca letter on this in the Pembrokeshire Record Office, HDX, 343/1.
119. The essential background to all of this is I. G. Jones, *Explorations and Explanations* (Llandysul, 1981), and K. O. Morgan, *Wales in British Politics, 1868–1922* (Cardiff, rev. edn. of 1970).
120. For a little background on this, see NLW, Cilgwyn, 34–5, and 44, letter books, and the comments and quotations on E. C. L. Fitzwilliams, in R. Colyer, 'The Gentry and the County in Nineteenth-Century Cardiganshire', *Welsh History Review*, 10: 4 (1981), 514, D. W. Howell, *Land and People in Nineteenth Century Wales* (London, 1977), p. 64, and J. H. Davies, op. cit. n. 2 above, p. 84.
121. The standard first-hand accounts of these evictions are Hansard, vol. 197 (1869), cols. 1294–325; PP 1868–9, Reports on Parliamentary and Municipal Elections; and PP 1896, XXXIV, Report on the Land, and vol. iii. Compare the cautious and revised account in D. W. Howell, op. cit. n. 120 above, pp. 64–5.
122. PP 1895, XL, Report on the Land, vol. iii. qu. 44, 276.

Sources

PRIMARY MATERIAL

Many sources were used for the book, and detailed information on them is given in the footnotes. The main sources are:

British Library

Francis Place, Additional MSS on the Working Men's Association and the General Convention.
Peel MSS.

Cambridge University Library (CUL)

Graham Papers (Microfilm).
Peel Papers (Microfilm).
Windsor Castle Papers (Microfilm).

Carmarthen Public Library

Census returns of 1841 and 1851 for Abernant, Llanarthne, Llanfihangel-ar-arth, Llannon, Llangadog, St Clears, and Tre-lech a'r Betws.

Carmarthen Record Office (CRO)

Carmarthen Borough Records.
Carmarthenshire County and District Roads Board Minutes.
Census Returns (typed abstracts) of 1851: Carmarthen and Llanelli.
Manuscript Collections: Aberglasney (Evans), Brigstocke, Castell Gorfod, Cawdor, Coedmore, Cwmgwili, Cynghordy, Davies-Evans, Derwydd, Dynevor, Glasbrook, John Francis (I), Plas Llanstephan, and Taliaris.
Other Collections: Acc. 4282A, 4916, Bryn Myrddin, CDX/22, Museum 303, 368, and 432.
Parish Registers: Meidrim, Llanarthne, Llanddarog, Llangendeirne, and St Ishmael's.
Quarter Session Records.
Poor Law Records.
Turnpike Trust Records: Main, Carmarthen and Newcastle, Kidwelly, Llandovery and Lampeter, and other trusts.
Vestry Minutes: Llanarthne, Llanddeusant, Llangadog, Llangunnor, Merthyr, Newchurch, and Tre-lech a'r Betws.

Sources

Ceredigion Record Office
Census (typed abstracts) of 1851: Cardigan.
Poor Law Records.
Parish Registers (typed abstracts): Llechryd and Troedyraur.

National Library of Scotland (NLS)
MSS 2843 and 2844 (Brown correspondence).

National Library of Wales (NLW)
Manuscript Collections: Ashburnham, Behrens, Bronwydd, Cilgwyn, Clynfiew, Crosswood, Cwmgwili, Derry Ormond, Dolaucothi, Edwinsford, Eaton Evans and Williams, Falcondale, Glansefin, Harpton Court, D. T. Jones, Llidiardau, Llwyngwair, Lucas, Mynachty, Nanteos, Nevill, Noyadd Trefawr, Orielton, Penpont, Picton Castle, Slebech, D. P. Williams, and William and Williams.
Other Records: Cardiganshire Board of Guardian Records, Cardiganshire County Roads Board, Cardiganshire Quarter Sessions, Calendar of the Diary of L. W. Dillwyn, Church of Wales (St Davids), MSS 1398B, 2114C, 3141–9F (Poor Law Vols.), 3294E, 6244B, 12368E, 14005C, 14536E, 14590E, 16582C, and 21209C.
Parish Records and Vestry Minutes: Cenarth, Llandyfriog, Llanfihangel-ar-arth, Llanfyrnach, Manordeifi, and Penboyr.

Pembrokeshire Record Office
Census (typed abstracts) of 1851: Haverfordwest and Pembroke.
Calendars of Prisoners for Assizes.
Other records: HDX/343.
Poor Law Records.
Prison Register, Haverfordwest.
Quarter Session Records.
Turnike Trust Records.

Public Record Office (PRO)
Assize Rolls, 71/7, 8, 72/1, 73/2, 73/3, 76/1, 77/1, and 77/2.
Home Office Letters and Papers (HO): 40/40, 46, 51; 41/13, 14, 17, 18; 43/54; 45/265, 347A, 453, 454, 642, 642A, 642B, 642C, 642D, 1431, 1611, 1809; and 52/35–8, 43, 44.
Ministry of Health, Poor Law Union Papers: MH 12.

Welsh Folk Museum, St Fagans
MS 2249.

Sources

NEWSPAPERS AND JOURNALS

Annual Register, 1843.
Anti-Bread Tax Circular/League, 1843–4.
Blackwood's Magazine, 1843.
Cambrian, 1843–6.
Carmarthen Journal (*CJ*), 1815–69.
Charter, 1839.
Y Diwygiwr, 1843.
Edinburgh Review, 1843–4.
Fraser's Magazine, 1843–4.
Gentleman's Magazine, 1843–4.
Yr Haul, 1843.
Hereford Journal, 1843.
Hereford Times, 1843–4.
Morning Advertizer, 1843–4.
Morning Chronicle, 1843–4.
Morning Herald, 1843–4.
Nation, 1843–4.
Nonconformist, 1843–4.
Northern Star, 1839–43.
Patriot, 1843.
Planet, 1844.
Poor Man's Guardian, and Repealer's Friend, 1843.
Quarterly Review, 1843–4.
Red Dragon, 1883–7.
Seren Gomer, 1843.
Silurian, 1843.
Swansea Journal, 1843.
Tarian y Gweithiwr, 1886.
The Times, 1843–5.
Wales, 1912–13.
Welsh Gazette, 1900–1.
Welshman (*W*), 1832–69.

Index

Aberaeron 10, 11, 24, 80, 83, 133, 188, 225, 231, 274–5, 289, 297, 311, 323
Aberavon 16, 147
Abergwili 68, 119, 163, 226, 270, 296, 372
Abernant 23, 43, 48–9, 57–8, 73, 92, 104, 115–16, 126, 128, 165, 215–16, 218–19, 262, 278, 304, 314
Aberporth 11, 267, 301
Aberystwyth 1, 10–12, 18–25, 32, 44, 48, 51, 56, 71, 78, 113, 129, 132, 153–5, 163, 172, 178, 183, 193, 203, 207, 332, 350, 353, 361–3, 367, 375
Adams, Edward 10, 48, 51, 56, 69, 98, 123, 159, 174, 189, 229, 232, 243, 277, 294, 299, 306–7
Albert, Prince 346
Allen, Lancelot B. 90, 293
Allt Cunedda 65, 325–6, 341
Alltyrodyn 28, 47, 69, 89, 90, 94–5, 137, 194, 209, 211, 243, 250, 309, 377
Anthony, William 329, 338
Anti-Corn Law League 222, 320, 335
Anti-Tithe-Commutation and Anti-Poor Law Association 335

Begelly 15, 38, 71–2, 172, 190
Beynon, John 223, 373–4
Beynon, Rice P. 48, 73, 75, 90, 114, 178, 196, 207–8, 273, 300
Bishop, Charles 92, 126, 362, 374
Blackpool weir 292
Blue Boar gate 209, 244
Bolgoed 228
Bosanquet, Judge 152
Botalog 241, 244
Bowen, George 292, 299
Bowen, James 153, 299
Bowen, Dr John 31, 91, 107, 130, 132, 151, 193, 361
Bowen, Michael 49, 214, 219
Bowen, Samuel 215

Brackenbury, Augustus 51, 153, 159
Brechfa 176, 283, 287, 294, 298
Brecon 82, 169, 197, 204, 208, 225, 295, 350, 358
Breconshire 30, 203, 226, 241
Brigstocke, Reverend Augustus 194, 224, 293
Brigstocke, William O. 78, 373–4
Bronfelin 226
Bronwydd 70, 85, 94, 223, 290, 299, 319, 342
Brown, Major-General George 236, 258, 296, 299, 350–1
Bryn Cwmllynfell 336
Brynlloi gate 239
Builth 181, 226, 241–2
Bullin, Thomas 139, 200, 204–7
Bwlchyclawdd 138, 142, 211, 230

Canaston Bridge 176, 210
Capel Iwan 272, 303, 311
Cardiff 9, 18, 112, 216, 229, 238, 242, 295, 339
Cardigan 7–9, 11, 13, 18, 22, 24, 27, 31–3, 40–1, 49, 56, 88, 90, 101, 116, 132, 145–6, 153–6, 163, 167, 179–80, 185–6, 192, 196–7, 217, 221–3, 226, 240, 244, 254, 267, 269, 274–5, 287, 290–2, 295, 299, 323, 349, 351, 353, 355, 362, 366–70, 375
Lower District turnpike trust 204, 229
Cardiganshire 3, 9–10, 13–14, 26, 29, 42, 44–9, 52–5, 59, 61–2, 67–70, 81, 83–4, 88, 91, 99–100, 104, 110–11, 126–7, 129, 133–4, 138, 140, 146, 148, 152–3, 156, 159, 169–70, 173, 182, 186, 189, 200, 203, 222, 224, 226, 231, 249, 252, 258, 262, 265, 269, 278, 280, 284, 286, 295, 304, 310, 312, 319, 327, 331, 333–4, 337, 351–4, 371, 374, 377

Index

Carmarthen 1, 4, 7–11, 13–23, 25–7, 30–6, 39–40, 42, 44, 49, 51, 56, 63, 67, 71, 78, 82, 87–8, 90–1, 100, 105, 107, 112–24, 129–30, 133–9, 144–7, 151, 155–6, 161, 163–72, 178–87, 191–7, 199, 200, 204, 207–8, 210–30, 238–9, 242–3, 246–7, 254, 262–5, 270–3, 282–5, 289, 294, 296, 298–300, 307, 317, 323, 329–33, 335, 338–9, 347, 350, 360–7, 372
and Brechfa turnpike trust 139
and Lampeter turnpike trust 141, 204, 213
and Newcastle turnpike trust 139, 204, 213, 217
Carmarthen Journal 39, 83, 85–6, 94, 112, 123, 171, 177, 186, 193, 206, 213, 217, 235, 254, 270, 275, 287, 293, 303, 313, 317–19, 329, 336, 359, 372
Carmarthenshire 3, 5, 9, 13, 17, 19, 21, 23, 29, 42–3, 47–51, 54, 56, 59, 61, 66–70, 73, 77, 87, 89, 91, 94, 99, 111, 114, 118, 120, 123, 133, 135, 137–8, 143–8, 152–3, 156–9, 164, 170–3, 176, 179, 182, 187, 197, 199, 203–4, 213, 217, 221–2, 224, 227, 231, 238, 244, 249, 262, 264, 277–8, 286, 289, 291, 294, 306, 308, 310, 314, 319, 321–2, 327, 333, 335, 337–8, 340, 352, 354–60, 366–8
Castell Rhingyll 221, 348
Castle Malgwyn 5, 13, 112, 171, 292, 306, 358
Cawdor, earl of 5–6, 13, 17, 19, 29, 47, 54, 56, 59–61, 83, 85, 87, 89, 97, 107, 125, 138, 148–9, 169, 171, 246, 289, 291, 305, 352, 354–5, 359, 361, 374
Ceffyl Pren 196–8, 252, 259, 267, 286, 297, 312, 370, 372–3
Cefn Coed yr Arllwyd 325, 331
Cefn Llanddewi 241
Cenarth 164, 250, 273, 292, 303
Chadwick, Edwin 33, 91, 129, 275
Chambers, William 14, 30, 42, 69, 90, 91, 167, 178, 235, 274, 277, 292, 296, 299–302, 309, 325, 328, 361, 373
Charles, Isaac 219
Chartism 96, 206, 222, 230, 247, 307,
316, 320, 323, 329, 331, 337–41, 344–7, 357
Child, James Mark 15, 61, 69, 72–4, 90, 133, 167, 170, 178, 190, 206, 325, 328, 335
Cilgerran 7, 22, 123–4, 192, 197, 292
Ciliau Aeron 18, 22, 48
Cilrhedyn 23, 106, 175, 184, 196, 289, 304, 315, 325, 336
Cilycwm 21–2, 240, 267, 315
Cilymaenllwyd 23, 72, 325, 340
Clive, Robert 347
Clydau 23, 44, 53, 170, 180, 278
Clynfiew 60, 148, 198, 252, 325, 367
Coedmore 60, 97, 148, 292
Colbys 47, 58–9, 70, 83, 148, 211, 245, 283
Commercial Inn bar 207
Cooke, Thomas 56, 61, 97, 149, 232–3, 243, 263, 286, 294, 306–7
County and Borough Police Act of 1856 352, 354
Cresswell, Judge 238
Cripps, William 347
Croesllwyd gate 137–8, 224
Crosswood 46, 61, 175
Currie, William Pitt 56, 190–1
Cwm Cile 66, 228
Cwmdwr 240
Cwmgwili 277
Cwmifor 324, 330
Cwmmawr 15–16, 247, 314
Cynwyl Elfed 39, 49, 75, 98, 120, 135–8, 143, 157, 212, 214, 259, 262, 287, 306, 324, 329

Danyrallt 124, 171, 277, 360
Davies, Bankes 219, 265, 300
Davies, David, of Carmarthen 91, 130, 138, 178, 195–6, 191, 215–16, 300
Davies, David (Dai Cantwr) 147, 233, 235, 238, 302, 308, 313
Davies, David Saunders 47, 59, 80, 88, 94, 125, 129, 144, 186, 194, 213, 293, 360
Davies, Dr Enoch 94, 377
Davies, Reverend Henry Lewis 93, 272, 316, 358, 363
Davies, Howell 49, 75, 287, 306
Davies, John Lloyd 28–9, 69, 89–90, 95, 98, 124, 135, 137, 139, 141,

Index

149, 159, 194, 209, 211, 213, 218, 223, 250, 309, 323
Davies, William, of Pantyfen 92, 213, 246
Day, William 94, 129–31, 242
de Rutzen, Baron 10, 47, 51, 59, 97, 125, 174, 190, 277, 284, 292, 306, 355, 358
Devonald, Mrs 285
Devynnock 226
Diwygiwr 33, 81, 86, 124, 321, 332, 335
Dolauhirion 240, 243–4
Dolhaidd 69, 91, 94, 97, 130, 194, 273, 289, 291, 305, 323, 348, 374
Downes, Joseph 24, 232, 234, 264, 357
Dynevor, Lord 19, 47, 59, 68–9, 83, 89, 207, 227, 231, 277, 297, 300, 305, 359, 361

Education Commission of 1847 27, 37, 41, 77, 79, 81, 319, 358
Edwardes, William 90, 245, 327, 342, 354
Edwards, John 127, 279, 294, 309
Efailwen 198–9, 204, 206, 247, 337
Eglwyswrw 84, 262, 270, 294
Ellis, George 141, 347
England 4, 17, 18, 21, 26–7, 45, 48–9, 57, 62, 70–1, 83, 93, 104, 123, 159, 195, 316, 340, 357, 364, 370
Evans, David, of Penlan 212, 245, 248
Evans, Reverend Eleazar 83, 123, 278, 304, 313
Evans, Frances 219
Evans, Henry Tobit 256
Evans, Reverend John 90, 114, 208, 363
Evans, Lewis 48, 70, 73, 90–1, 99, 126, 149, 178, 181, 218, 245, 258, 321, 323, 325, 328
Evans, Stephen 77, 106, 130, 143, 227, 246, 249, 329, 341

Factory Education Bill of 1843 93, 332
Felingigfran weir 292
Ferryside 11, 177, 179, 307, 366
Ffairfach 138, 199
Fforest 20, 207
Ffynone 47, 59, 70, 148, 153, 211, 245
Fishguard 3, 11, 24, 140, 153, 160, 190, 230, 239, 243, 247, 249, 267, 303, 366, 373
turnpike trust 141, 213
Forster, W. E. 359
Foster, Thomas Campbell 9, 27, 63, 83, 96–100, 104, 108, 129, 149, 157, 195, 225–6, 236, 244, 250, 256, 270, 273, 276, 287, 291, 305, 320–1, 324, 329–33, 345, 364
Freeman, Captain 353
Furnace gate 139, 233

Gelliwernen 279, 294, 309, 312
Gladstone, William E. 377
Glamorgan 4, 14, 17, 24, 26, 28, 30, 32, 34, 43, 66, 147, 164, 168, 203, 226, 228, 238, 303, 335
Glangwili 47, 89, 93, 137, 194, 211, 243, 290, 299, 357
gate 138, 180, 214, 230
Glanrhydw 48, 89, 142, 176, 300, 348
Glansefin 47, 59, 70, 72, 95, 227, 352, 377
Gogerddan 13, 47, 60, 71, 88, 94, 353
Golden Grove 5, 10, 47, 61, 174, 227, 277
Goulbourn, Judge 151
Gower, Abel L, 5, 13, 112, 197, 258, 306, 328, 358
Graham, Sir James 45, 62, 69, 74, 88, 93, 98, 146, 154, 183, 209, 211, 216, 221, 224, 236, 246, 257–8, 267, 276, 279, 305–6, 314, 316, 320, 332, 346–50, 353, 358
Gravelle, David 92, 124–5, 145, 329, 334–5, 343
Griffiths, George Davies 153, 245
Griffiths, Walter 92, 131
Guest, Charlotte 242
Guest, Josiah John 164
Gurney, Judge 238
Gwendraeth foundry and ironworks 14, 227, 309, 314
Gwernant 72, 194, 223, 229, 227, 325, 373
Gwynedd, Ieuan 357
Gwynfe 23, 73, 90, 96, 99, 124, 129, 138, 146, 198, 262, 325, 334, 354
Gwynne, Alban L. 245, 274, 299

Hafod 42, 47, 51, 68, 112
Hall, Edward Crompton Lloyd 45, 77, 79, 97–8, 100, 154, 157, 187, 211,

Hall, Edward Crompton Lloyd (*cont.*):
  218, 221–3, 244, 247, 249, 251,
  258, 264, 268, 281, 292–3, 308,
  315–16, 341, 364, 376
Hall, Thomas J. 347
Hankey, Colonel 159, 351, 353
Harford, John S. 30, 59, 68, 149, 354
Harries, David Lloyd 90, 373
Harries, Henry Lloyd 31, 280
Harries, John 49, 214–15, 218, 220
Harris, Daniel 265, 309, 318
Harvey, John 56
Haul, Yr 74, 319, 333
Haverfordwest 1, 7, 11–12, 15, 19, 24,
  31–2, 36, 40, 42–3, 48, 56, 71, 82,
  88, 91, 99, 107, 110, 114, 117,
  127, 130, 134, 138, 140, 155–6,
  159, 161–4, 169, 180, 182, 184,
  193, 204, 210, 224, 229–30, 242,
  247, 272, 323, 332, 337, 362–6,
  375
Hendy 15, 234–6, 253
Highmead 176, 194
Hill, William 338
Hopkins, William 34, 247
Howell, David 310
Howells, David, farmer 209
Howells, Frances 283
Howells, Phillip 283, 341
Howells, Thomas 209, 245
Hugh, Jenkin 234, 250
Hugh, John, Rebeccaite 235, 238
Hughes, John 32, 61, 148
Hughes, John, Rebeccaite 235, 238,
  242, 247
Hughes, Reverend John 126, 278
Hughes, William G. 90, 220
Hunter, Dr 101

Ireland 10, 13, 17–18, 20, 24, 26, 41,
  48, 68, 94, 98, 104, 254, 279, 288,
  319, 328, 342–3, 346, 349, 358,
  370
Ivorites 33, 114, 196, 333, 364

James, James, bugle-player 216
James, Reverend James W. 84, 239,
  300, 302–3, 311–12
Jenkins, John 17, 329, 335
Jenkins, Reverend John 218
Jenkins, William 323, 337
John, Martha 310
Johnes, John 364

Johnes, Thomas 112
Jones, Benjamin 298, 312, 317
Jones, David, Llanwrda 281
Jones, David Pontarddulais 235, 238
Jones, Eleanor 244
Jones, Evan 298
Jones, George 240, 296
Jones, Griffith 231, 153
Jones, John, 'Captain' 232
Jones, John, Danygarn 281
Jones, John, informer 53, 228, 238,
  243
Jones, John, labourer 216
Jones, John MP 71, 88–9, 159, 171,
  207, 245, 286
Jones, Reverend John 304, 357
Jones, John (Shoni Sgubor Fawr) 147,
  233, 235, 238, 249, 302, 308–9,
  317
Jones, Jonathan 216, 219
Jones, Joseph 289
Jones, Walter D. 259, 300–1, 307, 333
Joshua, David 214, 230, 252–3

Kensington, Lord 68
Kidwelly 14–16, 20, 24, 52, 56, 65,
  138, 193, 195, 210, 226, 229, 247,
  290, 329, 340, 372
  turnpike trust 136–7, 140, 204,
  226, 229, 306
Knighton 241

Lampeter 4, 9, 18, 28, 30, 51, 56, 81–
  2, 91–2, 110, 113, 132, 138–9,
  153, 155, 164, 204, 213, 223, 226,
  231, 240, 244, 323, 325, 338, 353,
  355, 363
Land Commission, report of 1896
  369, 376, 377
Laugharne 4, 63, 117, 170, 196, 285
Laws, Edward 75
Leach, Reverend Francis G. 127
Leach, Henry 91, 130, 135, 151, 162,
  351–2, 361
Lewes, Reverend Benjamin 90, 106–
  7, 130, 174, 223, 304
Lewes, Major Price 18
Lewis, Daniel 34, 228, 247
Lewis, David, labourer 216
Lewis, Howell 216
Lewis, John 92, 246, 297, 330–1
Lewis, John, fisherman 219
Lewis, John, of Minwear House 191

Index

Lewis, Jonathan 216
Lewis, Lewis 73, 90, 96, 124, 138, 325, 354
Lewis, Shadrach 170, 180–1
Lewis, Thomas Frankland 347
Lewis, William, of Clynfiew 148, 252, 325
LEX 99, 329
Lingen, Ralph R. W. 79
Lisburne, earl of 10, 13, 19, 29, 46, 54, 83
Llanarth 58, 92, 126, 132, 174, 286, 325, 328, 331, 337, 373
Llanarthne 7, 59–60, 80, 84, 92, 114, 128, 140, 162, 164, 173, 193, 266, 291, 292, 339, 355
Llanboidy 7, 23, 92, 96, 114, 141, 156–7, 166, 168, 182, 236, 333, 337
Llanddarog 35, 63, 125, 137–8, 210, 263, 286, 341
Llandeilo 1, 20, 24, 39, 48, 54, 80, 89, 127, 138, 155, 159, 162, 169, 181, 188, 197–9, 213, 224, 229, 231, 238, 278, 280, 294, 296–7, 324, 355, 363, 366, 368, 372–3, 375
  and Llangadog turnpike trust 204
  Walk gate 231
Llandeilo-yr-ynys 136, 160, 213, 226–7
Llandovery 1, 6, 7, 9, 19, 24, 31, 41, 62, 69, 126, 134–6, 139, 145, 194, 197, 199, 204, 207, 225, 229, 231, 240–1, 248, 250, 253, 263, 265, 273, 280, 287, 296, 299, 303, 358, 362, 373
Llandybie 10, 52, 138, 291, 304
Llandyfaelog 48, 96, 140–1, 147, 334–5
Llandysul 18, 65, 79, 116, 154, 164, 210, 212, 230, 259, 262, 267, 281, 323, 339, 363, 366, 377
Llanedi 23, 143, 191, 355
Llanegwad 35, 54, 92, 172, 213, 272, 282, 288, 355
Llanelli 7, 9, 14–22, 24, 32, 36, 40, 42–3, 54, 56, 79, 83–4, 88–93, 109, 112, 123–7, 138, 143, 146–7, 164, 167, 178, 186, 192, 200, 233, 235–6, 239, 245, 250, 262–3, 272, 274, 277, 279, 290, 294–6, 301–2, 309–10, 313, 319, 332, 353, 361, 363, 367–9, 372, 375

Llanfihangel-ar-arth 35, 65, 92, 100, 124, 133, 180, 212, 252, 265, 309
Llanfynydd 33, 39, 157, 169, 179, 183, 200, 283, 341
Llanfyrnach, 54, 62, 191, 282
Llangadog 24, 34–5, 52, 68, 92, 96, 100, 114, 124, 156, 180, 198, 231–2, 239–40, 246, 248, 288, 295, 297, 309, 324–5, 334, 339, 360, 365
Llangeler 19, 48, 78, 169, 211, 268, 271, 296, 353, 360, 372
Llangendeirne 10, 40, 92, 164, 295, 311, 372
Llangrannog 11, 83, 123, 128, 278, 294, 304
Llangunllo 269, 282, 303
Llangunnor 92, 124, 128
Llangyfelach 24, 107, 328, 338
Llanllawddog 265, 289
Llannon 15, 23, 69, 94, 109, 112, 117, 125, 138, 143, 145, 182, 226, 235–16, 239, 242, 245, 247, 279–80, 291, 309, 314, 321, 324, 336, 341
Llanon 226, 231
Llanpumpsaint 119, 296
Llansadwrn 23, 54, 303, 322, 330
Llanstephan 54, 89, 116–17
Llanwrda 108, 240, 281, 310
Llanwrtyd Wells 11, 13, 22
Llanybyther 43, 287
Llawhaden 211, 292, 330
Llechryd 22, 123, 179, 196–7, 292–3, 327–8, 374
Llewellin, Reverend Llewellin 89, 91, 353, 363
Lloyd, Edward P. 47, 59, 72–3, 93, 352
Lloyd, James R. Lewes 69, 91, 94, 97, 129–30, 273, 289, 291, 305, 323, 325, 348, 374
Lloyd, John W. 124, 171, 277
Lloyd, Thomas, of Bronwydd 70, 85, 94, 223, 290, 299, 319, 342, 361
Lloyd, Thomas, of Coedmore 97, 292, 328
Llwyngwair 70, 175, 292, 299
Llyn Llech Owen 321, 325–6
Llysnewydd 48, 69, 89, 223, 299, 325, 360
London 4, 14, 26, 33, 41–2, 69, 71, 124, 129, 131, 155, 189, 208, 242, 258, 279, 301, 321, 349–50, 365, 370

420 Index

Love, Colonel 28, 89, 98, 121, 154–5, 159, 198, 223–4, 231, 236, 239, 244, 251, 259, 273, 289, 297–9, 317, 320, 340, 344, 347, 350–1, 353
Luke, Daniel 206

Maesgwynne 47, 72, 207, 256, 325, 354, 360, 377
  gate 206
Maesoland gate 137, 209
Main turnpike trust 139, 204, 207, 210
Mainwaring 291
Manordeifi 280, 295, 316
Mansel, Sir John 14, 89
Mansel's Arms 138, 295
Martin, Reverend Thomas 130, 142, 151
Mathry 156, 169, 262
McKiernin, Francis 234, 295
Mechanics institutes 32, 363
Meidrim 23, 164, 217–18, 221, 262, 286, 294
Meinciau 226–7
Mermaid gate 207–8
Merthyr Tydfil 9, 17, 43, 112, 118, 147, 163–4, 169, 192, 223, 310, 313, 337–40, 344, 351, 367
Middleton Hall 48, 51, 61, 69, 98, 123, 159, 174, 227, 229, 243, 277, 286, 306–7
Milford 15, 18, 24–5, 155, 162–3, 239, 366, 372
  turnpike trust 204
Milford, Lord 170
Molloy, Pat 256
Monmouthshire 14, 28, 34, 43, 147, 163, 168, 172, 203, 227, 249
Montgomeryshire 249, 344
Morgan, Dr Charles 264
Morgans, of Cwm Cile 66, 228
Morning Advertiser 225
Morning Chronicle 25, 75, 100, 225, 231, 245, 252
Morning Herald 6, 77, 199, 225, 294, 328
Morris, Caleb 321
Morris, David 32, 87–8, 146, 155, 357
Morris, Reverend Ebenezer 17, 20, 80, 84, 123, 125, 178, 303, 363
Morris, Lewis 223

Nanteos 13, 46, 57, 61, 68, 88, 148, 175
Napier, Captain 228, 235, 238, 344
Narberth 7, 9, 23, 31, 121, 124, 130, 138, 168, 175, 189, 191, 194, 207–8, 210, 223–4, 226, 229, 239, 267, 271–3, 275, 282, 302, 323–4, 332, 351, 375
  Agricultural Association 9
Nathan, Reverend Henry 303
Nation 331, 343, 355
Neath 16, 18–19, 117, 172
Nevern, river 11, 292, 374
Nevill, Richard Janion 14, 16, 89–90, 167, 367
Newbridge on Wye 241
Newcastle, Duke of 42, 47, 51
Newcastle Emlyn 45, 62, 69, 130, 134, 138, 140, 145, 154, 168, 174, 179, 187, 210, 217, 221–6, 243–7, 250, 253, 258–9, 262, 265–7, 270, 274, 280, 284, 292, 299–300, 305, 308, 314–15, 323–6, 339
Newchurch 23, 91, 196, 214, 218–19, 296
New Inn 231, 295
Newman, Charles 14, 30, 314
Newport, Pembrokeshire 7, 118, 168, 184, 262, 275, 290, 292, 333
Newport, Rising of 28, 310, 333, 337, 345–6, 357, 369
New Quay 18, 203, 366
Night Poaching Act of 1844 176
Nonconformist 321, 332, 337
Northern Star 337

O'Brien, Bronterre 338–9
O'Connell, Daniel 319, 343
O'Connor, Feargus 338–9
Oddfellows 114, 333, 364
Orielton 47, 88, 284
Owen, Hugh 15, 17, 90
Owen, Sir John 47, 86, 89, 322
Owen, William 247

Pantycendy 48, 70, 73, 90–1, 99, 126, 149, 181, 218, 223, 245, 321
Parlby, Major 220
Parry-Jones, D. 160
Peel, Sir Robert 48, 98, 145–6, 228, 236, 276, 279, 320, 334, 346
Peel, William 51, 89, 277, 358
Pell, Captain 245

Index

Pembrey 16, 20, 24, 40, 55–6, 72, 112, 184, 204, 290, 295, 299, 308, 314, 336, 340–1, 373
Pembroke 4, 18–19, 24–5, 42, 75, 78, 87, 113, 117, 124, 132, 134, 139, 155, 162, 170, 183, 191, 275, 335, 362
Dock 156, 209, 223, 245, 350, 355, 363
Farmers' Club 5, 359
Pembrokeshire 3–4, 9, 15, 19, 22, 24, 26, 29, 37–8, 42–9, 61–3, 68–74, 84, 88–91, 99, 101, 109, 112–13, 118, 123, 128, 135, 143, 146, 148, 152, 161–2, 167, 170, 173, 178, 180, 184–5, 197, 199, 203–4, 210, 224, 240, 262, 278, 285–6, 290, 292–4, 319, 327, 330, 333–4, 351–2, 355, 357, 359, 363, 366–7, 373
Pembrokeshire Herald 86
Penblewin bar 207
Penboyr 23, 106, 164, 246, 276
Penbryn 126, 128, 154, 275, 278
Pencader 169, 177, 192
Pencrwcybalog 325
Pen Das Eithin 92, 325, 331, 336
Pensarn gate 138, 142
Pentre, estate 47, 88, 94
Penygarn gate 209, 213, 289, 254
Peterwell 30, 68, 354
Philip, William 206
Philipps, Grismond 277
Philipps, John Henry 90, 299, 358
Philipps, Sir Richard 14, 47, 88, 125, 138, 176
Phillip, Phillip 92, 97, 124, 128, 140, 282, 341
Phillip, William 282
Phillips, George 234
Phillips, John Walter 126, 277
Phillips, Thomas 75, 156, 368
Phillips, Thomas, of Topsail 184, 238, 308
Picton Castle 14, 47, 69, 88, 174, 176
Plaindealings gate 138, 211
Plasbach bar 240
Police 12, 41, 43, 73, 147, 151, 154–9, 164, 167, 177, 179, 187, 204, 206, 209, 213–16, 227–8, 232, 236, 239, 241–2, 255, 257, 280, 282, 285, 293–6, 299, 306, 328, 333, 348–54

Pontardawe 328, 336
Pontarddulais 15, 24, 34, 53, 139, 204, 207, 210, 226–9, 235–6, 239, 244, 247, 250, 254, 268, 292, 301, 325
Pontarllechau 138, 209, 231–2, 253
Pontyates 15, 240, 249
Pontyberem 14–15, 20, 224, 227, 236, 242, 249, 311, 314, 325, 340
Poor Law Amendment Act of 1834 73, 91, 103, 128–30, 134, 181, 183, 328, 334, 348
Porthyrhyd 39, 138, 224, 226, 232–3, 238–9, 243, 245, 249, 253, 264, 295–6, 297, 315
Powell, Timothy 48, 90, 114, 135, 207–8, 221, 300
Powell, Walter R. H. 47, 72, 80, 96, 207, 256, 325, 335, 354, 361, 377
Powell, William E. 13, 47, 54, 88–9, 148, 319, 376
Prendergast gate 138, 210, 226, 249
Preseli 12, 272, 206, 284
Price, John Lloyd 47, 89, 93, 137, 194, 211, 298, 357
Pritchard, Charles A. 59, 126, 174, 299, 303
Prothero, Reverend David 84, 290, 304
Pryse, Pryse 13, 17, 47, 53, 55, 70, 83, 85, 88, 94, 353
Pugh, David 89, 127, 148, 277, 356, 377
Pwlltrap gate 208, 273

Quarterly Review 159

Radnorshire 200, 203–4, 229, 241, 249, 296
Read, Clare Sewell 4, 6
Rechabites 333, 362
Rees, David 33, 83–4, 93, 97, 124, 245, 323, 332–3, 341, 359, 362–3, 375
Rees, John 58, 126, 132, 174, 286, 325, 328, 331
Rees, John Hughes 72, 90, 299, 325, 340, 351
Rees, Mary 52, 291
Rees, Thomas, of Carnabwth 206, 247
Rees, Thomas, of Swansea 370
Reform Act of 1832 88, 322
of 1867 376

Index

Reform crisis 8, 34, 88, 93, 155, 322
Rhandirmwyn 13, 263
Rhayader 241–2, 244
Rhydodyn 10, 47, 123, 174, 361
Rhydypandy 56, 226, 228, 248
Rice, Walter 69, 126
Richard, Henry 156, 333, 345, 368, 370, 374
Robeston Wathen 84, 210, 239, 300, 302
Roch, George 130, 151
Royal Agricultural Society 5
Royal Oak gate 138
Russell, Lord John 123, 150, 349, 352

St Clears 10, 23, 48, 54–6, 63, 73, 75–6, 90, 94, 107, 114, 121, 128, 135, 137, 139, 154, 196, 200, 204, 207–10, 213, 219, 221, 223–4, 239, 244, 247, 253, 262, 264, 267, 273, 280, 295, 300, 302, 322, 337, 355, 363
St Davids, Bishop of 125, 358, 361
St Davids, diocese of 83, 355
St Davids College, Lampeter 82, 89
St Dogmaels 174, 179, 187, 268, 355
Sandy gate 139, 233, 295
Saunders, John E. 89, 135, 138, 141–2, 176, 300, 306, 322, 348, 351
Scotch Cattle 147, 182, 227, 249, 263, 340
Scotland 4, 17, 20, 27, 47, 57, 342–3, 370
Scott, Captain 352, 372
Scott, Major Rochfort 223
Sealyham 48, 90, 245, 325, 327, 354
Seaton, Reverend William 110
Seren Gomer 81, 86, 133, 355
Slebech Park 10, 47, 51, 56, 59, 97, 148, 174, 190, 277, 284, 286
Slocombe, James 30, 309, 314
Soldiers 1, 3, 12, 25, 147, 156, 162, 177, 204, 208–9, 213, 216, 219–21, 223, 225, 227, 229, 231, 236, 242, 245, 251, 254, 286, 293, 295–9, 303, 305–6, 333, 350–1
South Wales Railway Bill of 1845 365
Turnpike Act of 1844 347
Stacey, Edmund H. 32, 98, 130, 215, 300
Stanley, Lord 258
Steynton gate 239
Swann, William B. 183, 206, 337

Swansea 7, 9, 14, 16, 19, 22, 24, 39, 41, 112, 139, 144, 146, 169, 172, 192, 229, 233–5, 263, 301, 319, 324, 338, 340, 344, 365, 367, 370
turnpike trust 204, 225
Swansea Journal 225, 231, 301, 322
Swing, Captain 8, 343, 346

Taf Bridge gate 203
Taliaris 5, 48, 51, 277, 358
Talog 49, 214–16, 219–20, 247, 250
Tavernspite 171
turnpike trust 204
Teifi river 3, 4, 11, 47, 52, 65, 78, 82, 97, 127, 168, 175, 195, 210–12, 221, 236, 262, 265, 282, 292, 311, 317, 331, 336, 372, 374
Tenby 12, 24, 40, 52, 160, 167, 183, 338, 361, 366
Thomas, David of Llanllawddog 265, 309
Thomas, Evan, Porthyrhyd constable 232, 245, 297
Thomas, John, of Cwmmawr 16, 314
Thomas, Margaret 264
Thomas, Mary, of Llanelli 109, 127, 143
Thomas, Michael 99, 129, 146, 321, 328, 334, 341, 377
Thomas, Rees Goring 58, 80, 89–90, 125, 127, 129, 135, 138, 223, 256, 279, 299, 316, 322, 325, 336, 351–2, 363, 377
Thomas, Thomas Emlyn 329, 331, 341
Thomas, Thomas, murderer 169, 270
Thomas, Thomas, Pantycerrig 298–9, 317
Thomas, Thomas, Porthyrhyd 245–6
Thomas, Thomas, shopkeeper 215–18
Thomas, William, Llannon 321, 331, 341
Three Commotts turnpike trust 137, 139, 204, 226, 229
Three Crosses, 226, 253
Times 58, 364
Tithe Commutation Act of 1836 84, 125, 130, 327
Tobias, Thomas 92, 97
Trawsmawr 138, 199, 215–16
Tregaron 3, 9, 18, 114, 132, 145, 164

Index

Tre-lech a'r Betws 23, 49, 52, 73, 84, 90–2, 115, 117, 120, 123, 129, 156, 194, 198–9, 214, 219, 266–7, 290, 327, 336
Trevaughan gate 209–10
Trevor, George Rice 69, 88, 123, 141, 194, 207–8, 213, 220, 223–4, 238, 344, 250, 274, 300, 322, 350, 352, 359, 376
Troedyraur 5, 8, 93, 153, 269, 272, 299, 316, 339, 358
Tycoch bar 234
Tywi river 3–4, 11, 25, 136, 175, 210, 226–7, 292, 374

Vagrancy 147–8, 162–3, 310
Vale of Towy Agricultural Society 359
Vaughan, Griffith 228–9
Vaughan, Herbert 221–2, 250, 286, 299, 306, 353
Verwig 116, 138, 240
Victoria, Queen 214, 228, 230, 242, 258, 276, 316, 323–4, 346
Vincent, Henry 363
Volunteers 361

Wagner, S.P.R. 280, 316
Walsh, Sir John 241, 296
Walters, Jane 158, 243, 258
Water Street gate 138, 199, 214, 216, 219, 247
Waunystradfeiris gate 232, 240
Wellington, Duke of 258, 305, 336, 342, 350
Welshman 14, 29, 33, 69, 86–7, 106, 163, 174, 179, 192, 221, 255, 286, 310, 320, 344
Whitehouse gate 138
Whitland 145, 204, 210, 230, 244, 247, 288, 292
turnpike trust 139, 141, 204–7, 210, 213
William, Thomas 288–9
Williams, Bridget, 273, 310
Williams, Professor David 242, 256
Williams, David, of Bronmeurig 117, 120
Williams, Edward Lloyd 72, 126, 194, 223–4, 229, 244, 277, 293, 325–8, 353, 373
Williams, Reverend Erasmus 30, 69, 128, 352
Williams, Reverend Herbert 80, 84, 140, 355
Williams, Hugh 98, 132–3, 195, 247, 316, 320, 323, 329, 331, 335, 337, 341, 345
Williams, Sir James 47, 98, 123, 138, 277
Williams, John 31, 280, 285
Williams, Lady, of Rhydodyn 117, 361
Williams, R.B. 56, 148, 246
Williams, Richard 232–3, 238
Williams, Sarah 234–5
Williams, Thomas, auctioneer 108, 280–1
Williams, Thomas, blacksmith 96, 321
Williams, Thomas, informer 230
Williams, William, MP 157, 357
Williams, William, of Water Street gate 199
Wolfscastle 327, 333
Wrigley, C. J. 274¬5

Yelverton, Miss 288
Ystrad 51, 71, 88

Zacharius, William 238, 243